罗里波文集

模型论与计算复杂度

李仲来／主编

北京师范大学出版集团
BEIJING NORMAL UNIVERSITY PUBLISHING GROUP
北京师范大学出版社

2013 · 北京

图书在版编目(CIP)数据

模型论与计算复杂度——罗里波文集／罗里波著，李仲来主编.—北京：北京师范大学出版社，2013.12
(北京师范大学数学家文库)
ISBN 978-7-303-15890-4

Ⅰ.①模… Ⅱ.①罗… ②李… Ⅲ.①模型论－文集②线性复杂度－文集 Ⅳ.①O141.4-53 ②O157.4-53

中国版本图书馆CIP数据核字（2013）第011584号

营销中心电话 010-58802181 58805532
北师大出版社高等教育分社网 http://gaojiao.bnup.com
电子信箱 gaojiao@bnupg.com

出版发行：北京师范大学出版社 www.bnup.com
北京新街口外大街19号
邮政编码：100875
印　　刷：北京京师印务有限公司
经　　销：全国新华书店
开　　本：155 mm × 235 mm
印　　张：21
插　　页：4
字　　数：325千字
版　　次：2013年12月第1版
印　　次：2013年12月第1次印刷
定　　价：52.00元

策划编辑：岳昌庆　　责任编辑：岳昌庆
美术编辑：王齐云　　装帧设计：王齐云
责任校对：李　菡　　责任印制：孙文凯

◀ 1984 年在 Michigan 大学的博士毕业照。

▶ 1987 年南京大学的一个博士生的答辩会上的照片：左起（下同）：莫绍揆、丁德成、李未、罗里波。

◀ 1984 年在 Michigan 大学的博士毕业典礼之后和教授们的合影：罗里波的博士导师之一 R. Lyndon 教授、罗里波、M. Reade 教授。

◀ 1988 年罗里波的博士导师之一 Y. Gurevich 来华访问，在明十三陵的合影。

▶ 1987 年德国 Freiburg 大学 D. Ebbinghaus 教授夫妇来华访问，在天安门城楼上的合影。

◀ 1990 年，作为首批访问台湾的大陆学者得到李焕的接见：罗里波、工作人员、李焕。

▶ 1988 年，和侯宝林在第 7 届全国人民代表大会住所的合影。

▶ 1985 年在 Mississippi 大学数学系讲课的照片。

▶ 1981 年 10 月，数学系首届（1978 级）硕士毕业生合影。
第 1 排：刘绍学 *、孙永生 *、张禾瑞 *、王世强 *、严士健 *；
第 2 排：陈木法、罗里波、罗承忠 *、李英民 *、张英伯、罗俊波；
第 3 排：唐守正、王成德、郑小谷、孙晓岚、王昆扬、程汉生、沈复兴。
（有 * 的是教师）

◀ 1983 年王世强先生访问 Michigan 大学在研究生院 Rackham Hall 前面的合影。

▶ 1987 年出席《中国科学》和《科学通报》的编委会，会间休息时和杨乐院士的合影。

◀ 2010 年 10 月家人聚会。前排：葛月明、罗里波；后排：孔伟成、罗宁、罗南、Lance Winn。

自　序

这本文集收集了我所发表过的数学方面的主要文章，其中有研究论文 20 篇，科普文章 2 篇，指导研究生的论文 3 篇以及历史上重要文献的中译文 1 篇. 我开始做数学研究工作于 1956 年，第一篇研究论文《有限结合系与有限群(I)》是在王世强教授的指导之下并和他合作写出来的. 由于当时国内的数学研究论文较少，这篇文章得以在 1956 年举行的全国第一届数学论文宣读会上报告并且发表在《数学进展》上. 我的第 2～3 篇研究论文《强不可接近基数上 $P(K)$ 的插入定理》和《关于代数系统自同构群的一个问题》1980 年发表在《数学学报》上. 这两篇文章作者的署名单位是内蒙古锡林浩特中学，投稿时我是该校英语教研组的一名教师.《关于代数系统自同构群的一个问题》由于解决了 Birkhoff 的《格论》中所提出的一个问题，后来在该书的第 3 版中得到了引用作为参考文献.

经过我的导师王世强的多方努力，我于 1978 年由内蒙古返回北京师范大学做数理逻辑方向的硕士研究生. 这时由于年龄较大，在同一批研究生中我得了一个雅号，叫作“大师兄”. 这个名字给了我一股压力，使我觉得什么事都要做得快一点. 我用较快的速度完成了研究生的毕业论文两篇，它们是《模型的并、积与齐次模型》和《自由群内方程的讨论》. 这两篇文章后来分别发表在《北京师范大学学报(自然科学版)》和《中国科学》. 除了毕业论文之外我还写了 3 篇文章《可换群中无限生成元直和项消去条件的探讨》发表在《数学学报》上；另外两篇《On the Number of Countable Homogeneous Models》解决了模型论里面的 Vaught 猜想中的一个情形，《The τ—Theory for Free Groups is Undecidable》解决了 Huber—Dyson 所提出的一个问题. 由于这些问题比较重要，我和王先生商量看能不能投到美国的《The Journal of Symbolic Logic》去发表. 王先生说可以试一试. 当时的条件比较差，并且为了节省邮费，我用一种很薄的纸在老式的机械打字机上打印出底稿，寄到了美国.《The Journal of Symbolic Logic》发表了这两篇文章.《On the Number of Countable Homogeneous Models》和《模型的并、积与齐次模型》后来得到了美国 Chang 的书《Model Theory》第 3 版的引用. 英国的 Hodges 的书《Model Theory》引用了前一篇.

这时数学系又有了选派出国访问学者的任务，领导建议我去参加考试，我考取了一个名额. 在联系赴美访问的过程中，美国 Michigan 大学数学系的 Lyndon 教授又将我的身份由访问学者改成了他的一个以教学助教名义攻读博士学位的研究生. 从 1981 年到 1984 年里的三年间，我用最短的时间修够了必要的学分，通过了资格考试并且撰写了博士论文. 我的毕业论文的题目《On the Computational Complexity of the Theory of Abelian Groups》是 Michigan 大学计算机系的 Gurevich 教授提出的，他也就成为我的另一位博士导师. 这篇文章后来发表在荷兰的《Annals of Pure and Applied Logic》杂志上.

1984 年取得博士学位以后，我先后在美国 Mississippi 大学数学系、北京师范大学数学系、美国 Mississippi 大学计算机系、美国 Wichita 州立大学计算机系和日本名古屋商科大学计算机系工作. 这一期间我发表了《The Computational Complexity of Positive Formulas in Real Addition》(《名古屋商科大学论文集》)和《Functions and Functionals on Finite Systems》(美国《The Journal of Symbolic Logic》).

1995 年我回到北京师范大学数学系工作. 我受邀做了一个关于计算

机科学发展的报告，并在报告后应《数学通报》的请求将其发表，题目是《计算机科学发展漫谈》. 这是一篇科普文章，后来我把它放在我的博客上，也有一些点击率. 这段时间我写了两篇研究论文，《多个一元关系上的 Vaught 猜想》发表在《数学学报》上，《P-Time Algorithms in Number Theory》发表在《南京大学学报》上.

1999 年我在北京师范大学数学系退休. 退休之后应广东女子职业技术学院的邀请去帮他们筹建计算机系. 这一期间我写了研究论文《无原子布氏代数的计算复杂性》，发表在《数学研究》. 另外和该学院的同事合作写了一篇科普文章《利用计算机计算古典数论问题》，发表在《数学通报》. 我的同事在文章的有关程序运行上做了很多工作. 国内有些中学将这篇文章列为数学小组的必读材料.

2005 年我应邀到贵州民族学院数学与计算机科学系担任首席教授. 在这期间我写了研究论文《The Theory of Finite Models without Equal Sign》，发表在《Acta Mathematica Sinica》. 又应邀参加在广州中山大学举办的第四届逻辑与认知国际会议，在会上作了题为《Remove Infinity from Computer Science》的报告. 会后这篇文章收入了该会议的文集.

2007 年我应邀到石家庄经济学院的信息工程学院担任特聘教授. 在这期间我写了 4 篇研究论文《康托尔实数的局限性》《The Stability of Turing Machines in Computing Real Functions》《非良基集合论模型悖论》《Computing Functions on Real Numbers with Stable ω—Turing Machines》，其中前两篇发表在《数学研究》，第 3 篇发表在《北京师范大学学报(自然科学版)》，第 4 篇发表在《前沿科学》.

回国后，在北京师范大学数学系我招收过 4 个硕士研究生，其中于丽荣的毕业论文是《New Theorems in Non-Standard Number Theory》，她和我联名将这篇文章发表在《Acta Mathematica Sinica》. 我参与指导过沈复兴教授的研究生的论文《完全二叉树的量词消去》，他们和我联名将该文发表在《数学学报》. 我参与指导过贵州大学李祥教授的博士生李志敏的毕业论文《完全二叉树理论的计算复杂度》，他们和我联名将该文发表在《数学学报》上.

我翻译的重要文献，Turing 的著名论文《On Computable Numbers, with an Application to the 'Entscheidungs Problem'》，也收在这本文集

之中.

纵观我的研究工作,先是在代数方面,后来转到数理逻辑,又转到计算复杂性. 近年来我的文章主要是想说明数学基础中的问题,我希望将来能有人在这方面做一些工作.

感谢北京师范大学数学科学学院和北京师范大学出版社的大力支持和帮助.感谢我的导师王世强教授对我的教导和极大的帮助,感谢本丛书主编李仲来教授所做的很多工作和大力的帮助.

罗里波

2012 年 3 月

目 录

Contents

数学进展,1957,3(2):268～270

有限结合系与有限群(Ⅰ)[①]

Finite Systems and Finite Groups

在群的各种著名性质中,哪些是为群所特有的?哪些还有其他类似系统能适合?这是一类很有趣的问题.本文只限于讨论有限结合系及Lagrange定理.结果如下:

设S为一具有Lagrange性质(定义见下)的有限结合系(以下简称L系),则

1. 当S之元数n为奇数时,S为群.

2. 当S之元数n为偶数时,若S只含1个幂等元,则除$n=2$时是一个特例外,S为群.

当元数为偶数时,含多个幂等元的L系是存在的,本文暂不讨论.

定义 设S为一非空有限集合,在其中定义了一种适合结合律的二元运算"·",则称S(对"·"言)为一有限结合系.若T为S的子集,对"·"封闭,则称T为S的子系.设S含n元,若S的每一子系其元数皆为n的因数,则称S具有Lagrange性质,此种S简称为L系.

引理1 设S为有限结合系,$a\in S$,则存在正整数r使a^r为一幂等元e(特知:每一有限结合系皆含有幂等元).

证 在无限序列$a,a^2,a^3,\cdots$中必有相同者,设$a^p=a^{p+q}(p,q\geqslant 1)$,取$m$使$r=mq>p$,则

$$(a^r)^2=a^{mq-p}a^{p+mq}=\cdots=a^{mq-p}a^p=a^r.$$

① 本文与王世强合作.

定理 1　设 S 为一 $n(>2)$ 元结合系，不含 $n-1$ 元子系，只含 1 个幂等元 e，则 S 为群.

证　(1) S 中任一元 a 皆能表为另二元之积：$a=bc(b,c\neq a)$. 因否则易见 $S-\{a\}$ 为 $n-1$ 元子系.

(2) 设 $a\in S$，考虑子系 $(a)=\{a,a^2,a^3,\cdots\}$ 及 $(a)'=\{a^2,a^3,a^4,\cdots\}$.

若 $a\in(a)'$，则有 $S\geqslant 2$ 使 $a^S=a$，因而易见 (a) 为一循环群.

若 $a\notin(a)'$，则易见 (a) 不为群，称此种 (a) 为 N 型子系.

(3) 若有 $a_1\in S$ 使 (a_1) 为 N 型.

(i) 设 $a_1=a_2b_2(a_2\neq a_1)$.

若 (a_2) 为群，则必以幂等元 e 为单位元. 又由引理 1 知有 $r\geqslant 1$ 使 $a_1^r=e$，所以

$$a_1=a_2b_2=ea_2b_2=ea_1=a_1^{r+1}\in(a_1)',$$

这与 (a_1) 为 N 型之假设不合. 故 (a_2) 为 N 型.

(ii) 再设 $a_2=a_3b_3(a_3\neq a_2)$，

仿上知 (a_3) 为 N 型.

今证 $a_3\neq a_1$：若 $a_3=a_1$，则

$$a_1=a_2b_2=a_3b_3b_2=a_1b(b=b_3b_2),$$

设 $b^t=e$，则 $a_1=a_1b=a_1b^2=\cdots=a_1b^t=a_1^{r+1}\in(a_1)'$，这与 (a_1) 为 N 型之假设不合.

(iii) 再设 $a_3=a_4b_4(a_4\neq a_3)$，

仿上知 (a_4) 为 N 型，且 $a_4\neq a_2,a_1$.

如此继续，可得 S 中无限多互异之元 $a_1,a_2,a_3,\cdots$ 与 S 之有限性不合.

故对每 $a\in S$，(a) 皆为群，以 e 为单位元. 由此即易见 S 自身为群.

定理 2　设 S 为一 L 系，只含 1 个幂等元. 则除一 2 元特例外，S 为群.

证　(1) 当 S 之元数 $n\leqslant 2$ 时，除群外易见只有如下一个只含 1 个幂等元的 L 系：

$\cdot$	e	a
e	e	e
a	e	e

(2)当$n>2$时，由 Lagrange 性质知S不含$n-1$元子系，故由定理1知S为群.

引理2 设有限结合系S无2元子系，则S只含1个幂等元.

证 若S之幂等元数$h\geqslant 2$. 设a,b为S之二不同的幂等元. 令$(ab)^p=c$，$(ba)^q=d$，其中c,d为幂等元，则c,d至少有一$\neq a,b$，否则易见$\{a,b\}$为2元子系(例如若$(ab)^p=a$，则$ab=(ab)^pb=(ab)^p=a$，等等). 不妨设$c\neq a,b$.

令T为S中所有适合$ax=x$之元x所成之集合，则易见T为S之子系，因而T无2元子系. 又易见$a,c\in T$而$b\notin T$(因若$ab=b$，则$c=(ab)^p=b^p=b$，与上不合). 所以T所含幂等元数$k\geqslant 2$而$<h$.

同样，可得T之一子系U，无2元子系，所含幂等元数$l\geqslant 2$而$<k$.

如此继续，有限次后显然可得出矛盾.

定理3 设S为一奇数元L系，则S为群.

证 当S之元数$n=1$时显然为群. 当$n>2$时，由引理2知S只含1个幂等元，再由定理1即知S为群.

注 含多个幂等元的偶数元L系是存在的，例如：

$\cdot$	a	b
a	a	a
b	b	b

,

$\cdot$	a	b	c	d
a	a	a	c	c
b	b	b	d	d
c	a	a	c	c
d	b	b	d	d

此种L系之构造本文暂不讨论.

Abstract The following results are obtained:

Let S be a finite associative system in which Lagrange theorem holds (i. e. the order of every subsystem of S devides the order of S).

1. If the order n of S is odd, then S is a group.

2. If the order n of S is even, and S contains only one idempotent element, then S is a group with but one exception when $n=2$.

数学学报,1980,23(2):177～182

强不可接近基数上 $P(K)$ 的插入定理[①]

The Interpolation Theorem of $P(K)$ on a Strongly Inaccessible Cardinal Number

在《符号逻辑》杂志 1975 年 6 月份第 40 卷第 2 期上 Friedman H. 选编了 102 个数理逻辑问题. 其中第 24 个问题是关于无穷正规基数 K 上的命题演算 $P(K)$ 是否遵守插入定理的问题. 问题后面还引述了 Friedman H. 自己的两个结果:当 K 是共尾数为 ω 的强极限基数的后继基数时,$P(K)$ 遵守插入定理. 当 K 是一个共尾数大于 ω 的基数的后继基数时,$P(K)$ 不遵守插入定理. 余下的主要情形就是强不可接近基数上的 $P(K)$ 是否遵守插入定理的问题(非正规极限基数的 $P(K)$ 不讨论). 本文证明了这种 $P(K)$ 是遵守插入定理的. (并且于文中引理 8 还证明了 $P(K)$ 遵守无穷分配律). 在承认广义连续统假设的前提下,这就完全解决了上述的第 24 个问题(否则只是部分解决).

定义 1 设 K 是强不可接近基数,因此它也就是正规基数. 下面定义 $P(K)$ 的式子:

(1)选取 K 个互不相同的命题记号

$$\{s_0, s_1, \cdots, s_\alpha, \cdots\} = \sum.$$

$s \in \sum$,称 s 为 1 阶式子,又叫作原子式子.

(2)设 $\alpha < K$,如果对于所有小于 α 的 β,$P(K)$ 的 β 阶式子已有定义,下面定义 $P(K)$ 的第 α 阶式子.

(i)设 $\alpha = \beta + 1$,如果 α 是 β 阶式子,那么 $\sim(a)$ 是 α 阶式子.

① 收稿日期:1977-07-27.

如果$a_0,a_1,\cdots,a_r,\cdots$(个数小于$K$)中每一个$a_r$的阶数均$\leqslant\beta$,并且至少有一个$a_r$的阶数等于$\beta$,那么

$$a_0\wedge a_1\wedge\cdots\wedge a_r\wedge\cdots$$

是一个α阶式子.

(ii)设α是极限序数,如果

$$a_0,a_1,\cdots,a_r,\cdots(\text{个数小于 }K)$$

中每一个a_r的阶数均$<\alpha$,并且

$$\sup(a_r\text{ 的阶数})=\alpha,$$

那么

$$a_0\wedge a_1\wedge\cdots\wedge a_r\wedge\cdots$$

是一个α阶式子.

(3)$P(K)$的式子是上述(1)(2)两项所定义的全体$\alpha(<K)$阶式子的总和.

引理 1 由超限归纳法易见每一个$P(K)$式子有唯一确定的阶数(并且不会出现无限个"$\sim$"记号连用的情况,例如$\cdots\sim\sim a$).

定义 2 $A\subseteq\sum$,φ是一个$P(K)$式子,以下定义由A产生的φ的赋值φ_A:

(1)对于1阶式子s,如果$s\in A$,那么$s_A=t$. 如果$s\notin A$,那么$s_A=f$.

(2)对于任意$\alpha<K$,如果全体阶数小于α的式子ψ,ψ_A已有定义,下面定义α阶式子φ的赋值φ_A.

(i)如果$\varphi\equiv\sim(a)$,其中a是阶数小于α的式子,由归纳假设a_A已有定义,如果$a_A=t$,那么定义$\varphi_A=f$. 如果$a_A=f$,那么定义$\varphi_A=t$("$\equiv$"表示式子全同).

(ii)如果$\varphi\equiv(a_0\wedge a_1\wedge\cdots\wedge a_\beta\wedge\cdots)$其中各个$a_\beta$是阶数小于$\alpha$的式子,由归纳假设各个$(a_\beta)_A$已有定义. 如果对各个$\beta$均有$(a_\beta)_A=t$,那么定义$\varphi_A=t$. 否则,即在有一个$(a_\beta)_A=f$时,定义$\varphi_A=f$.

定义 3 如果$\varphi_A=t$,那么记作$A\models\varphi$,此时又称A是φ的一个模型,或者φ在A上成立.

定义 4 如果对于任意$A\subseteq\sum$均有$A\models\varphi$,那么记作$\models\varphi$. 此时又称φ是一个恒真式.

定义 5 简记号:

$(a_0 \vee a_1 \vee \cdots \vee a_\alpha \vee \cdots) \equiv \sim(\sim a_0 \wedge \sim a_1 \wedge \cdots \wedge \sim a_\alpha \wedge \cdots)$.

$(a \to b) \equiv (\sim a \vee b)$.

$(a \leftrightarrow b) \equiv ((a \to b) \wedge (b \to a))$.

引理 2 对于任意 $A \subseteq \Sigma$，$\varphi \in P(K)$，φ_A 有确定的值 t 或 f（证略）.

引理 3 以下 4 条与普通命题演算类似可得证：

（1）$(a_0 \wedge a_1 \wedge \cdots \wedge a_\alpha \wedge \cdots)_A = t$ 的充分必要条件是每一个 a_α 均满足

$$(a_\alpha)_A = t.$$

$(a_0 \wedge a_1 \wedge \cdots \wedge a_\alpha \wedge \cdots)_A = f$ 的充分必要条件是有一个 a_α 满足 $(a_\alpha)_A = f$.

（2）$(a_0 \vee a_1 \vee \cdots \vee a_\alpha \vee \cdots)_A = t$ 的充分必要条件是有一个 a_α 均满足

$$(a_\alpha)_A = t.$$

$(a_0 \vee a_1 \vee \cdots \vee a_\alpha \vee \cdots)_A = f$ 的充分必要条件是每一个 a_α 均满足

$$(a_\alpha)_A = f.$$

（3）$(a \to b)_A = t$ 的充分必要条件是 $a_A = f$ 或 $b_A = t$ 二者至少有一成立.

$(a \to b)_A = f$ 的充分必要条件是 $a_A = t$ 与 $b_A = f$ 同时成立.

（4）$(a \leftrightarrow b)_A = t$ 的充分必要条件是 $a_A = b_A$.

$(a \leftrightarrow b)_A = f$ 的充分必要条件是 $a_A \neq b_A$.

定义 6 如果每一个 φ 的模型都是 ψ 的模型，记作 $\varphi \models \psi$.

引理 4 $\varphi \models \psi$ 成立的充分必要条件是 $\models \varphi \to \psi$.

证 （1）设 $\varphi \models \psi$，由定义 6 对于任意 $A \subseteq \Sigma$，如果 $\varphi_A = t$，那么 $\psi_A = t$. 再由引理 3 第（3）项 $(\varphi \to \psi)_A = t$；如果 $\varphi_A = f$，由引理 3 第（3）项也得 $(\varphi \to \psi)_A = t$. 由 A 的任意性知 $\models \varphi \to \psi$.

（2）反过来，设 $\models \varphi \to \psi$，而 $\varphi_A = t$. 假如 $\psi_A = f$，由引理 3 第（3）项 $(\varphi \to \psi)_A = f$. 但由 $\models \varphi \to \psi$ 的定义知 $(\varphi \to \psi)_A = t$，矛盾. 因此不可能 $\psi_A = f$. 即有 $\psi_A = t$. 由 A 的任意性知 $\varphi \models \psi$.

引理 5 $\varphi \models \psi_0 \wedge \psi_1 \wedge \cdots \wedge \psi_\alpha \wedge \cdots$ 成立的充分必要条件是对于所有 α 均有 $\varphi \models \psi_\alpha$.

证 设 $\varphi \models \psi_0 \wedge \psi_1 \wedge \cdots \wedge \psi_\alpha \wedge \cdots$ 对于任意 $A \subseteq \Sigma$，如果 $\varphi_A = t$，那么 $(\psi_0 \wedge \psi_1 \wedge \cdots \wedge \psi_\alpha \wedge \cdots)_A = t$. 由引理 3 第（1）项对每个 α 均有 $(\psi_\alpha)_A = t$，再由 A 的任意性知 $\varphi \models \psi_\alpha$.

反过来，如果 $\varphi \models \psi_\alpha$ 对于所有 α 均成立，A 是 φ 的一个模型，我们有

$\varphi_A=t$. 但 φ 的模型也是 ψ_α 的模型，因此 $(\psi_\alpha)_A=t$. 由 α 的任意性得 $(\psi_0\wedge\psi_1\wedge\cdots\wedge\psi_\alpha\wedge\cdots)_A=t$. 再由 A 的任意性得 $\varphi\models\psi_0\wedge\psi_1\wedge\cdots\wedge\psi_\alpha\wedge\cdots$

引理 6 假设 $\varphi(a_0,a_1,\cdots,a_\alpha,\cdots)$ 是含有 $a_0,a_1,\cdots,a_\alpha,\cdots$ 作为部分式子的式子. 并且 $\models a_0\leftrightarrow b_0,\cdots,\models a_\alpha\leftrightarrow b_\alpha,\cdots$ 那么

$$\models\varphi(a_0,a_1,\cdots,a_\alpha,\cdots)\leftrightarrow\varphi(b_0,b_1,\cdots,b_\alpha,\cdots).$$

证 对 φ 的阶数用超限归纳法. (1) 设 φ 为 1 阶式子，易见引理结论成立. (2) 下面假设对任意阶数小于 α 的式子 ψ，引理的结论已经成立，证明 α 阶式子 φ 的情形.

(i) 设 $\varphi(a_0,a_1,\cdots,a_\beta,\cdots)\equiv\sim\psi(a_0,a_1,\cdots,a_\beta,\cdots)$. ψ 是阶数小于 α 的式子. 由归纳假设

$$\models\psi(a_0,a_1,\cdots,a_\beta,\cdots)\leftrightarrow\psi(b_0,b_1,\cdots,b_\beta,\cdots).$$

利用引理 3 第(4)项，对于任意 $A\subseteq\Sigma$：

$$(\psi(a_0,a_1,\cdots,a_\beta,\cdots))_A=(\psi(b_0,b_1,\cdots,b_\beta,\cdots))_A.$$

$$(\sim\psi(a_0,a_1,\cdots,a_\beta,\cdots))_A=(\sim\psi(b_0,b_1,\cdots,b_\beta,\cdots))_A.$$

$$\models\sim\psi(a_0,a_1,\cdots,a_\beta,\cdots)\leftrightarrow\sim\psi(b_0,b_1,\cdots,b_\beta,\cdots).$$

$$\models\varphi(a_0,a_1,\cdots,a_\beta,\cdots)\leftrightarrow\varphi(b_0,b_1,\cdots,b_\beta,\cdots).$$

(ii) 设 $\varphi(a_0,a_1,\cdots,a_\beta,\cdots)\equiv\bigwedge_i\psi_i(a_0,a_1,\cdots,a_\beta,\cdots)$. 其中各 ψ_i 是阶数小于 α 的式子，由归纳假设

$$\models\psi_i(a_0,a_1,\cdots,a_\beta,\cdots)\leftrightarrow\psi_i(b_0,b_1,\cdots,b_\beta,\cdots).$$

设 $(\varphi(a_0,a_1,\cdots,a_\beta,\cdots))_A=t$，由引理 3 对每一个 i 均有 $(\psi_i(a_0,a_1,\cdots,a_\beta,\cdots))_A=t$. 再由引理 3 第(4)项知 $(\psi_i(b_0,b_1,\cdots,b_\beta,\cdots))_A=t$. 由 i 的任意性及引理 5

$$\bigwedge_i(\psi_i(b_0,b_1,\cdots,b_\beta,\cdots))_A=t.\quad(\varphi(b_0,b_1,\cdots,b_\beta,\cdots))_A=t.$$

再由 A 的任意性得

$$\models\varphi(a_0,a_1,\cdots,a_\beta,\cdots)\rightarrow\varphi(b_0,b_1,\cdots,b_\beta,\cdots).$$

类似地可以得到

$$\models\varphi(b_0,b_1,\cdots,b_\beta,\cdots)\rightarrow\varphi(a_0,a_1,\cdots,a_\beta,\cdots).$$

合并这两条，我们得到

$$\models\varphi(a_0,a_1,\cdots,a_\beta,\cdots)\leftrightarrow\varphi(b_0,b_1,\cdots,b_\beta,\cdots).$$

引理 7 $\alpha,\beta<K$，那么 $\alpha^\beta<K$.

证 只讨论 β 为无限的情况即可(β 有限时显然). 如果 $\alpha\leqslant\beta$，那么

$\alpha^\beta \leqslant 2^\beta$. 如果 $\alpha > \beta$，那么 $\alpha^\beta \leqslant (\alpha^\alpha)^\beta = \alpha^{\alpha\beta} = \alpha^\alpha = 2^\alpha$. 因为 K 是强不可接近基数，所以 $2^\alpha, 2^\beta < K$.

定义 7 如果式子 φ 可以写成下面形式：

$$\varphi \equiv (a_{00} \vee a_{01} \vee \cdots \vee a_{0j0} \vee \cdots) \wedge (a_{10} \vee a_{11} \vee \cdots \vee a_{1j1} \vee \cdots) \wedge \cdots \wedge (a_{i0} \vee a_{i1} \vee \cdots \vee a_{iji} \vee \cdots) \wedge \cdots$$

其中每一个 a_{ii} 都是命题记号 s 或 $\sim s$ 形式，那么 φ 称为一个布氏展开式.

引理 8 设 $\varphi \equiv (\bigvee_{0 \leqslant i < \alpha} \bigwedge_{0 \leqslant j < \beta_i} a_{ij})$,

$$\psi \equiv (\bigwedge_{p \in \pi} \bigvee_{0 \leqslant i < \alpha} a_{ip(i)}).$$

其中 p 跑过全体 $\pi = (\beta_0 \times \beta_1 \times \cdots \times \beta_i \times \cdots)$ 的元. 我们有 $\models \varphi \leftrightarrow \psi$（无穷分配律）.

证 首先证明 $|\pi| = |\beta_0 \times \beta_1 \times \cdots \times \beta_i \times \cdots| < K$. 因为共有 α 个 β_i，$\alpha < K$，K 是强不可接近基数，它不可能为 α 个比它小的序数的极限，所以 $\sup \beta_i = \beta < K$. 而由引理 7

$$|\beta_0 \times \beta_1 \times \cdots \times \beta_i \times \cdots| \leqslant \beta^\alpha < K.$$

（附注：在 K 不是强不可接近基数，而是其他正规基数时，可以 $\beta^\alpha \geqslant K$.）

如果 A 是 φ 的模型，那么 $\varphi_A = t$. 由引理 3 第(2)项可以找到一个 l，使

$$(\bigwedge_{0 \leqslant j < \beta_l} a_{lj})_A = t.$$

由引理 3 第(1)项

$$(a_{l_0})_A = t, \cdots, (a_{lj})_A = t, \cdots$$

因此在每一个 $\bigvee_{0 \leqslant i < \alpha} a_{ip(i)}$ 中均有 $(a_{lp(l)})_A = t$（对任意 $p \in \pi$）. 由引理 3 第(1)项

$$(\bigwedge_{p \in \pi} \bigvee_{0 \leqslant i < \alpha} a_{ip(i)})_A = t.$$

即 $\psi_A = t$. 由 A 的任意性知 $\varphi \models \psi$.

反之，我们证明如果 A 不是 φ 的模型，那么 A 不是 ψ 的模型. 因为由 $\varphi_A = f$，即对各 i 均有 $(\bigwedge_{0 \leqslant i < \beta_i} a_{ij})_A = f$，由引理 3 第(1)项可以找到一个 $(a_{il})_A = f$. 令 $p(i) = l$. $p(i)$ 对每一个 i 均可选取，所以 $p \in \pi$. 而

$$(\bigvee_{1 \leqslant i < \alpha} a_{ip(i)})_A = f.$$

由引理 3 第(1)项

$$(\bigwedge_{p\in\pi}\bigvee_{1\leqslant i<\alpha}a_{ip(i)})_A=f.$$

即 $\psi_A=f$. 因此每一个 ψ 的模型都是 φ 的模型. 即得 $\psi\models\varphi$. 综上我们可以得到 $\models\varphi\leftrightarrow\psi$.

引理 9 对于任意式子 φ 均有一个布氏展开式 ψ，满足 $\models\varphi\leftrightarrow\psi$. 并且每一个 ψ 中出现的命题记号都在 φ 中出现.

证 对式子 φ 的阶数 λ 实行超限归纳法. (1) $\lambda=1$ 时命题显然成立. (2)以下假设阶数小于 λ 的式子均已能展开，现在证明 λ 阶的情形.

(i)如果 $\varphi\equiv(a_0\wedge a_1\wedge\cdots\wedge a_\beta\wedge\cdots)$(共 r 项). 因为各 a_β 是阶数小于 λ 的式子，由归纳假设可以找到布氏展开式满足

$$\models a_\beta\leftrightarrow(\bigwedge_{0\leqslant i<\delta(\beta)}\ \bigvee_{0\leqslant j<\varepsilon(\beta,i)}a_{ij}^{(\beta)}).$$

因此令

$$\psi\equiv(\bigwedge_{0\leqslant\beta<r}\ \bigwedge_{0\leqslant i<\delta(\beta)}\ \bigvee_{0\leqslant j<\varepsilon(\beta,i)}a_{ij}^{(\beta)}).$$

ψ 是布氏展开式，总项数 $<K$，并且其中每一个命题记号都在 φ 中出现. 再由引理 6 $\models\varphi\leftrightarrow\psi$.

(ii)如果

$$\varphi\equiv\sim a\equiv\sim(a_0\wedge a_1\wedge\cdots\wedge a_\beta\wedge\cdots),$$

利用赋值可以证明

$$\models\varphi\leftrightarrow(\sim a_0\vee\sim a_1\vee\cdots\vee\sim a_\beta\vee\cdots).$$

并且所出现的命题记号集不变. 由归纳假设($\sim a_\beta$ 的阶数小于 λ)每一个 $\sim a_\beta$ 均可展开，即有

$$\models(\sim a_\beta)\leftrightarrow(\bigwedge_{0\leqslant i<\delta(\beta)}\ \bigvee_{0\leqslant j<\varepsilon(\beta,i)}a_{ij}^{(\beta)}).$$

右端各命题记号均出现于 $\sim a_\beta$，因此也出现于 φ. 由引理 8 有

$$\psi\equiv\bigwedge_{p\in\pi}\Big[\bigvee_{0\leqslant\beta<\alpha}(\bigvee_{0\leqslant j<\varepsilon(\beta,p(i))}a_{p(\beta)j}^{(\beta)})\Big],$$

满足 $\models\varphi\leftrightarrow\psi$，而 ψ 中出现的命题记号均出现于 φ.

(iii)如果 $\varphi\equiv\sim\sim a$，那么 φ 的展开式与 a 的展开式全同.

引理 10 任意一个不是恒真的布氏展开式

$$\varphi\equiv\bigwedge_{1\leqslant i<\alpha}a_i,\quad a\equiv\bigvee_{1\leqslant j<\beta_i}b_{ij},(1\leqslant i<\alpha)$$

可以化简为一个布氏展开式 ψ，其中每一个 a_i 中不同时出现某命题记号

的 s 与 $\sim s$. 并且 $\models\varphi\leftrightarrow\psi$.

证 先将 φ 中同时含有某 s 与 $\sim s$ 的 a_i 全部集中起来，组成部分式子 T(T 是恒真式). 因为 φ 不是恒真式，T 不可能就是 φ. 所以有 $\varphi\equiv\psi\wedge T$，其中 ψ 的 a_j 再不同时含有某 s 与 $\sim s$. 对于任意 A，由引理 3 第(1)项如果 $\varphi_A=t$，那么能推出 $\psi_A=t$. 在 $\varphi_A=f$ 时能推出 $\psi_A=f$，或 $T_A=f$ 至少有一个成立，但 $T_A\neq f$，只可能 $\psi_A=f$. 因此对任意 A 均有 $\varphi_A=\psi_A$. 再由引理 3 第(4)项得 $\models\varphi\leftrightarrow\psi$.

定理 1 如果 $\varphi\models\psi$，那么(1)$\models\psi$，或者(2)$\models\sim\varphi$，或者(3)可以找到一个式子 θ 满足 $\varphi\models\theta,\theta\models\psi$，并且 θ 中每一个命题记号均同时出现于 φ,ψ.

证 设 φ,ψ 不满足(1)(2). 将 ψ 展开为布氏化简型(如引理 10 所说的)：

$$\psi'\equiv\bigwedge_{0\leqslant i<\alpha}\ \bigvee_{0\leqslant j<\beta(i)}a_{ij}.$$

对于每一个 $\bigvee\limits_{0\leqslant j<\beta(i)}a_{ij}$，将其分为两部分

$$\Big(\bigvee_{0\leqslant j<\gamma(i)}b_{ij}\Big)\vee\Big(\bigvee_{0\leqslant j<\delta(i)}c_{ij}\Big),$$

其中每一个 b_{ij} 的命题记号均出现于 φ，每一个 c_{ij} 的命题记号均不出现于 φ. 我们证明

$$\varphi\models\bigvee_{0\leqslant j<\gamma(i)}b_{ij}.\qquad\text{①}$$

如果(Ⅰ)式不成立，即有一个 A，$\varphi_A=t$，而 $\Big(\bigvee\limits_{0\leqslant j<\gamma(i)}b_{ij}\Big)_A=f$. 设 φ 中全体出现的命题记号组成集为 $\Delta\subseteq\sum$，可以取 $A\subseteq\Delta$(否则考虑 $A'=A\cap\Delta$，在 A' 上仍有 $\varphi_{A'}=t$，$\Big(\bigvee\limits_{0\leqslant j<\gamma(i)}b_{ij}\Big)_{A'}=f$). 因为各 c_{ij} 的命题记号均不出现于 Δ，也就是不出现于 A. 又不会同时出现某 s 与 $\sim s$. 如果 s 出现为 c_{ij}，我们令 $s\notin B$，如果 $\sim s$ 出现为 c_{ij}，我们令 $s\in B$. (B 由出现有 $\sim s\equiv c_{ij}$ 的全体 s 组成)令 $C=A\cup B$. 因为新加入的 B 的元不出现于 φ，所以有 $\varphi_C=\varphi_A=t$. 而

$$\Big[\Big(\bigvee_{0\leqslant j<\gamma(i)}b_{ij}\Big)\vee\Big(\bigvee_{0\leqslant j<\delta(i)}c_{ij}\Big)\Big]_C=f.$$

(例如 $\vee c_{ij}=(s_0\vee s_1\vee\cdots\vee\sim s_2\vee\sim s_3\vee\cdots)$，那么 $B=\{s_2,s_3,\cdots\}$. A 中各命题记号不出现于 c_{ij}，所以

$$(\vee c_{ij})_C=(\vee c_{ij})_{A\cup B}=(\vee c_{ij})_B=f).$$

与 $\varphi \models \bigvee_{0\leqslant j<\beta(i)} a_{ij}$ 矛盾. 因此①成立.

对每一个 $0\leqslant i<\alpha$,取出上述的 $\bigvee_{0\leqslant j<\gamma(i)} b_{ij}$ 作成 $\theta=\bigwedge_{0\leqslant i<\alpha}\bigvee_{0\leqslant j<\gamma(i)} b_{ij}$.

我们证明 $\varphi\models\theta$, $\varphi\models\psi$.

(1) $\varphi\models\theta$ 的证明:由①及引理 5

$$\varphi\models\bigwedge_{0\leqslant i<\alpha}\bigvee_{0\leqslant j<\gamma(i)} b_{ij}\text{,即 }\varphi\models\theta.$$

(2) $\theta\models\psi$ 的证明:(由 $a\models a\vee b$:)

$$\Big(\bigvee_{0\leqslant j<\gamma(i)} b_{ij}\Big)\models\Big(\bigvee_{0\leqslant j<\gamma(i)} b_{ij}\Big)\vee\Big(\bigvee_{0\leqslant j<\delta(i)} c_{ij}\Big).$$

(再由 $a\models b$ 能推出 $a\wedge c\models b$:)

$$\bigwedge_{0\leqslant i<\alpha}\bigvee_{0\leqslant j<\gamma(i)} b_{ij}\models\Big(\bigvee_{0\leqslant j<\beta(i)} a_{ij}\Big).$$

再由引理 5

$$\Big(\bigwedge_{0\leqslant i<\alpha}\bigvee_{0\leqslant j<\gamma(i)} b_{ij}\Big)\models\Big(\bigwedge_{0\leqslant i<\alpha}\bigvee_{0\leqslant j<\beta(i)} a_{ij}\Big).$$

这就是 $\theta\models\psi'$. 但由引理 10 $\models\psi'\rightarrow\psi$. 故 $\theta\models\psi$. 而 θ 中出现的命题记号均同时出现于 φ 和 ψ.(定理 1 证完)

参考文献

[1] Fridman H. One hundred and two problems in mathematical logic. The journal of symbolic logic, 1975, 40(2): 113－129.

[2] Fridman H. Uncountable propositional logic. Stanford University. 1968 (unpublished).

[3] Chang C C and Keisler H J. Model Theory. North-Holland Publishing Company, Amsterdam, 1973.

数学学报,1980,23(4):500～505

关于代数系统自同构群的一个问题①

A Problem on the Group of Automorphisms on an Algebraic System

在 Birkhoff 的《格论》第 2 版(1948)中有一个未解决问题如下(见于该书第 ix 页,代数前言,习题 10(b)):

对于一切正整数 n,求定出最小可能的 $f(n)$,使得当任一有限群 G 的元数不超过 n 时,恒存在元数不超过 $f(n)$ 的代数系统 A 以 G 为自同构群.

作者曾与王世强共同考虑此问题,在[1]中王给出了两个较弱的类似问题的解答. 本文利用[1]的结果进一步给出了上述 Birkhoff 原问题的解答.(在[1]的末尾,王曾提到作者本文的结果. 当时作者未将证明整理成文. 近因见 Math. Reviews, Vol. 35,(1968),5381 的评论注意到此事并感到困惑,故将证明整理出来. 这个证明比 1963 年时已有所简化.)

下面的引理 1 是对上述问题的解答. 而引理 2 是证明引理 1 所需要的群论引理之一. 这个引理对群论自身看来也是有意义的.(从内容的基本性来看,这个引理也可能不是新的结果,但作者未找到出处.)

引理 1 G 为一 n 元群,如果 G 同构于对称群 S_m(m 不一定等于 n)的一个子群 H(下面有时也说 G 可以嵌入于 S_m),那么有一个 m 元代数系统 A,A 以 H 为自同构群.

证 和[1]引理 1 类似.

引理 2 N 是 G 的正常子群,N 有 n 元,G 有 g 元,$l=g/n$. 如果 N

① 收稿日期:1977-12-22.

可以嵌入于 $S_m(m<n)$ 中，那么 G 可以嵌入于 $S_{ml}(ml<n)$ 中.

在下面的证明中几次用到[3]中，p. 155 定理Ⅺ，现抄译如下：

"G 是一个 g 元有限群. H 是 G 的子群（H 有 h 元，$g/h=m$），那么 G 有一个$(k,1)$同态，其同态象是（S_m 的一个子群）m 级可迁置换群. 其中 k 是 H 中某个 G 的正常子群的元数. 当 H 不包含 G 的一元以上正常子群时，同态实为同构."

引理 2 的证明 按[3]上抄定理的证明办法任意取定 N 的一组右陪集代表元 $\tau_1,\tau_2,\cdots,\tau_l$.（但现在题设 N 是 G 的正常子群，故它们同时也是左陪集代表元.）由[3]p. 155 知以 $\tau_1,\tau_2,\cdots,\tau_l$ 为置换对象可构造一个 l 级置换群 H，并且该处给出了一个 G 到 H 的同态对应. 由 N 的正常性可证该同态对应的核即为 N 自身.

现在形式地写出 ml 个记号，作为以下进行置换的对象.

$$1\tau_1,\cdots,m\tau_1,\cdots,1\tau_l,\cdots,m\tau_l.$$

按本引理的题设将 N 嵌入于 S_m 中，即每一个 N 的元均看作一个 m 级置换在 $1,2,\cdots,m$ 中进行.

规定如下三个函数 ϕ,ψ,χ.

$$\tau_i\tau_j=\phi(i,j)\tau_{\chi(i,j)},$$

其中

$$\phi(i,j)\varepsilon N,1\leqslant i,j\leqslant l,1\leqslant\chi(i,j)\leqslant l.$$

$$\tau_i a=\psi(i,a)\tau_i,$$

其中

$$a\varepsilon N,\psi(i,a)\varepsilon N,\quad 1\leqslant i\leqslant l.$$

G 中每一个元都可唯一地表示为 $a\tau_i(a\varepsilon N,1\leqslant i\leqslant l)$ 形状，将其映射到下列的置换：（以 * 表示由 G 到 S_{ml} 的一个对应.）

$$G\ni a\tau_i\xrightarrow{*}(a\tau_i)^*=:$$

$$\begin{pmatrix} 1\tau_1 & 2\tau_1 & \cdots & m\tau_1 & 1\tau_2 & 2\tau_2 & \cdots & m\tau_2 & \cdots \\ x_{11}\tau_{y_1} & x_{12}\tau_{y_1} & \cdots & x_{1m}\tau_{y_1} & x_{21}\tau_{y_2} & x_{22}\tau_{y_2} & \cdots & x_{2m}\tau_{y_2} & \cdots \end{pmatrix}. \text{①}$$

其中

$$\begin{cases} x_1 = \begin{pmatrix} 1 & 2 & \cdots & m \\ x_{11} & x_{12} & \cdots & x_{1m} \end{pmatrix} =: \psi(1,a)\phi(1,i), \\ x_2 = \begin{pmatrix} 1 & 2 & \cdots & m \\ x_{21} & x_{22} & \cdots & x_{2m} \end{pmatrix} =: \psi(2,a)\phi(2,i), \cdots \\ y_1 =: \chi(1,i), y_2 =: \chi(2,i), \cdots \end{cases}$$

由①，对任意 $b\tau_j$ 应有如下两种置换表示式：

$$G \ni b\tau_j \xrightarrow{*} (b\tau_j)^* =: \begin{pmatrix} 1\tau_1 & \cdots & m\tau_1 & 1\tau_2 & \cdots & m\tau_2 & \cdots \\ u'_{11}\tau_{v'_1} & \cdots & u'_{1m}\tau_{v'_1} & u'_{21}\tau_{v'_2} & \cdots & u'_{2m}\tau_{v'_2} & \cdots \end{pmatrix}$$

$$= \begin{pmatrix} x_{11}\tau_{y_1} & \cdots & x_{1m}\tau_{y_1} & x_{21}\tau_{y_2} & \cdots & x_{2m}\tau_{y_2} & \cdots \\ u_{11}\tau_{v_1} & \cdots & u_{1m}\tau_{v_1} & u_{21}\tau_{v_2} & \cdots & u_{2m}\tau_{v_2} & \cdots \end{pmatrix}.$$

前一个置换所满足的条件为

$$\begin{cases} \begin{pmatrix} 1 & 2 & \cdots & m \\ u'_{11} & u'_{12} & \cdots & u'_{1m} \end{pmatrix} = u'_1 = \psi(1,b)\phi(1,j), \\ \begin{pmatrix} 1 & 2 & \cdots & m \\ u'_{21} & u'_{22} & \cdots & u'_{2m} \end{pmatrix} = u'_2 = \psi(2,b)\phi(2,j), \cdots, \\ v'_1 = \chi(1,j), v'_2 = \chi(2,j), \cdots \end{cases}$$

后一个置换所满足的条件为

$$\begin{cases} \begin{pmatrix} x_{11} & x_{12} & \cdots & x_{1m} \\ u_{11} & u_{12} & \cdots & u_{1m} \end{pmatrix} = u'_{y_1} = \begin{pmatrix} 1 & 2 & \cdots & m \\ u'_{y_1 1} & u'_{y_1 2} & \cdots & u'_{y_1 m} \end{pmatrix}, \\ v_1 = v'_{y_1} = \chi(y_1, j); \\ \begin{pmatrix} x_{21} & x_{22} & \cdots & x_{2m} \\ u_{21} & u_{22} & \cdots & u_{2m} \end{pmatrix} = u'_{y_2} = \begin{pmatrix} 1 & 2 & \cdots & m \\ u'_{y_2 1} & u'_{y_2 2} & \cdots & u'_{y_2 m} \end{pmatrix}, \\ v_2 = v'_{y_2} = \chi(y_2, j); \cdots \end{cases}$$

$$\therefore \quad u_1 =: \begin{pmatrix} 1 & 2 & \cdots & m \\ u_{11} & u_{12} & \cdots & u_{1m} \end{pmatrix} = x_1 u'_{y_1} = \psi(1,a)\phi(1,i)\psi(y_1,b)\phi(y_1,j)$$

$$= \psi(1,a)\phi(1,i)\psi(\chi(1,i),b)\phi(\chi(1,i),j). \qquad ②$$

$$u_2 =: \begin{pmatrix} 1 & 2 & \cdots & m \\ u_{21} & u_{22} & \cdots & u_{2m} \end{pmatrix} = x_2 u'_{y_2} = \psi(2,a)\phi(2,i)\psi(y_2,b)\phi(y_2,j)$$

$$= \psi(2,a)\phi(2,i)\psi(\chi(2,i),b)\phi(\chi(2,i),j). \qquad ③$$

$\cdots$

$$v_1=\chi(y_1,j)=\chi(\chi(1,i),j),\qquad ④$$

$$v_2=\chi(y_2,j)=\chi(\chi(2,i),j),\qquad ⑤$$

$$\cdots$$

由二置换乘积的定义：

$$\therefore\ (a\tau_i)^*\cdot(b\tau_j)^*=\begin{pmatrix}1\tau_1 & \cdots & m\tau_1 & 1\tau_2 & \cdots & m\tau_2 & \cdots\\ u_{11}\tau_{v_1} & \cdots & u_{1m}\tau_{v_1} & u_{21}\tau_{v_2} & \cdots & u_{2m}\tau_{v_2} & \cdots\end{pmatrix}.\qquad ⑥$$

又由 ϕ,ψ 和 χ 的定义：

$$a\tau_i b\tau_j=a(\tau_i b)\tau_j=a\psi(i,b)\tau_i\tau_j=a\psi(i,b)\phi(i,j)\tau_{\chi(i,j)}=c\tau_k.$$

则

$$((a\tau_i)\cdot(b\tau_j))^*=(c\tau_k)^*=\begin{pmatrix}1\tau_1 & \cdots & m\tau_1 & 1\tau_2 & \cdots & m\tau_2 & \cdots\\ r_{11}\tau_{s_1} & \cdots & r_{1m}\tau_{s_1} & r_{21}\tau_{s_2} & \cdots & r_{2m}\tau_{s_2} & \cdots\end{pmatrix}.\qquad ⑦$$

其中

$$\begin{cases} r_1=\begin{pmatrix}1 & 2 & \cdots & m\\ r_{11} & r_{12} & \cdots & r_{1m}\end{pmatrix}=\psi(1,c)\phi(1,k)\\ \quad =\psi(1,a\psi(i,b)\phi(i,j))\phi(1,\chi(i,j)). & ⑧\\ r_2=\begin{pmatrix}1 & 2 & \cdots & m\\ r_{21} & r_{22} & \cdots & r_{2m}\end{pmatrix}=\psi(2,c)\phi(2,k)\\ \quad =\psi(2,a\psi(i,b)\phi(i,j))\phi(2,\chi(i,j)). & ⑨\\ \quad\cdots \end{cases}$$

$$\begin{cases} s_1=\chi(1,k)=\chi(1,\chi(i,j)). & ⑩\\ s_2=\chi(2,k)=\chi(2,\chi(i,j)). & ⑪\\ \quad\cdots \end{cases}$$

$$\begin{aligned}&\tau_1 a\tau_i b\tau_j\\ =&(\tau_1 a)\tau_i b\tau_j\\ =&\psi(1,a)(\tau_1\tau_i)b\tau_j\quad(\text{由 }\phi,\psi,\chi\text{ 的定义，下同})\\ =&\psi(1,a)\phi(1,i)(\tau_{\chi(1,i)}b)\tau_j\\ =&\psi(1,a)\phi(1,i)\psi(\chi(1,i),b)\tau_{\chi(1,i)}\tau_j\end{aligned}$$

$$=\psi(1,a)\phi(1,i)\psi(\chi(1,i),b)\phi(\chi(1,i),j)\tau_{\chi(\chi(1,i),j)}$$
$$=u_1\tau_{v_1}(由②④)(u_1\varepsilon N)$$
$$=\tau_1 a(\tau_i b)\tau_j(从第一个式子用结合律)$$
$$=\tau_1 a\psi(i,b)\tau_i\tau_j$$
$$=\tau_1(a\psi(i,b)\phi(i,j))\tau_{\chi(i,j)}=\psi(1,a\psi(i,b)\phi(i,j))\tau_1\tau_{\chi(i,j)}$$
$$=\psi(1,a\psi(i,b)\phi(i,j))\phi(1,\chi(i,j))\tau_{\chi(1,\chi(i,j))}$$
$$=r_1\tau_{s_1}(由⑧⑩)(r_1\varepsilon N).$$

所以

$$u_1=r_1,\quad v_1=s_1. \qquad ⑫$$

同理,通过计算 $\tau_2 a\tau_i b\tau_j,\tau_3 a\tau_i b\tau_j,\cdots,\tau_l a\tau_i b\tau_j$ 可得

$$u_2=r_2,\quad v_2=s_2. \qquad ⑬$$
$$u_3=r_3,\quad v_3=s_3. \qquad ⑭$$
$$\cdots$$
$$u_l=r_l,\quad v_l=s_l. \qquad ⑮$$

所以⑥与⑦右端相等,从而知对应“ * ”保持乘法运算.

下证对于 G 的不同元,所求出的代表置换也不相同. 设 $a\tau_i\neq b\tau_j$.

(1)当 $i\neq j$ 时,由[3]中证明 G 与 H 的同态过程知各 $\tau_{\chi(k,i)}$ 不全同于各 $\tau_{\chi(k,j)}(k=1,2,\cdots,l)$.(因否则在所说的同态中易见 G 的两个不同元 τ_i,τ_j 将对应于 H 中的同一个置换

$$\begin{pmatrix}\tau_1 & \tau_2 & \cdots & \tau_l\\ \tau_{\chi(1,i)} & \tau_{\chi(2,i)} & \cdots & \tau_{\chi(l,i)}\end{pmatrix},$$

因而 $\tau_i\tau_j^{-1}$ 应在同态对应的核 N 中,得到了 $i=j$ 与上面矛盾.)故由①所规定的置换也不相等.

(2)当 $i=j$ 时,$a\neq b$. 故

$$\psi(1,a)\neq\psi(1,b).$$

(如 $\psi(1,a)=\psi(1,b)$ 则 $\tau_1 a=\psi(1,a)\tau_1=\psi(1,b)\tau_1=\tau_1 b$,从而推得 $a=b$,与上面矛盾.)故

$$\psi(1,a)\phi(1,i)\neq\psi(1,b)\phi(1,j),$$
$$y_1=\chi(1,i)=\chi(1,j).\quad(因\ i=j)$$

由此二式可知由 $a\tau_i$ 与 $b\tau_j$ 依①所对应的两个置换也不相等. 这样我们证明了①所规定的对应是由 G 到 S_{ml} 内的一个同构对应,即 G 可以嵌

入于 S_{ml} 之内.

引理 3 p 为一素数,G 中有一个 p 元子群 H 不是正常子群,则 G 可以嵌入于 S_l 之内,这里 $l=g/p$. g 为 G 的元素个数.

证 由前面抄译的[3]中 p. 155 定理Ⅺ知 G 与一个 l 级置换群 H 同态,而 H 不是 G 的正常子群,也不可能包含 G 的一个一元以上的正常子群,故同态实为同构.

引理 4 设群 G 的任何素元子群均为正常子群,$p,q,r,\cdots$ 为 G 的元数 n 的不同素因子,取 $a,b,c,\cdots$ 各生成 p 元,q 元,r 元,…的子群(由[3], p. 58,定理ⅩⅢ及 p. 68,定理ⅩⅥ知可以找到这样的 $a,b,c,\cdots$.),则$\{a,b,c,\cdots\}$生成 G 的一个可换正常子群.(不一定是在中心之内.)

证 由[3]中 p. 51 定理Ⅺ,两个除单位元以外无其他公共元的正常子群,它们的元之间有乘法可换性,故各 $a^i,b^j,c^k,\cdots$ 互相可换,即生成可换子群,其正常性也易见.

引理 5 设 $n=p_1^{r_1}p_2^{r_2}\cdots p_k^{r_k}(k\geqslant 2)$ 是 n 的欧几里得分解,令

$$l=(p_1+p_2+\cdots+p_k)p_1^{r_1-1}p_2^{r_2-1}\cdots p_k^{r_k-1}.$$

则存在素数幂 p^m 满足 $l\leqslant p^m\leqslant n$.

证 不妨设 $p_1<p_2<\cdots<p_k(k\geqslant 2)$. 容易看出

$$\frac{l}{n}=\frac{p_1+p_2+\cdots+p_k}{p_1\cdot p_2\cdot\cdots\cdot p_k}\leqslant\frac{p_1+p_2+\cdots+p_{k-1}}{p_1\cdot p_2\cdot\cdots\cdot p_{k-1}}\leqslant\cdots$$

$$\leqslant\frac{p_1+p_2}{p_1p_2}=\frac{1}{p_1}+\frac{1}{p_2}\leqslant\frac{1}{2}+\frac{1}{3}=\frac{5}{6}.$$

所以有 $\frac{6}{5}l\leqslant n$.

(i)当 $l\geqslant 25$ 时,根据 Nagura 在[4]中的结果,在 l 与 $\frac{6}{5}l$ 之间恒存在一个素数,所以本引理成立.

(ii)当整数 $l=2,3,4,\cdots,24$ 时,由验证可知虽然在闭区间 $\left[l,\frac{6}{5}l\right]$ 中不一定存在素数,但素数幂却总是存在的,所以本引理的结论仍然成立.

引理 6 对于元数为 p^m(p 素数)的循环群而言,以循环群 C 为自同构群的代数系统其最小可能的元素为 $k=p^m$ 自己.

证 见于[1].

定理 1 $f(n)=$不超过 n 的素数幂中的最大者.

证 (1)若 $n=p^m$(p 素),任取一个元素个数不超过 n 的群 G,设其元数为 n_1,由群嵌入对称群的基本定理(见于[3],p. 55,定理Ⅻ)知 G 可以嵌入于 S_{n_1} 之中,而由引理 1 可知存在元数为 $n_1(\leqslant n)$ 的代数系统以 G 为自同构群. 由此可见 $f(n)\leqslant n$. 但由引理 6 又可知 $f(n)\geqslant n$. 故 $f(n)=n$. 并且 n 本身就是不超过 n 的最大素数幂.

(2)若 $n=p_1^{r_1}p_2^{r_2}\cdots p_k^{r_k}$($p_1,p_2,\cdots,p_k$ 为互异素数),$k\geqslant 2$,各 $r_i\geqslant 1$,为 n 的欧几里得分解. 设不超过 n 的最大素数幂为 p^m. 由引理 5 知

$$l=(p_1+p_2+\cdots+p_k)p_1^{r_1-1}p_2^{r_2-1}\cdots p_k^{r_k-1}\leqslant p^m<n.$$

(Ⅰ)对于任意一个 n 元群 G.

(i)如果 G 有一个 p_i 元子群不是正常子群,由引理 3 知 G 可以嵌入于 S_t 之中,$t=n/p_i$. $p_i\geqslant 2$. 由数论中已证明 Bertrand 假设(参看 Hardy and Wright《The Theory of Numbers》第三版(1954),p. 343,定理 418)可知在 t 及 n 之间有素数存在,设其为 p',$t\leqslant p'\leqslant n$. 由此易见 G 可以嵌入于

$$S_t\subseteq S_{p'}\subseteq S_{p^m}$$

之中.(可以有一些被置换的符号不动.)

(ii)如果 G 的每一个 p_i 元子群都是正常子群,由引理 4 知 G 有一个可换正常子群 N,N 的元素个数是 $p_1p_2\cdots p_k$. 由[1]引理 5 知 N 可以嵌入于 $S_{p_1+p_2+\cdots+p_k}$ 之中. 故由本文引理 2 知 G 可以嵌入于 S_l,此处 l 如引理 5 所示,并且由引理 5 可知满足 $l\leqslant p^m\leqslant n$. 故 G 可以嵌入于 S_{p^m}.

(Ⅱ)对于元数 n_1 适合 $p^m<n_1<n$ 的任一群 G_1,由类似的讨论知 G_1 可以嵌入于 S_{p^m} 之中:

将 n_1 作欧几里得分解 $n_1=q_1^{s_1}q_2^{s_2}\cdots q_u^{s_u}$,$s_i\geqslant 1$. 由 $n_1>p^m$,知 $u\geqslant 2$. 与(i)(ii)类似地讨论可知:

(i)如果 G_1 有一个 q_j 元的子群不是 G_1 的正常子群,那么 G_1 可以嵌入于 S_{p^m} 之中.

(ii)如果 G_1 的每一个 q_j 元的子群都是 G_1 的正常子群($j=1,2,\cdots,u$),那么 G_1 可以嵌入于 S_{l_1} 之中,此处

$$l_1=(q_1+q_2+\cdots+q_u)q_1^{s_1-1}q_2^{s_2-1}\cdots q_u^{s_u-1}.$$

由引理 5 的证明可知 $l_1\leqslant\frac{5}{6}n_1$. 从而

$$l_1 \leqslant \frac{5}{6} n_1 \leqslant \frac{5}{6} n \leqslant p^m.$$

故 G_1 可以嵌入于 S_{p^m} 之中.

(Ⅲ)对于元数 $n_1 \leqslant p^m$ 的任一群 G_1 显见也能嵌入于 S_{p^m} 之中.

由(Ⅰ)～(Ⅲ)及引理 1 可知

$$f(n) \leqslant p^m.$$

又由引理 6 可知

$$f(n) \geqslant p^m.$$

故得

$$f(n) = p^m. \quad \text{(定理 1 证完)}$$

参考文献

[1] 王世强. 关于代数系统自同构群的一个注记. 数学进展，1964，7(2)：213—218.

[2] Birkhoff G. Lattice Theory，2nd-printing. New York，1948.

[3] Carmicheal RD. Introduction to the Theory of Groups of Finite Order. Boston，Ginn and Co. ，1937.

[4] Nagura，J. On the Interval Containing at Least One Prime Number. Proc. Japan Acad. ，1952，28(4)：177—181(文摘见于 Math. Reviews，1953，14(4)：355).

北京师范大学学报(自然科学版),1980,(3－4):31～39

模型的并、积与齐次模型①

The Union, Product of Models and Homogeneous Models

这是我们研究齐次模型的第二篇文章. 在第一篇文章里我们证明了:(1)任意可数理论 T 的可数模齐次型数 $h_T(\aleph_0)\leqslant\aleph_0$,或者 $h_T(\aleph_0)=2^{\aleph_0}$(不用 GCH). (2)如果可数完备理论 T 的每一个可数模型都是齐次的,那么 $h_T(\aleph_0)\neq n$,$(1<n<\omega)$. 对非齐次的理论 T,[2]习题 5.1.16 提出了如下的未解决问题:

对什么样有限数 n,存在一个可数的理论 T_n,使得对任意无穷基数 α,T_n 恰好有 n 个数为 α 的齐次模型?

原来在[1]中说已经正面解决了这个问题,但是容易验证其所给出的例子是错误的,在[2]的上述习题中说明了该问题对 $n=1,2$ 和 3 是有解的. 我们通过对模型的并与积的研究,将这个问题有解的情况推进到 $n=2^r3^s$,从而使该问题中 n 的有解数达到无穷多个.

本文采用[2]的记号和述语,为了节省篇幅,我们略去了一部分预理和定理的证明,对此,下面我们不再一一注出.

§ 1. 模型的并

我们给出一个方法来将任意两个不含有函数记号的模型并成一个模型,所用到的一个式子关于一个一元关系的相对化式子请参阅[3],本节和下一节所用到的语言和理论都不一定是可数的. 本节的预理和定理的

① 收稿日期:1980-01-23.

证明较之下一节要简单一些，我们把它们都略去了. 有兴趣的读者可以照下一节将其补出.

定义 设 L,L' 都是不含有函数记号的语言. $L\cap L'=\varnothing$, T,T' 各是 L,L' 内的理论，令 $\overline{L}=L\cup L'\cup\{P,P'\}$ (P,P' 是两个新的一元关系). $\overline{L}$ 内的理论 $\overline{T}$ 的公理是以下 8 条所规定的句子：

(1) T^P.

(2) $T'^{P'}$.

(3) $P(x)\vee P'(x)$.

(4) $\neg(P(x)\wedge P'(x))$.

(5) $P(c)$, $(c\varepsilon L)$.

(6) $P'(c')$, $(c'\varepsilon L')$.

R,R' 是关系记号.

(7) $R(x_1x_2\cdots x_m)\rightarrow P(x_1)\wedge P(x_2)\wedge\cdots\wedge P(x_m)$, $(R\varepsilon L)$.

(8) $R'(y_1y_2\cdots y_n)\rightarrow P(y_1)\wedge P(y_2)\wedge\cdots\wedge P'(y_n)$, $(R'\varepsilon L')$.

我们说理论 $\overline{T}$ 是 T 和 T' 的并，记作 $\overline{T}=T\vee T'$.

定义 设 $\overline{T}=T\vee T'$. 如果 $\overline{\mathscr{U}}$, $\mathscr{U}$ 和 $\mathscr{U}'$ 各是 $\overline{T}$, T 和 T' 的模型，并且分别局限于对 L,L' 的关系，常数的解释时，$\overline{\mathscr{U}}$ 中满足 P 的全体元素与 $\mathscr{U}$ 同构；$\overline{\mathscr{U}}$ 中满足 P' 的全体元素与 $\mathscr{U}'$ 同构，那么 $\overline{\mathscr{U}}$ 叫作 $\mathscr{U}$ 与 $\mathscr{U}'$ 的并模型，记作 $\overline{\mathscr{U}}=\mathscr{U}\vee\mathscr{U}'$.

预理 1.1 设 $\overline{\mathscr{U}}=\mathscr{U}\vee\mathscr{U}'$, $\overline{\mathscr{B}}=\mathscr{B}\vee\mathscr{B}'$, 那么 $\overline{\mathscr{U}}\cong\overline{\mathscr{B}}$，当且仅当，$\mathscr{U}\cong\mathscr{B}$ 并且 $\mathscr{U}'\cong\mathscr{B}'$.

定义 L 的理论 T 叫作满足对于基本式子集 B 的量词消去性质，当且仅当，L 的任意式子都 T 等价于 B 中式子的布氏组合，

预理 1.2 设 T,T' 各是 L,L' 的理论，$\overline{T}=T\vee T'$, B 是以下 3 条所规定的 $\overline{L}$ 的式子集：

(1) $P(x)$, $P'(y)$.

(2) $P(x_1)\wedge P(x_2)\wedge\cdots\wedge P(x_m)\wedge\varphi^P(x_1x_2\cdots x_m)$, $(\varphi\varepsilon L)$.

(3) $P'(y_1)\wedge P'(y_2)\wedge\cdots\wedge P'(y_n)\wedge\psi^{P'}(y_1y_2\cdots y_n)$, $(\psi\varepsilon L')$.

(其中 φ,ψ 各是只含 L,L' 的符号所构成的式子，并且不含有除括号所记的变数之外的其他变数.) 那么 $\overline{T}$ 对基本式子集 B 有量词消去性质.

定理 1.3 假设 $\overline{\mathscr{U}}=\mathscr{U}\vee\mathscr{U}'$.

(1) $\overline{\mathscr{U}}$ 是 α 齐次模型，当且仅当 $\mathscr{U}$，$\mathscr{U}'$ 都是 α 齐次模型.

(2) $|A|=\alpha$. $\overline{\mathscr{U}}$ 是齐次模型，当且仅当 $\mathscr{U}$，$\mathscr{U}'$ 都是 α 齐次模型，$|A|$ 和 $|A'|$ 都 $\leqslant\alpha$，并且其中至少有一个 $=\alpha$.

§2. 模型的积

模型的积研究得较多的是直积，超积，这些都是同一种语言的模型的积. 这一节里，我们要引进一种完全不同的积，它可以适用于语言不同的只含关系记号的两个模型. 首先我们给出一个式子 φ 对一个二元等价关系 $\sim$（或 $\wr$）的相对化式子 $\varphi^{\sim}$（或 $\varphi^{\wr}$）的定义，然后研究积模型的性质.

定义 (1) 在 φ 的子式子 $(\exists x)[\cdots\cdots]$ 中量词 $(\exists x)$ 的作用范围内，如果另外还有对这个子式子来说是自由的变数，那么在 $(\exists x)$ 之后，方括号之前，加上关于二元等价关系 $\sim$（或 $\wr$ 与之类似），使其变为

$$(\exists x)((\bigwedge_{1\leqslant i\leqslant m} x\sim x_i)\wedge[\cdots\cdots]).$$

(2) 类似地当子式子是 $(\forall x)[\cdots\cdots]$ 时变为

$$(\forall x)((\bigwedge_{1\leqslant i\leqslant m} x\sim x_i)\rightarrow[\cdots\cdots]).$$

(3) 对式子 φ 的每一个含有量词的子式子都作了上述添加之后，所得的结果叫作 φ 对 $\sim$ 的相对化式子，记作 $\varphi^{\sim}$（类似地 $\varphi^{\wr}$）.

定义 设 L，L' 是两个只含关系记号的语言，$L\cap L'=\varnothing$. T，T' 各是 L，L' 的理论. 设 $\overline{L}=l\cup L'\cup\{\sim,\wr\}$，$L$ 的理论 $\overline{T}$ 由以下 6 类公理及其所推出的式子组成.

(1) $T^{\sim}$，$(T')^{\wr}$.

(2) $\sim$，$\wr$ 是等价关系.

(3) 对 L 的 m 元关系 R 和 L' 的 n 元关系 R'.

$$R(x_1x_2\cdots x_m)\rightarrow(\bigwedge_{1\leqslant i,j\leqslant m} x_i\sim x_j),$$

$$R'(y_1y_2\cdots y_n)\rightarrow(\bigwedge_{1\leqslant i,j\leqslant n} y_i\ \wr\ y_j).$$

(4) $R\varepsilon L$，$R'\varepsilon L'$ 如 (3).

$$R(x_1x_2\cdots x_m)\wedge(\bigwedge_{1\leqslant i\leqslant m} x_i\ \wr\ y_i)\wedge(\bigwedge_{1\leqslant i,j\leqslant m} y_i\sim y_j)\rightarrow R(y_1y_2\cdots y_m),$$

$$R'(x_1x_2\cdots x_n)\wedge(\bigwedge_{1\leqslant i\leqslant n} x_i\sim z_i)\wedge(\bigwedge_{1\leqslant i,j\leqslant n} z_i\ \wr\ z_j)\rightarrow R'(z_1z_2\cdots z_n).$$

(5) $(\forall xy)(\exists z)(x\sim z\wedge y\ \wr\ z)$.

(6) $x \sim y \wedge x \wr y \rightarrow x \equiv y$.

我们把 $\overline{T}$ 叫作是 T 和 T' 的积,记作 $\overline{T}=T \cdot T'$.

定义 如果 $\overline{\mathscr{U}}$, $\mathscr{U}$ 和 $\mathscr{U}'$ 各是 $\overline{T}$, T 和 T' 的模型,并且 $\overline{\mathscr{U}}$ 中有一个 $\sim$ 等价类在 L 的关系解释之下与 $\mathscr{U}$ 同构, $\overline{\mathscr{U}}$ 中有一个 $\wr$ 等价类在 L' 的关系解释之下与 $\mathscr{U}'$ 同构,那么 $\overline{\mathscr{U}}$ 叫作是 $\mathscr{U}$ 和 $\mathscr{U}'$ 的积模型,记作 $\overline{\mathscr{U}}=\mathscr{U} \cdot \mathscr{U}'$.

预理 2.1 设 $\overline{\mathscr{U}}=\mathscr{U} \cdot \mathscr{U}'$, $\overline{\mathscr{B}}=\mathscr{B} \cdot \mathscr{B}'$. $\overline{\mathscr{U}} \cong \overline{\mathscr{B}}$,当且仅当, $\mathscr{U} \cong \mathscr{B}$ 并且 $\mathscr{U}' \cong \mathscr{B}'$.

预理 2.2 设 $\overline{L}$ 的式子集 B 是由以下二类式子所组成

(1)
$$(\exists y_1 y_2 \cdots y_m)[(\bigwedge_{1 \leqslant i \leqslant m} x_i \wr y_i) \wedge (\bigwedge_{1 \leqslant i, j \leqslant m} y_i \sim y_j) \wedge \varphi^{\sim}(y_1 y_2 \cdots y_m)],$$

其中 φ 是一个 L 式子

(2)
$$(\exists z_1 z_2 \cdots z_m)[(\bigwedge_{1 \leqslant i \leqslant n} x_i \sim z_i) \wedge (\bigwedge_{1 \leqslant i, j \leqslant n} z_i \wr z_j) \wedge \psi^{\sim}(z_1 z_2 \cdots z_n)],$$

其中 ψ 是一个 L' 的式子.

那么 $\overline{T}$ 对基本式子集 B 有量词消去性质.

证 (1)每一个 $\sim$ 等价类的元素叫作是在同一条横线上的点集. 每一个 $\wr$ 等价类叫作是在同一条竖线上的点集.

(2) $l^{\sim}$ 是一条横线, $a \varepsilon \overline{A}$. 在 $l^{\sim}$ 内存在唯一的一个点 b 使得 $b \wr a$, b 叫作 a 在 $l^{\sim}$ 内的射影. 类似地定义点 a 到竖线 $l^{\sim}$ 内的射影.

(3)射影是横线间(竖线间)对语言 L(对 L')的同构对应.

(4)元子式子可以用基本式子来表示. 例如

$$x \sim y \leftrightarrow (\exists uv)[(u \sim x \wedge v \sim y) \wedge (u \wr v) \wedge (u \equiv v)],$$

$$R(x_1 x_2 \cdots x_n) \leftrightarrow (\bigwedge_{1 \leqslant i, j \leqslant n} x_i \sim x_j) \wedge (\exists y_1 y_2 \cdots y_n) [(\bigwedge_{1 \leqslant i \leqslant n} x_i \wr y_i) \wedge (\bigwedge_{1 \leqslant i, j \leqslant n} y_i \sim y_j) \wedge R^{\sim}(y_1 y_2 \cdots y_n)].$$

(5)基本式子的否定式可以化为基本式子.

$$\neg (\exists y_1 y_2 \cdots y_n)[(\bigwedge_{1 \leqslant i \leqslant n} x_i \wr y_i) \wedge (\bigwedge_{1 \leqslant i, j \leqslant n} y_i \sim y_j) \wedge \varphi^{\sim}(y_1 y_2 \cdots y_n)]$$
$$\leftrightarrow (\exists y_1 y_2 \cdots y_n)[(\bigwedge_{1 \leqslant i \leqslant n} x_i \wr y_i) \wedge (\bigwedge_{1 \leqslant i, j \leqslant n} y_i \sim y_j) \wedge \neg \varphi^{\sim}(y_1 y_2 \cdots y_n)].$$

(6)由[2], §1.5 知只需考虑含有 x 的基本式子的合取式如何消去前面加上的量词 $(\exists x)$. B 中含有 x 的基本式子有以下二类:

$$(6.1)\qquad (\exists y_0y_1\cdots y_m)[x \wr y_0 \wedge (\bigwedge_{1\leqslant i\leqslant m} u_i \wr y_i)\wedge (\bigwedge_{1\leqslant i,j\leqslant m} y_i\sim y_j)\wedge \varphi_1^{\sim}(y_0y_1\cdots y_m)],$$

φ_1 是一个 L 式子.

$$(6.2)\qquad (\exists z_0z_1\cdots z_m)[x\sim z_0 \wedge (\bigwedge_{1\leqslant i\leqslant n} v_i\sim z_i)\wedge (\bigwedge_{0\leqslant i,j\leqslant n} z_i \wr z_j)\wedge \varphi_2^{\wr}(z_0z_1\cdots z_n)],$$

φ_2 是一个 L'式子.

(7)两个(6.1)型式子的合取式可以合并成为一个(6.1)型式子.

(8)当 φ 只含一个(6.1)型式子时,(Ex)φ 可消去(Ex)如下:

$$(\exists x)(\exists y_0y_1\cdots y_m)[x \wr y_0 \wedge (\bigwedge_{1\leqslant i\leqslant m} u_i \wr y_i)\wedge (\bigwedge_{1\leqslant i,j\leqslant m} y_i\sim y_j)\wedge \varphi_1^{\sim}(y_0y_1\cdots y_m)]$$
$$\leftrightarrow (\exists y_1y_2\cdots y_m)[(\bigwedge_{1\leqslant i\leqslant m} u_i \wr y_i)\wedge (\bigwedge_{1\leqslant i,j\leqslant m} y_i\sim y_j)\wedge (\exists y_0)[(\bigwedge_{1\leqslant i,j\leqslant m} y_i\sim y_j)\wedge \varphi^{\sim}(y_0y_1\cdots y_m)]],$$

右端恰好是一个(6.1)型式子.

(9)当 φ 含有(6.1),(6.2)型式子各一个时可按如下办法消去($\exists x$)φ 中的($\exists x$):

$$(\exists x)[(\exists y_0y_1\cdots y_m)[x \wr y_0 \wedge (\bigwedge_{1\leqslant i\leqslant m} u_i \wr y_i)\wedge (\bigwedge_{1\leqslant i,j\leqslant m} y_i\sim y_j)\wedge \varphi_1^{\sim}(y_0y_1\cdots y_m)]\wedge$$
$$(\exists z_0z_1\cdots z_n)[x\sim z_0 \wedge (\bigwedge_{1\leqslant i\leqslant n} v_i\sim z_i)\wedge (\bigwedge_{1\leqslant i,j\leqslant n} z_i \wr z_j)\wedge \psi_2^{\wr}(z_0z_1\cdots z_n)]]$$
$$\leftrightarrow (\exists y_0y_1\cdots y_m)[(\bigwedge_{1\leqslant i\leqslant m} u_i \wr y_i)\wedge (\bigwedge_{1\leqslant i,j\leqslant m} y_i\sim y_j)\wedge (\exists y_0)[(\bigwedge_{1\leqslant i\leqslant m} y_0\sim y_i)\wedge \psi_1^{\sim}(y_0y_1\cdots y_m)]]\wedge$$
$$(\exists z_0z_1\cdots z_n)[(\bigwedge_{1\leqslant i\leqslant n} v_i\sim z_i)\wedge (\bigwedge_{1\leqslant i,j\leqslant n} z_i \wr z_j)\wedge (\exists z_0)](\bigwedge_{1\leqslant i\leqslant n} z_0 \wr y_i)\wedge \psi_2^{\wr}(z_0z_1\cdots z_n),$$

右端恰好是(6.1)(6.2)型式子各一个的合取式.

(10)其他未写出的情形均与上面类似,所以 $\overline{T}$ 对 B 的量词消去性质成立.

预理 2.3 设 $\overline{\mathscr{U}}=\mathscr{U}\cdot\mathscr{U}'$ 各是 $\overline{T}=T\cdot T'$ 的模型.不妨设 $\mathscr{U}$ 是 $\overline{\mathscr{U}}$ 的

一条横线，$\mathscr{U}'$是$\overline{\mathscr{U}}$的一条竖线. 对任意$\bar{x}$，$\bar{y}\varepsilon^{\xi}\bar{A}$，它们在$\mathscr{U}$，$\mathscr{U}'$内的射影各是$x$，$y\varepsilon^{\xi}A$，$x'$，$y'\varepsilon^{\xi}A'$. 那么以下二条等价：

(1) $$(\overline{\mathscr{U}},\bar{x}_\eta)_{\eta<\xi}\equiv_L(\overline{\mathscr{U}},\bar{y}_\eta)_{\eta<\xi},$$

(2) $$(\overline{\mathscr{U}},\bar{x}_\eta)_{\eta<\xi}\equiv_L(\overline{\mathscr{U}},\bar{y}_\eta)_{\eta<\xi}，并且$$
$$(\mathscr{U}',x'_\eta)_{\eta<\xi}\equiv_{L'}(\mathscr{U}',y'_\eta)_{\eta<\xi}.$$

定理 2.4 (1)$\overline{\mathscr{U}}=\mathscr{U}\cdot\mathscr{U}'$是$\alpha$齐次模型，当且仅当$\mathscr{U}$和$\mathscr{U}'$都是$\alpha$齐次模型.

(2)$|A|=\alpha$，$\overline{\mathscr{U}}=\mathscr{U}\cdot\mathscr{U}$. $\overline{\mathscr{U}}$是齐次模型，当且仅当$\mathscr{U}$和$\mathscr{U}'$都是α齐次模型，$|A|$和$|A'|$都$\leqslant\alpha$，并且至少有一个$=\alpha$.

预理 2.3 和定理 2.4 的证明可以利用预理 2.2 顺利地写出.

§3. α齐次稠密无端有序集

稠密无端有序集，以下简记作DLN集，是构造模型的一个工具. 特别是α齐次DLN集，可以用来构造多种齐次模型. 我们在这一节里专门研究DLA集的性质及其构造. 这一部分的内容在代数的观点上看来也是有意义的.

定义 设A是DLN集，A_1，$A_2\subseteq A$. f是A_1到A_2上的保持序关系的一一对应，那么f叫作A的一个局部同构.

定义 设A是DLN集，A叫作是α齐次的，当且仅当，对任意A的局部同构f：$A_1\to A_2$，(其中$|A_1|$，$|A_2|<\alpha$)和任意$a\varepsilon A$，存在$b\varepsilon A$使得$g=f\cup\{(ab)\}$也是A的局部同构.

定义 A是DLN集，A叫作是α不凝聚的，当且仅当A满足以下两条：

(1)对任意A的上升列

$$a_0<a_1<\cdots<a_\eta<\cdots,(\eta<\varepsilon<\alpha)$$

不存在$a\varepsilon A$使得$a=\sup_{\eta<\xi}\{a_\eta\}$.

(2)对任意A的下降列

$$b_0>b_1>\cdots>b_\eta>\cdots,(\eta<\xi<\alpha)$$

不存在$b\varepsilon A$使得$b=\inf_{\eta<\xi}\{b_\eta\}$.

定义 设A是DLN集，A叫作是α松的，当且仅当，对任意X，$Y\subseteq A$，$|X|$和$|Y|$都小于α，$X<Y$(即对任意$x\varepsilon X$，$y\varepsilon Y$都有$x<y$，下同)存在

$a\varepsilon A$ 使得 $X<a<Y$（X,Y 中可以有一个是空集.）

预理 3.1 设 A 是势为 α 的 DLN 集，以下(1)能推出(2)，(1)和(3)等价.

(1)A 是 α 齐次的.

(2)A 是 α 不凝聚的.

(3)A 是 α 松的.

证 (1)⇒(2)：设 A 不是 α 不凝聚的，那么存在上升列（存在下降列的情形类似）$a_0<a_1<\cdots<a_\eta<\cdots$，$(\eta<\xi<\alpha)$ 和 $a\varepsilon A$，使得 $\sup_{\eta<\xi}\{a_\eta\}=a$，任取 $d>a$，$f:(a_0,a_1,\cdots,a_\eta,\cdots,d)\rightarrow(a_0,a_1,\cdots,a_\eta,\cdots,a)$ 是 A 的局部同构，但是 A 中不存在 x 使得 $g:(a_0,a_1,\cdots,a_\eta,\cdots,d,a)\rightarrow(a_0,a_1,\cdots,a_\eta,\cdots,a,x)$ 作成局部同构，所以 A 不是 α 齐次的.

(1)与(2)⇒(3)：设有 $X,Y\subset A$，$|X|$ 和 $|Y|$ 都小于 α，$X<Y$. 以下分九种情形来讨论，我们只写出其中三种小情形，其他几种我们只列出标题，详细证明可类似地作出.

①设 X,Y 都 $\neq\varnothing$，存在 $x=\sup X$，$y=\inf Y$. 由(2)知 $x\varepsilon X$，$y\varepsilon Y$，所以 $x<y$. 又因 A 是稠密的，所以存在 $x<a<y$. 即有 $X<a<Y$.

②设 X,Y 都 $\neq\varnothing$，存在 $x=\sup X$，不存在 $\inf Y$. 在 Y 中尽可能选取 $y_0>y_1>\cdots>y_\eta>\cdots(\eta<\xi)$，假设对任意 $a>x$，都存在 η 使得 $a>y_\eta>x$，那么 $\inf\{y_\eta\}=x$，矛盾于(2)，所以存在 $a>x$，$a>y_\eta$（对所有 $\eta<\xi$），即有

$$X<a<Y.$$

③设 X,Y 都 $\neq\varnothing$，不存在 $\sup X$，存在 $\inf Y$.

④设 X,Y 都 $\neq\varnothing$，$\sup X$ 和 $\inf Y$ 都不存在. 仿上尽可能选取 $x_0<x_1<\cdots<x_\eta<\cdots$，$(\eta<\xi<\alpha)$；$y_0>y_1>\cdots>y_{\eta'}>\cdots$，$(\eta'<\xi'<\alpha)$. 其中 $x_\eta\varepsilon X$，$y_{\eta'}\varepsilon Y$. 再任取 $b_0<c<d_0\varepsilon A$，在 b_0 与 c 之间选取 $b_0<b_1<\cdots<b_\eta<\cdots(<c)$，$(\eta<\xi)$. 这样选取的 b_η 在 ξ 步之内不会停止，否则有 $\zeta<\xi$ 使得 $\sup_{\eta<\zeta}b_\eta=c$，矛盾于(2). 类似地可以选取 $d_0>d_1>\cdots>d_{\eta'}>\cdots(>c)$ $(\eta'<\xi')$. 考虑局部同构 $f:(b_\eta,d_{\eta'})_{\eta<\xi,\eta'<\xi'}\rightarrow(x_\eta,y_{\eta'})_{\eta<\xi,\eta'<\xi'}$. 由 A 的 α 齐次性知存在 a 使得 $g:(b_\eta,c,d_{\eta'})_{\eta<\xi,\eta'<\xi'}\rightarrow(x_\eta,a,y_{\eta'})_{\eta<\xi,\eta'<\xi'}$，也是局部同构. 对任意 $x\varepsilon X$，$y\varepsilon Y$，存在 $x_\eta>x$，$y_{\eta'}<y$. 所以 $x<x_\eta<a<y_{\eta'}<y$，$X<a<Y$.

⑤设 $Y=\varnothing$，$x=\sup X$，易见有 $a>X$.

⑥设 $Y=\varnothing$，$\sup X$ 不存在. 在 X 中尽可能选取 $x_0<x_1<\cdots<x_\eta<\cdots$，$(\eta<\xi<\alpha)$. 任取 $b_0<c\varepsilon A$，仿④再选取 $b_0<b_1<\cdots<b_\eta<\cdots(<c)$，$(\eta<\xi)$. 存在 $a\varepsilon A$ 使得 $g:(b_\eta,c)_{\eta<\xi}\to(x_\eta,a)_{\eta<\xi}$是局部同构即有 $X<a$.

⑦，⑧$X=\varnothing$的两种情形.

(3)⇒(1)，设 $f:(a_0,a_1,\cdots,a_\eta,\cdots)\to(b_0,b_1,\cdots,b_\eta,\cdots)$，$(\eta<\xi<\alpha)$是局部同构，对任意 $a_\xi\varepsilon A$，如果 a_ξ 是某 a_η，那么令 $b_\xi=b_\eta$，$f=g$，g 就是由 f 扩充而得的局部同构. 下设 $a_\xi\neq$各 $a_\eta(\eta<\xi)$. 令 $X=\{a_\eta:\eta<\xi,a_\eta<a_\xi\}$，$Y=\{a_\eta:\eta<\xi,a_\eta>a_\xi\}$，$X'=\{b_\eta:a_\eta\varepsilon X\}$，和 $Y'=\{b_\eta:a_\eta\varepsilon Y\}$. 得到 $X<a_\xi<Y$. f 是局部同构，其象集也有同样关系. $X'=f(X)<f(Y)=Y'$. 由 A 的 α 可松性知存在 b_ξ 使得 $X'<b_\xi<Y'$，$g:(a_0,a_1,\cdots,a_\eta,\cdots,a_\xi)\to(b_0,b_1,\cdots,b_\eta,\cdots,b_\xi)$是 A 的局部同构. 所以 A 是α 齐次的.

定理 3.2(GCH) 对每一个无穷正规基数 α，存在势为 α 的α 齐次 DLN 集.

证 用超限归纳法定义如下递增的有序集列($A\subset B$ 除集合包含之外，次序也保持，并集也保持原来的次序).

(1)$A_0=1$ 元有序集.

(2)如果 A_ξ 中任意(严格)单调的势都$<\alpha$，并且$|A_\xi|\leqslant\alpha$. 定义 A_ξ 的一个 D-分划$[X,Y]$为满足 $X\cup Y=A_\xi$，$X\cap Y=\varnothing$，$X<Y$(X,Y 中允许有一个$=\varnothing$)的 A_ξ 子集对. 对每一个 A_ξ 的 D-分划引进一个新的元 $a_{[X,Y]}$. 令

$A_{\xi+1}=A_\xi\cup\{a_{[X,Y]}:[X,Y]$是 A_ξ 的 D-分划$\}$. 定义 $A_{\xi+1}$的“$<$”关系如下：

①A_ξ 中的元保留原来的关系.

②$X<a_{[X,Y]}<Y$.

③传递性.

容易验证 $A_{\xi+1}$是一个有序集并且满足

④$A_\xi\subset A_{\xi+1}$.

⑤A_ξ 稠密性：对任意 $a<b\varepsilon A_{\xi+1}$，存在 $c\varepsilon A_\xi$ 使得 $a<c<b$.

⑥$A_{\xi+1}$中每一个单调的势$<\alpha$.

⑦对任意 $a\varepsilon A_{\xi+1}$($A_{\xi+1}$的左端元除外)，可以选取一个势$<\alpha$ 的 A_ξ 的

单调列 $a_0<a_1<\cdots<a_\zeta<\cdots$,$(\zeta<\eta)$,使得 $\sup_{\zeta<\eta}\{a_\zeta\}=a$. 易见对任意 $a\neq a'$,如上所选取的单调列不可能全同. 而 A_ξ 的势 $<\alpha$ 的单调上升列至多有 $\sum_{\eta<\alpha}|A_\xi^\eta|\leqslant\alpha$,(GCH). 所以 $|A_{\xi+1}|\leqslant\alpha$.

(3)当 ξ 是 $<\alpha$ 的极限序数时,我们令 $A_\xi=\bigcup_{\eta<\xi}A_\eta$.

①A_ξ 的单调列可以看成 A_η 的单调列的并,所以势仍 $<\alpha$.

②$|A_\xi|=\sum|A_\eta|\leqslant\eta\alpha=\alpha$.

(4)令 $A_\alpha=\bigcup_{\xi<\alpha}A_\xi$. 容易验证 A_α 是 DLN 集,$|A_\alpha|=\alpha$,对任意 X,$Y\subset A_\alpha$,$|X|$ 和 $|Y|$ 都 $<\alpha$,$X<Y$. 易见有 $X,Y\subset$ 某 A_ξ,$(\xi<\alpha)$. 在 $A_{\xi+1}$ 中存在 $a<X<b<Y<c$. 再由预理 3.1 知 A_α 是 α 齐次的 DLN 集.

定理 3.3 任意两个势为 α 的 α 齐次 DLN 集互相同构.

证 用 Cantor 过来过去的方法和预理 3.1 可得证.

定理 3.4(GCH) 对任意奇异基数 α 和任基数 $\omega\leqslant\beta<\alpha$,存在一系列 DLN 集 $B_\omega\subset B_{\omega_1}\subset\cdots\subset B_\beta\subset\cdots$,($\beta$ 是 $<\alpha$ 的基数). 使得每一个 B_β 是 β^+ 齐次的 DLN 集.(这时 $B=\bigcup_{\beta<\alpha}B_\beta$ 叫作 α 弱齐次 DLN 集.)

证 在定理 3.2 的证明中选取 A_{β^+} 作为 B_β. 所得序列 $\{B_\beta:\beta$ 是 $<\alpha$ 的基数$\}$ 符合要求.

定理 3.5 任意势 $<\alpha$ 的 DLN 集不可能是 α 齐次的(在 α 是奇异基数时,不是弱齐次的).

证 在 α 是正规基数时,假设 $|A|<\alpha$,A 是 DLN 集. 任取 $a_0<c\in A$,尽可能选取 $a_0<a_1<\cdots<a_\eta<\cdots(<c)$,直至不能再选. 因为 $|A|<\alpha$,这个元列的势也 $<\alpha$. 而 $\sup\{a_\eta\}=c$(否则仍可继续选取). 所以 A 不是 α 不凝聚的,再由预理 3.1 知 A 不是 α 齐次的.

(在 α 是奇异基数时,假设 $|A|<\alpha$,A 是 DLN 集. 因为 α 是极限基数,存在基数 β 使得 $|A|<\beta<\alpha$. $\beta^+<\alpha$ 是正规基数. A 不可能含有势为 β^+ 的子 DLN 集,所以 A 不可能含有 β^+ 齐次的子集. A 不可能是 α 弱齐次的.)

§4. 齐次模型的例子

例 4.1(GCH.) 设 T_{1h} 是 DLN 集的理论,对于任意基数 α,T 恰好有一个势为 α 的齐次(弱齐次)模型.

证 由[2]定理 1.5.3 知 T_{1h} 对元子式子集 B 有量词消去性质. 因此

对任意 T 模型 $\mathscr{U}$，$(\mathscr{U},a_\eta)_{\eta<\xi}\equiv(\mathscr{U},b_\eta)_{\eta<\xi}$ 成立的充分必要条件是 $f:a_\eta\to b_\eta(\eta<\xi)$ 是局部同构. 再利用§3各预理、定理知 T_{1h} 有一个 α 齐次（弱齐次）模型. 由[1]定理 A 知 T_{1h} 的齐次（弱齐次）模型数是一个不增的函数. 所以 T 恰好有一个势为 α 的齐次（弱齐次）模型.

例 4.2(GCH.) 设 T_{2h} 是 Ehrenfeucht 集理论（即[3]p. 318 的 T_3）那么对每一个无穷基数 α，T_{2h} 恰好有两个势为 α 的齐次（弱齐次）模型.

证 (1) T_{2h} 是在 T_{1h} 的基础之上加入无限多个常数 $c_0<c_1<\cdots<c_n<\cdots(n<\omega)$ 而得. 由 T_{1h} 对其元子式集的量词消去性质可以导出 T_{2h} 对其元子式集的量词消去性质.

(2)可验 T_{2h} 恰好有两个可数齐次模型.

(3)选取 $\aleph_0$ 个势为 α 的 α 齐次（弱齐次）DLN 集 $A_0,A_1,\cdots,A_\omega$ 和元素 $e_0,e_1,\cdots$ 构造模型 $\mathscr{U}$，$\mathscr{B}$ 如下：

$\mathscr{U}:A_0<e_0<A_1<e_1<\cdots$

$\mathscr{B}:A_0<e_0<A_1<e_1<\cdots A_\omega$.

以 e_i 解释 c_i，利用§3各定理可以验证 $\mathscr{U}$，$\mathscr{B}$ 都是势为 α 的 T_{2h} 齐次（弱齐次）模型. 再由[1]定理 A 知 T_{2h} 恰好有两个势为 α 的齐次（弱齐次）模型.

例 4.3 设 T_{3h} 是离散有一端点的有序集理论. 对任意无穷基数 α，T_{3h} 有 3 个 α 齐次（弱齐次）模型. 其中 1 个势为 α，另外两个势为 $\aleph_0$.

证 (1) $\mathscr{U}=\langle\omega,<\rangle$ 是 α 齐次（弱齐次）的.

(2) $\mathscr{B}=\langle\omega\oplus Z,<\rangle$（其中 Z 代表整数有序集，$\oplus$ 是有序和）是 α 齐次（弱齐次）的.

(3)取一个势为 α 的 α 齐次（弱齐次）DLN 集 D，作 $\mathscr{C}=\langle\omega\oplus D\otimes Z,<\rangle$，（其中 $\otimes$ 是有序积）. 利用§3各预理、定理可以验证 $\mathscr{C}$ 是 α 齐次的.

例 4.4 设 $T_{3H}=T_{3h}\cdot T_{1h}$，对任意无穷基数 α，T_{3H} 恰好有三个势为 α 的齐次（弱齐次）模型.

证 由例 4.1，4.3 及§2的定理知 T_{3H} 存在三个势为 α 的齐次模型. 再由[1]定理 A 知 T_{3H} 恰好有三个势为 α 的齐次模型（弱齐次类似）.

例 4.5 $n=2^r3^s$ 时，用互不相交的语言构造 r 个相当于 T_{2h} 的理论，s 个相当于 T_{3H} 的理论 $T_{2h}^1,T_{2h}^2,\cdots,T_{2h}^r,T_{3H}^1,T_{3H}^2,\cdots,T_{3H}^s$. 令

$$T_{nh}=((\cdots(T_{2h}^1\vee T_{2h}^2)\vee\cdots)\vee T_{2h}^r)\vee((\cdots(T_{3H}^1\vee T_{3H}^2)\vee\cdots)\vee T_{3H}^s)$$

那么对任意无穷基数 α，T_{nh} 恰好有 n 个势为 α 的齐次（弱齐次）模型.

证 反复运用例 4.2，例 4.3，定理 1.3 和[1]定理 A 可得证.

本节各例中所用到理论 T_{1h}，T_{2h}，T_{3h}，T_{3H} 和 T_{nh}（$n=2^r3^s$）的完备性可利用量词消去法证出.

参考文献

[1] H J Keisler and M D Morley. On the number of homogeneous models of a given power. Israel J. of Math，1967，5：73—78.

[2] C C Chang and H J Keisler. Theory of models. North-Holland Publishing Co.，1973.

[3] R L Vaught. Denumerable models of complete theories，Infinitistic method. Warsaw，1961，303—321.

Abstract This is the 2nd paper of a series of papers in studying the number of homogeneous models. In the lst paper we have proved：(1) For a countable theory T，either $h_T(\aleph_0)\leqslant\aleph_0$ or $h_T(\aleph_0)=2^{\aleph_0}$. (2) If all models of a complete countable theory T are ω-homogeneous，then $h_T(\aleph_0)\neq n(1<n<\omega)$.

In [2] Ex. 5. 1. 16，there is a relevant open problem. For which finite n does a complete countable theory T exist such that for every infinite cardinal α，T has exactly n nonisomorphic (weakly) homogeneous models of power α? (For the cases $n=1,2,3$ such theories are known to exist.)

In this paper，we study the union and product of models，and improve the above cases of existence for the problem to any $n=2^r3^s$.

中国科学，1981，25A(2)：161～170

自由群内方程的讨论①

A Discussion on the Equations in Free Groups

摘要 本文讨论了三种自由群方程.对非蜕化的一元方程，证明它没有变数解，并对 $AxBx^{-1}=1$ 的短解的消去式作了详细的讨论，对方程 $PxQyRx^{-1}Sy^{-1}=1$，证明了它的有解性是有限可判定的，它的全部解可以归入有限个递归解集合.对以上方程有变数解的条件和变数解的形式，文中也给出了较为完整的结果.

Lyndon[1]给出了自由群一元方程的解法.经过 Lorenc 和 Appel 的精确化[2,3]，自由群一元方程的解法已逐渐完善.但是，自由群二元方程的解法至今尚未有很好的结果.Lorenc[4]给出了无系数的二元方程的解法.关于有系数的二元方程，Appel 给出了一个特殊的二元方程，证明它的解不可能写成任何有限参数形式，得到否定性的结果[5].本文给出了 Appel 所提供的这一类自由群二元方程的有解判别法.虽然这一类方程一般地说来不是有限参数解的，但本文给出了一种解的递归形式，并且证明了任意一个这类方程如果它是有解的，那么可以给出它的一组含有全部解的有限个递归形式的解.对于这类方程有变数解的条件和变数解的形式，本文也给出了较为完整的结果.我们所采用的记号和术语均出自文献[6].

我们主要讨论以下三种自由群方程：

$$C_1x^{\varepsilon_1}C_2x^{\varepsilon_2}\cdots C_tx^{\varepsilon_t}=1, \qquad ①$$

① 收稿日期：1979-11-19，收压缩稿日期：1980-12-19.

其中 $\varepsilon_i=\pm 1(i=1,2,\cdots,t)$. 令 $M=2\mathrm{Max}\{|C_i|\}$,

$$AxBx^{-1}=1, \qquad ②$$

$$PxQyRx^{-1}Sy^{-1}=1, \qquad ③$$

命题 1 如果 $A\neq 1$,那么 $|A^k|<|A^{k+1}|(k\geqslant 0)$ (见文献[6],p. 9).

命题 2 如果存在 m, $x=v^m$ 是方程①的解(本命题中允许某些 $C_i=1$,同时 $\varepsilon_{i-1}\neq\varepsilon_i$,但在其他场合,我们要求这种 $C_i\neq 1$). 其中 v 是循环既约的, $|v|\leqslant M$,并且 $|v^m|>12N$,这里 $N=\sum_i|C_i|$. 那么 $x=v^n$ 是方程①的参数解.

证 对于 $t\leqslant 2$ 的情形,可以利用文献[4]的预理 1 经过对解的长度的讨论而证出. 在 $t>2$ 时,如果有一个 $C_i=v^{k_i}$,并且两侧 $\varepsilon_{i-1}\neq\varepsilon_i$,那么可以利用归纳法讨论较小的 t. 对其他情形也可利用文献[4]预理 1 来证明不可能发生.

命题 3 如果存在 m,方程①有解 $x=uv^mw$,其中 $|u|\leqslant l$, $|v|\leqslant M$, $|w|\leqslant p$, v 是循环既约的,并且 $|v^m|>12t(l+M+p)$. 那么 $x=uv^nw$ (n 任意)是方程①的参数解.

命题 4 方程①有参数解的充分必要条件是方程①有长度大于 $12t^2+133tM+10M$ 的解.

证 必要性明显. 现证充分性. 设方程①有长度大于 $12t^2+133tM+10M$ 的解 u,由文献[4]知,u 可以写成 $d\cdot z^m\cdot y^n\cdot f$ 形. 其中 $|d|$ 和 $|f|$ 小于 $2M$, z 和 y 是本原的并且长度小于 M. 并由文献[4]知,不可能有 $|z^m|>(t+6)M$, $|y^n|>(t+6)M$ 以及 $z\neq y^{\pm 1}$ 同时成立. 对于 $z=y^{\pm 1}$,我们可以将 y 和 z 合并. 在 $|z^m|\leqslant(t+6)M$ 时($|y^n|<(t+6)M$ 类似),我们有

$$|d\cdot z^m|\leqslant(t+8)M,$$

$$|dz^m|+|y^n|+|f|=|u|>12t^2+133tM+10M,$$

$$|y^n|>12t^2+133tM+10M-(t+8)M-2M$$
$$=12t(t+11M).$$

利用命题 3 知,$x=dz^my^nf$ (n 任意)是方程①的参数解.

命题 5 设方程①有参数解 $x=uv^nw$,那么可以选取 $x_1=u_1v_1^nw_1$ 使 v_1 是本原循环既约,并且 $u_1\neq u_2\cdot v_1^{\pm 1}$ 形,$w_1\neq v_1^{\pm 1}\cdot w_2$ 形.

命题 6 方程①非蜕化时没有非蜕化的变数解.

证 设 $f(c, v_1, \cdots, v_r)$ 是方程①的非蜕化的变数解. 令 $p=6M+r$, 取 p 个在 f 中不出现的互不相同的自由群生成元 $d_1, \cdots, d_p$. 由文献[4]知, 有

$$f(c, d_1, \cdots, d_{r-1}, d_r, \cdots, d_p)=a \cdot z^m \cdot y^n \cdot b.$$

其中 $|a|$ 和 $|b|<2M$, $|z|$ 和 $|y| \leqslant M$. 上式左端含有长度为 $6M+1$ 的不出现重复元的段, 而右端至多允许出现长度为 $6M$ 的不出现重复元的段, 故产生矛盾. 方程①不可能有这种解.

命题 7 如果方程②有解, 并且 $B \neq 1$, 那么它有一个参数解.

证 设 x_0 是方程②的解, 那么 $x=x_0 B^n$ 是方程②的解. 它就是方程②的参数解.

命题 8 a, b 都是循环既约的, $b=v^{-1} a v$, 那么存在 v_1, v_2 和整数 n 满足 $a=v_1 \cdot v_2, b=v_2 \cdot v_1$ 和 $v=v_1(v_1^{-1} v_2^{-1})^n$.

证 $v^{-1} a$ 和 $a v$ 两个乘积中有并且只有一个能相消, 再对 v 的长度作归纳法即可得证.

现在我们来讨论方程②的有解条件, 解的形式和解与系数的关系. 从现在起, 我们尽量稳定各个字母所代表的意义, 以便于各命题之间连用.

定义 1 如果 v_1, v_2 均 $\neq 1$, 那么我们把 $x_1=u v_1 w$ 叫作方程②的第一种短解, $x_2=u v_2^{-1} w$ 叫作方程②的第二种短解. 如果 $v_1 \neq 1, v_2=1$ (或 $v_1=1, v_2 \neq 1$), 那么 $x_3=u w$ 叫作方程②的第三种短解. 在 $v_1=v_2=1$ 时, $x_4=1$ 叫作方程②的第四种短解. 此外无其他短解.

命题 9 设方程②有解, 将 A, B 分别写成

$$A=u \cdot a^k \cdot u^{-1}, B=w^{-1} \cdot b^l \cdot w,$$

其中 $k, l \geqslant 0$, 并且 a, b 都是循环既约的和本原的. 我们有以下性质(1)～(7).

(1) $k=l$, 并且存在 v_1, v_2 满足

$$a=v_1 \cdot v_2, b^{-1}=v_2 \cdot v_1.$$

(2) $x=u v_1(v_1^{-1} v_2^{-1})^n w$ 是方程②的唯一(最全的)参数解.

(3)第一种短解的消去式如下:

$$A x_1 B x_1^{-1}=\left[_e(u \cdot v_1)\left[_c(v_2 \cdot(v_1 \cdot v_2)^{k-1})\left[_a u^{-1} u\right]_a\left[_b(v_1 \cdot w)(w^{-1} \cdot v_1^{-1})\right]_b\right.\right.$$

$$((v_2^{-1}\cdot v_1^{-1})^{k-1}\cdot v_2^{-1})]_c[_d wrw^{-1}]_d(v_1^{-1}\cdot u^{-1})]_e=1. \qquad ④$$

(4)第二种短解的消去式如下：

$$Ax_2Bx_2^{-1}=[_e u[_c((v_1\cdot v_2)^{k-1}\cdot v_1)[_a(v_2\cdot u^{-1})(u\cdot v_2^{-1})]_a[_b wrw^{-1}]_b$$
$$(v_1^{-1}\cdot(v_2^{-1}\cdot v_1^{-1})^{k-1})]_c[_d(v_2^{-1}\cdot w)(w^{-1}\cdot v_2)]_d u^{-1}]_e=1. \qquad ⑤$$

(5)设 $u=u_1\cdot r, w=r^{-1}\cdot w_1, r, r^{-1}$ 是 u, w 之间相消的最大可能部分，那么第三种短解的消去式如下.

$$Ax_3Bx_3^{-1}=[_e u_1[_c(r\cdot v_1^k\cdot r^{-1})[_a u_1u_1^{-1}]_a[_b w_1w_1^{-1}]_b\cdot$$
$$(r\cdot v_1^{-k}\cdot r^{-1})]_c[_d w_1w_1^{-1}]_d u_1^{-1}]_e=1. \qquad ⑥$$

(6)上述三个消去式中各不同方括号的元之间不会交叉相消，相消对应元唯一确定，并且 x 的元和 x^{-1} 的元不相消.

(7)假设 $A=A_1\cdot A_2\cdot A_3, x_1=X_1\cdot X_2, B=B_1\cdot B_2\cdot B_3, x_1^{-1}=X_4^{-1}\cdot X_3^{-1}$，能使

$$A_3X_1=X_2B_1=A_2B_2=B_3X_4^{-1}=A_1X_3^{-1}=1.$$

并且又知道 x_1 是方程(Ⅱ)的第二种短解，那么一定有

$$\left.\begin{aligned}&u^{-1}=A_3, u=X_1,\\&v_1\cdot w=X_2, w^{-1}\cdot v_1^{-1}=B_1,\\&v_2\cdot(v_1\cdot v_2)^{k-1}=A_2, (v_2^{-1}\cdot v_1^{-1})^{k-1}\cdot v_2^{-1}=B_2,\\&w=B_3, w^{-1}=X_4,\\&u\cdot v_1=A_1 \text{ 和 } v_1^{-1}\cdot u_1^{-1}=X_3^{-1}.\end{aligned}\right\} \qquad ⑦$$

关于第二种短解和第三种短解也有类似的等式. 这里不再列出但给以编号⑧和⑨.

证 利用命题 8 可验证(1)和(2).

由 $u\cdot v_1\cdot v_2\cdot u^{-1}$ 和 $w^{-1}\cdot v_1^{-1}\cdot v_2^{-1}\cdot w$ 以及它们的逆关系 $u\cdot v_2^{-1}\cdot v_1^{-1}\cdot u^{-1}$ 和 $w^{-1}\cdot v_2\cdot v_1\cdot w$ 可以证出(3)～(7).

命题 10 方程②的短解短于其他解.

定义 2 如果 (x_0, y_0) 是方程③的解，并且 x_0 是方程

$$(Sy_0^{-1}P)x(Qy_0R)x^{-1}=1 \qquad ⑩$$

的短解之一，那么 x_0 叫作 y_0 对方程③的一个短解.

如果 (x_0, y_0) 是方程③的解，并且 y_0 是方程

$$(Px_0Q)y(Rx_0^{-1}S)y^{-1}=1 \qquad ⑪$$

的短解之一，那么 y_0 叫作 x_0 对方程③的一个短解.

定义 3 如果(x_0,y_1)，(x_0,y_2)是方程③的解，y_1，y_2 是 x_0 对方程③的全部短解，并且 x_0 各是 y_1，y_2 对方程③的短解之一，那么(x_0,y_1)，(x_0,y_2)叫作方程③的一组同 x 型短解. 同 y 型短解组的定义类似. 以上两种短解组中的每一个解都叫作方程③的短解. 此外方程③再无其他短解.

命题 11 设(x_0,y_0)是方程③的解，那么存在 $x_1,x_2,\cdots,x_l$ 和 $y_1,y_2,\cdots,y_l$，使 y_i 是 x_{i-1} 对方程③的短解，x_i 是 y_i 对方程③的短解，并且(x_l,y_{l-1})，(x_l,y_l)是方程③的一个同 x 型短解组. 同 y 型短解组类似.

证 对 $m=|x_0|$ 作归纳法. $m=0$ 的情形易验. 假设命题对小于 m 的情形已经成立，现在来证明 $|x_0|=m$ 的情形. 设方程(e. 8)的短解是 a,b，

$$(Sa^{-1}P)x(QaR)x^{-1}=1 \text{ 和 } (Sb^{-1}P)x(QbR)x^{-1}=1$$

的短解分别是 c,d 和 e,f.

(1)如果 c,d 中有一个是 x_0，并且 e,f 中也有一个是 x_0，那么令 $x_1=x_2=x_0$，$y_1=a$，$y_2=b$. 这样得(x_2,y_1)，(x_2,y_2)是方程③的一组同 x 型短解.

(2)如果 c,d 均 $\neq x_0$，那么由命题 10 知 $|c|$，$|d|$ 均小于 m. 令 $y_1=a$，$x_1=c$. y_1，x_1 分别是 x_0，y_1 对方程③的短解，(x_1,y_1)是方程③的解，并且 $|x_1|<m$. 故由归纳法知命题成立.

命题 12 方程③能否有解是有限可判定的，其短解个数有限，可以有有限可计算方法将这些短解全部求出，并且短解中不含有变数.

证 由命题 11 知，只需讨论是否存在 y 型短解，它的求法，个数和组成方式. 设(x_0,y_0)，(x_1,y_0)是方程③的一组同 y 型短解.

1. 假设(P,x_0,Q)，(Q,y_0,R)，(R,x_0^{-1},S)和(S,y_0^{-1},P) 4 个三元组中有一个是奇异的，例如(P,x_0,Q)是奇异的，那么 $x_0=P_1^{-1}\cdot Q_1^{-1}$，其中 P_1，Q_1 各是 P，Q 的一部分. 所以 x_0 只有有限种可能. 将这些可能的 x_0 分别代入方程③得到相应的一元方程⑪，再利用命题 9 求出相应的短解或验算不存在这样的解. 又由命题 9 知，上述过程不会产生变数解.

下面情形 2～6 中，我们假设情形 1 所列的 4 个三元组都不是奇异的.

2. 假设(x_0,Q,y_0)，(y_0,R,x_0^{-1})，(x_0^{-1},S,y_0^{-1})和(y_0^{-1},P,x_0)都是奇异三元组. 所以有

$$Px_0Qy_0Rx_0^{-1}Sy_0^{-1}=(P_1\cdot P_2)(X_1\cdot X_2\cdot X_3)(Q_1\cdot Q_2)(Y_1\cdot Y_2\cdot Y_3)\cdot(R_1\cdot R_2)(X_6^{-1}\cdot X_5^{-1}\cdot X_4^{-1})(S_1\cdot S_2)(Y_6^{-1}\cdot Y_5^{-1}\cdot Y_4^{-1})=1, \quad ⑫$$

$$P_2X_1=X_3Q_1=Q_2Y_1=Y_3R_1=R_2X_6^{-1}=X_4^{-1}S_1=S_2Y_6^{-1}=Y_4^{-1}P_1=1.$$

相消之后得

$$X_2Y_2X_5^{-1}Y_5^{-1}=1. \quad ⑬$$

由命题 9 知，x_0 的元与 x_0^{-1} 的元不相消，y_0 的元与 y_0^{-1} 的元不相消，所以⑬式可以分解为

$$(X_7\cdot X_8)(Y_7\cdot Y_8)(X_{10}^{-1}\cdot X_9^{-1})(Y_{10}^{-1}\cdot Y_9^{-1})=1,\text{其中}$$

$$X_8Y_7=Y_8X_{10}^{-1}=X_9^{-1}Y_{10}^{-1}=Y_9^{-1}X_7=1.\text{所以}$$

$$Ax_0Bx_0^{-1}=(S_1Y_{10}^{-1}Y_9^{-1}P_2)(X_1X_7X_8X_3)(Q_1Y_7Y_8R_2)(X_6^{-1}X_{10}^{-1}X_9^{-1}X_4^{-1})=1. \quad ⑭$$

按照命题 9 中的括号 a,b,c,d 和 e 的顺序写出消去式为

$$Y_9^{-1}P_2X_1X_7=X_8X_3Q_1Y_7=1=Y_8R_2X_6^{-1}X_{10}^{-1}=X_9^{-1}X_4^{-1}S_1Y_{10}^{-1}.$$

可验证⑭式的四个乘积内是直分解. 由(e. 4)式得

$$v_2\cdot(v_1\cdot v_2)^{k-1}=A_2,(v_2^{-1}\cdot v_1^{-1})^{k-1}\cdot v_2^{-1}=B_2.$$

但由⑭式又可看出，$A_2=B_2=1$，所以 $v_1=v_2=1$. 这样(S,y_0^{-1},P)是奇异三元组，产生矛盾. 所以情形 2 不可能发生.

3. 假设(x_0,Q,y_0)不是奇异三元组，而(y_0,R,x_0^{-1})，(x_0^{-1},S,y_0^{-1})和(y_0^{-1},P,x_0)都是奇异三元组. 仿前陆续得下列式子：

$$Px_0Qy_0Rx_0^{-1}Sy_0^{-1}=(P_1\cdot P_2)(X_1\cdot X_2\cdot X_3)(Q_1\cdot Q_2\cdot Q_3)(Y_1\cdot Y_2\cdot Y_3)(R_1\cdot R_2)\cdot(X_6^{-1}\cdot X_5^{-1}\cdot X_4^{-1})(S_1\cdot S_2)(Y_6^{-1}\cdot Y_5^{-1}\cdot Y_4^{-1})=1, \quad ⑮$$

$$P_2X_1=X_3Q_1=Q_3Y_1=Y_3R_1=R_2X_6^{-1}=X_4^{-1}S_1=S_2Y_6^{-1}=Y_4^{-1}P_1=1,$$

$$(X_2\cdot Q_2\cdot Y_2)X_5^{-1}Y_5^{-1}=1,$$

$$(X_2\cdot Q_4\cdot Q_5\cdot Y_2)(X_{11}^{-1}\cdot X_{10}^{-1}\cdot X_9^{-1})(Y_{11}^{-1}\cdot Y_{10}^{-1}\cdot Y_9^{-1})=1,$$

$$Y_2X_{11}^{-1}=Q_5X_{10}^{-1}=X_9^{-1}Y_{11}^{-1}=Q_4Y_{10}^{-1}=X_2Y_9^{-1}=1,\text{和}$$

$$Ax_0Bx_0^{-1}=(S_1Y_{11}^{-1}Y_{10}^{-1}Y_9^{-1}Y_2)(X_1X_2X_3)(Q_1Q_4Q_5Y_2R_2)(X_6^{-1}X_{11}^{-1}X_{10}^{-1}X_9^{-1}X_4^{-1})$$

$$=[_e(S_1Y_{11}^{-1})[_cY_{10}^{-1}[_a(Y_9^{-1}Y_2)(X_1X_2)]_a[_bX_3Q_1]_bQ_4]_c$$
$$[_d(Q_5Y_2R_2)(X_6^{-1}X_{11}^{-1}X_{10}^{-1})]_d(X_9^{-1}X_4^{-1})]_e=1. \quad ⑯$$

以下分几种情形来计算 x_0, y_0 或 x_1.

(1)设 x_0 是⑯式相应方程的第一种短解. 与⑦式对比得 $B=Q_1Q_4Q_5Y_2R_2$. 所以

$$w^{-1}\cdot(v_1^{-1}\cdot v_2^{-1})^k=Q_1\cdot Q_4, w=Q_5\cdot Y_2\cdot R_2.$$

w^{-1}相当于 Q 的一部分 Q_6，即有

$$y_0=Q_3^{-1}Q_5^{-1}Q_6^{-1}R^{-1}.$$

(2)设 x_0 是⑯式相应方程的第二种短解. 与⑧式对比，得

$$w^{-1}\cdot v_1^{-1}\cdot(v_2^{-1}\cdot v_1^{-1})^{k-1}=Q_1\cdot Q_4, Q_5\cdot Y_2\cdot R_2=v_2^{-1}\cdot w.$$

再设 $k>1$，那么有 $w^{-1}=Q_6$, $v_2^{-1}=Q_7$，即有

$$y_0=Q_3^{-1}Q_5^{-1}Q_7Q_6^{-1}R^{-1}.$$

(3)各条件同(2)，但 $k=1$，由⑧式可得

$$w^{-1}=Q_1, v_1^{-1}=Q_4, Y_9^{-1}P_2=v_2\cdot u^{-1}\text{和}$$
$$Q_5Y_2R_2=v_2^{-1}w.$$

①假设 $|P_2|\geqslant|u^{-1}|$，那么 u^{-1}是 P_2 的一部分，$u^{-1}=p_3$. 我们得到

$$x_1=uv_1w=P_3^{-1}Q_4^{-1}Q_1^{-1}.$$

②假设 $|P_2|<|u^{-1}|$. 这时 v_2 全部由 Y_9^{-1} 的元所组成. 考虑以 x_1 去替换 x_0 所得的消去式如下：

$$[_euv_1[_cv_2[_au^{-1}u]_a[_bv_1wrw^{-1}v_1^{-1}]_bv_2^{-1}]_c[_dwrw^{-1}]_dv_1^{-1}u^{-1}]_e,$$

y_0 与 y_0^{-1} 的元也不应相消，所以 Y_2 完全落入 w 之中. 但前面已知 $w=Q^{-1}$，所以有 $Y_2=Q_6^{-1}$. 即得

$$y_0=Y_1Y_2Y_3=S_2^{-1}Q_6^{-1}R_1^{-1}.$$

(4)设 x_0 是⑯式相应方程的第三种短解，仿上也可以求出这个解.

4. 假设(x_0,Q,y_0)和(y_0,R,x_0^{-1})不是奇异三元组，而(x_0^{-1},S,y_0^{-1})和(y_0^{-1},P,x_0)是奇异三元组. 其分解式和消去式如下：

$$Px_0Qy_0Rx_0^{-1}Sy_0^{-1}=(P_1\cdot P_2)(X_1\cdot X_2\cdot X_3)(Q_1\cdot Q_2\cdot Q_3)$$
$$(Y_1\cdot Y_2\cdot Y_3)(R_1\cdot R_2\cdot R_3)(X_6^{-1}\cdot X_5^{-1}\cdot X_4^{-1})$$
$$(S_1\cdot S_2)(Y_6^{-1}\cdot Y_5^{-1}\cdot Y_4^{-1})=1 \quad ⑰$$

$$P_2X_1=X_3Q_1=Q_3Y_1=Y_3R_1=R_3X_6^{-1}=X_4^{-1}S_1=S_2Y_6^{-1}=Y_4P_1=1.$$

由此可以推出 Y_2 的元与 Y_5^{-1} 的元相消,与命题 9 的性质(5)矛盾. 所以这种情况不可能发生.

5. 假设(x_0,Q,y_0)和(x_0^{-1},S,y_0^{-1})不是奇异三元组,而(y_0,R,x_0^{-1})和(y_0^{-1},P,x_0)是奇异三元组. 首先我们有如下分解式和消去式:

$$Px_0Qy_0Rx_0^{-1}Sy_0^{-1}=(P_1\cdot P_2)(X_1\cdot X_2\cdot X_3)(Q_1\cdot Q_2\cdot Q_3)$$
$$(Y_1\cdot Y_2\cdot Y_3)(R_1\cdot R_2)(X_6^{-1}\cdot X_5^{-1}\cdot X_4^{-1})$$
$$(S_1\cdot S_2\cdot S_3)(Y_6^{-1}\cdot Y_5^{-1}\cdot Y_4^{-1})=1, \qquad ⑱$$

$$P_2X_1=X_3Q_1=Q_3Y_1=Y_3R_1=R_2X_6^{-1}=X_4^{-1}S_1=S_3Y_6^{-1}=Y_4^{-1}P_1=1$$

和

$$(X_2\cdot Q_2\cdot Y_2)(X_5^{-1}\cdot S_2\cdot Y_5^{-1})=1. \qquad ⑲$$

(1)假设 x_0 是方程

$$(Sy_0^{-1}P)x(Qy_0R)x^{-1}=1 \qquad ⑩$$

的第一种短解. 因为 x_0 与 x_0^{-1} 的元不相消,y_0 与 y_0^{-1} 的元不相消,⑲式的二括号对应直分解为

$$(X_2\cdot Q_4\cdot Q_5\cdot Y_2)(X_5^{-1}\cdot S_4\cdot S_5\cdot Y_5^{-1})=1.$$

其中,$Q_5Y_2X_5^{-1}S_4=S_5Y_5^{-1}X_2Q_4=1$. 再考虑

$$Ax_0BX_0^{-1}=(S_1S_4S_5Y_5^{-1}P_2)(X_1X_2X_3)(Q_1Q_4Q_5Y_2R_2)$$
$$(X_6^{-1}X_5^{-1}x_4^{-1})=1. \qquad ⑳$$

将⑳式与⑦式对比可以看出,在 $A=S_1S_4S_5Y_5^{-1}P_2$ 中与 x_0^{-1} 的元相消的元全部落在 S_1S_4 之内,$B=Q_1Q_4Q_5Y_2R_2$ 与 x_0 的元相消的元,全部落在 Q_1Q_4之内. 所以有 S 的两个部分 S_6,S_7 和 Q 的一个部分 Q_6满足

$$u=S_6,v_1=S_4,w^{-1}=Q_6.\ x_0=uv_1w=S_6S_7Q_6^{-1}.$$

(2)假设 x_0 是⑩式的第二种短解. 这种情形比较复杂,我们只列出其中主要式子.

(i)$X_2=X_7\cdot X_8$,$X_5^{-1}=X_{10}^{-1}\cdot X_9^{-1}$,其中

$$Y_5^{-1}X_7=S_2X_8=X_9^{-1}Q_2=X_{10}^{-1}Y_1=1.$$

$$Ax_0Bx_0^{-1}=(S_1S_2Y_5^{-1}P_2)(X_1X_7X_8X_3)(Q_1Q_2Y_2R_2)(X_6^{-1}X_{10}^{-1}X_9^{-1}X_4^{-1})=1, \qquad ㉑$$

$$S_2Y_5^{-1}P_2X_1X_7X_8=X_3Q_1=1=Q_2Y_2R_2X_6^{-1}X_{10}^{-1}X_9^{-1}=S_1X_4^{-1}.$$

仿照情形 2 可证本情形不会发生.

(ii)$X_2=X_7\cdot X_8$,$Q_2=Q_4\cdot Q_5$,其中

$$Y_5^{-1}X_7=S_5X_8=S_4Q_4=X_9^{-1}Q_5=X_{10}^{-1}Y_2=1,$$

$$Ax_0Bx_0^{-1}=(S_1S_4S_5Y_5^{-1}P_2)(X_1X_7X_8X_3)(Q_1Q_4Q_5Y_2R_2)(X_6^{-1}X_{10}^{-1}X_9^{-1}X_4^{-1})=1, \quad ㉒$$

$$S_5Y_5^{-1}P_2X_1X_7X_8=X_3Q_1=S_4Q_4=Q_5Y_2R_2X_6^{-1}X_{10}^{-1}X_9^{-1}=S_1X_4^{-1}=1.$$

$$u=S_2,w^{-1}=Q_1,(v_1\cdot v_2)^{k-1}v_1=S_4.$$

(iii) $Q_2=Q_4\cdot Q_5,X_5^{-1}=X_{10}^{-1}\cdot X_9^{-1}$,其中

$$Y_5^{-1}X_2=S_2Q_4=X_9^{-1}Q_5=X_{10}^{-1}Y_2=1,$$

$$Ax_0Bx_0^{-1}=(S_1S_2Y_5^{-1}P_2)(X_1X_2X_3)(Q_1Q_4Q_5Y_2R_2)(X_6^{-1}X_{10}^{-1}X_9^{-1}X_4^{-1})=1, \quad ㉓$$

$$Y_5^{-1}P_2X_1X_2=X_3Q_1=S_2Q_4=Q_5Y_2R_2X_6^{-1}X_{10}^{-1}X_9^{-1}=S_1X_4^{-1}=1,$$

$$u=S_1,w^{-1}=Q_1,v_1^{-1}\cdot(v_2^{-1}\cdot v_1^{-1})^{k-1}=Q_4.$$

(iv) $Q_2=Q_4\cdot Q_5\cdot Q_6,Y_5^{-1}=Y_{10}^{-1}\cdot Y_9^{-1},X_5^{-1}=X_{10}^{-1}\cdot X_9^{-1}$,其中

$$Y_9^{-1}X_2=Y_{10}^{-1}Q_4=S_4Q_5=X_9^{-1}Q_6=X_{10}^{-1}Y_2=1$$

$$Ax_0Bx_0^{-1}=(S_1S_2Y_{10}^{-1}Y_9^{-1}P_2)(X_1X_2X_3)(Q_1Q_4Q_5Q_6Y_2R_2)(X_6^{-1}X_{10}^{-1}X_9^{-1}X_4^{-1})=1, \quad ㉔$$

$$Y_9^{-1}P_2X_1X_2=X_3Q_1=S_2Y_{10}^{-1}Q_4Q_5=Q_6Y_2R_2X_6^{-1}X_{10}^{-1}X_9^{-1}=S_1X_4^{-1}=1,$$

$$u=S_1,w^{-1}=Q_1,v_1^{-1}\cdot(v_2^{-1}\cdot v_1^{-1})^{k-1}=Q_4\cdot Q_5.$$

(v) $X_2=X_7\cdot X_8,S_2=S_4\cdot S_5$,其中

$$Y_5^{-1}X_7=S_5X_8=S_4B_2=X_5^{-1}Y_2=1,$$

$$Ax_0Bx_0^{-1}=(S_1S_4S_5Y_5^{-1}P_2)(X_1X_7X_8X_3)(Q_1Q_2Y_2R_2)(X_6^{-1}X_5^{-1}X_4^{-1})=1. \quad ㉕$$

$$S_5Y_5^{-1}P_2X_1X_7X_8=X_3Q_1=S_4Q_2=Y_2R_2X_6^{-1}X_5^{-1}=S_1X_4^{-1}=1,$$

$$u=S_1,w^{-1}=Q_1,(v_1\cdot v_2)^{k-1}\cdot v_1=S_4.$$

(vi)不再作分解,其中

$$Y_5^{-1}X_2=S_2Q_2=X_5^{-1}Y_2=1,$$

$$Ax_0Bx_0^{-1}=(S_1S_2Y_5^{-1}P_2)(X_1X_2X_3)(Q_1Q_2Y_2R_2)(X_6^{-1}X_5^{-1}X_4^{-1})=1. \quad ㉖$$

$$Y_5^{-1}P_2X_1X_2=X_3Q_1=S_2Q_2=Y_2R_2X_6^{-1}X_5^{-1}=S_1X_4^{-1}=1,$$

$$u=S_1,w^{-1}=Q_1,(v_1\cdot v_2)^{k-1}\cdot v_1=S_2.$$

(vii) $Y_5^{-1}=Y_{10}^{-1}\cdot Y_9^{-1},Q_2=Q_4\cdot Q_5$,其中

$$Y_9^{-1}X_2=Y_{10}^{-1}Q_4=S_2Q_5=X_5^{-1}Y_2=1,$$

$$Ax_0Bx_0^{-1}=(S_1S_2Y_{10}^{-1}Y_9^{-1}P_2)(X_1X_2X_3)(Q_1Q_4Q_5Y_2R_2)(X_6^{-1}X_5^{-1}X_4^{-1})=1. \quad ㉗$$

$$Y_9^{-1}P_2X_1X_2=X_3Q_1=S_2Y_{10}^{-1}Q_4Q_5=Y_2R_2X_6^{-1}X_5^{-1}=S_1X_4^{-1}=1,$$

$$u=S_1, w^{-1}=Q_1, v_1^{-1}\cdot(v_2^{-1}\cdot v_1^{-1})^{k-1}=Q_4\cdot Q_5.$$

(viii) $X_2=X_7\cdot X_8, Y_2=Y_7\cdot Y_8, S_2=S_4\cdot S_5\cdot S_6$，其中

$$Y_5^{-1}X_7=S_6X_8=S_2Q_2=S_4Y_7=X_5^{-1}Y_8=1,$$

$$Ax_0Bx_0^{-1}=(S_1S_4S_5S_6Y_5^{-1}P_2)(X_1X_7X_8X_3)(Q_1Q_2Y_7Y_8R_2)(X_6^{-1}X_5^{-1}X_4^{-1})=1, \quad ㉘$$

$$S_6Y_5^{-1}P_2X_1X_7X_8=X_3Q_1=S_4S_5Q_2Y_7=Y_8R_2X_6^{-1}X_5^{-1}=S_1X_4^{-1}=1,$$

$$u=S_1, w^{-1}=Q_1, (v_1\cdot v_2)^{k-1}v_1=S_4\cdot S_5.$$

(ix) $Y_2=Y_7\cdot Y_8, S_2=S_4\cdot S_5$，其中

$$Y_5^{-1}X_2=S_5Q_2=S_4Y_7=X_5^{-1}Y_8=1,$$

$$Ax_0Bx_0^{-1}=(S_1S_4S_5Y_5^{-1}P_2)(X_1X_2X_3)(Q_1Q_2Y_7Y_8R_2)(X_6^{-1}X_5^{-1}X_4^{-1})=1, \quad ㉙$$

$$Y_5^{-1}P_2X_1X_2=X_3Q_1=S_4S_5Q_2Y_7=Y_8R_2X_6^{-1}X_5^{-1}=S_1X_4^{-1}=1,$$

$$u=S_1, w^{-1}=Q_1, (v_1\cdot v_2)^{k-1}v_1=S_4\cdot S_5.$$

(x) $Y_5^{-1}=Y_{10}^{-1}\cdot Y_9^{-1}, Q_2=Q_4\cdot Q_5, S_2=S_4\cdot S_5, Y_2=Y_7\cdot Y_8$，其中

$$Y_9^{-1}X_2=Y_{10}^{-1}Q_4=S_5Q_5=S_4Y_7=X_5^{-1}Y_8=1,$$

$$Ax_0Bx_0^{-1}=(S_1S_4S_5Y_{10}^{-1}Y_9^{-1}P_2)(X_1X_2X_3)(Q_1Q_4Q_5Y_7Y_8R_2)(X_6^{-1}X_5^{-1}X_4^{-1})=1, \quad ㉚$$

$$Y_9^{-1}P_2X_1X_2=X_3Q_1=S_4S_5Y_{10}^{-1}Q_4Q_5Y_7=Y_8R_2X_6^{-1}X_5^{-1}=S_1X_4^{-1}=1,$$

$$u=S_1, w^{-1}=Q_1, v_1\cdot(v_2\cdot v_1)^{k-1}=S_4\cdot S_5\cdot Y_{10}^{-1}=S_4\cdot S_5\cdot Q_4^{-1}.$$

(xi) $Y_5^{-1}=Y_{10}^{-1}\cdot Y_9^{-1}, Y_2=Y_7\cdot Y_8$，其中

$$Y_9^{-1}X_2=Y_{10}^{-1}Q_2=S_2Y_7=X_5^{-1}Y_8=1,$$

$$Ax_0Bx_0^{-1}=(S_1S_2Y_{10}^{-1}Y_9^{-1}P_2)(X_1X_2X_3)(Q_1Q_2Y_7Y_8R_2)(X_6^{-1}X_5^{-1}X_4^{-1})=1, \quad ㉛$$

$$Y_9^{-1}P_2X_1X_2=X_3Q_1=S_2Y_{10}^{-1}Q_2Y_7=Y_8R_2X_6^{-1}X_5^{-1}=S_1X_4^{-1}=1,$$

$$u=S_1, w^{-1}=Q_1, v_1\cdot(v_2\cdot v_1)^{k-1}=S_2\cdot Q_2^{-1}.$$

(3) x_0 是⑩式的第三种短解时仿照(2)可得证.

6. 只有一个奇异三元组和没有奇异三元组的两种情况均可引出矛盾. 命题 12 证毕.

命题 13　方程(Ⅲ)有变数解的充分必要条件是

$$SRQP=1 \quad ㉜$$

对方程③的任意变数解(x_0,y_0)必定存在$x_1,x_2,\cdots,x_l$和$y_1,y_2,\cdots,y_l$，其中y_i是x_{i-1}对方程③的短解，x_i是y_i对方程③的短解，并且(x_l,y_l)或者(x_{l-1},y_l)是下二解的特例，

$$(P^{-1}Q^{-1},g)\text{或}(f,Q^{-1}R^{-1}), \quad ㉝$$

其中f,g是变数.

证　条件(e. 29)的充分性明显，下证条件的必要性. 设(x_0,y_0)是方程③的变数解. 由命题 11 知有$x_1,x_2,\cdots,x_l$和$y_1,y_2,\cdots,y_l$，使y_i是x_{i-1}对方程③的短解，x_i是y_i对方程③的短解，并且(x_{l-1},y_l)，(x_l,y_l)是方程③的一组同y型短解. 由命题 12 知，(x_l,y_l)中不含有变数，因此(x_0,y_0)，(x_0,y_1)，$\cdots$，(x_{l-1},y_l)中有一个坐标含有变数，而另一个不含变数的. 设(x_k,y_{k+1})中x_k含有变数，而y_{k+1}中不含有变数(如系x_k不含有变数而y_k含有变数仿此可证). 考虑方程

$$(Sy_{k+1}^{-1}P)x(Qy_{k+1}R)x^{-1}=1 \quad ㉞$$

有变数解x_k. 由命题 6 知，它是蜕化方程，即有

$$Sy_{k+1}^{-1}P=Qy_{k+1}R=1,SRQP=1.$$

又这时方程㉞是$xx^{-1}=1$. 易见$(f,Q^{-1}R^{-1})$是方程③的解，而x_k是f的一个特例.

定义 4(解的递归形式).　设(a,b)是方程③的解，满足下面条件的元对集$\sum_{a,b}$叫作由(a,b)生成的解集合. 这里“元”是指自由群F与变数自由群X的自由积$F*X$中的元.

(1)$(a,b)\varepsilon\sum_{a,b}$.

(2)如果$(x_0,y_0)\varepsilon\sum_{a,b}$，$Rx_0^{-1}S\neq1$，设$C$是$Rx_0^{-1}S$的本原因子，那么

$$(x_0,y_0c^n)\varepsilon\sum_{a,b}.$$

(3)如果$(x_0,y_0)\varepsilon\sum_{a,b}$，$Qy_0R\neq1$，设$d$是$Qy_0R$的本原因子，那么

$$(x_0d^n,y_0)\varepsilon\sum_{a,b}.$$

(4)如果$(x_0,y_0)\varepsilon\sum_{a,b}$，$Rx_0^{-1}S=1$，那么$(x_0,g)\varepsilon\sum_{a,b}$，这里$g$是$F*X$中的任意元.

(5)如果$(x_0,y_0)\varepsilon\sum_{a,b}$，$Qy_0R=1$，那么$(f,y_0)\varepsilon\sum_{a,b}$，这里$f$是$F*X$的任意元.

(6) $\sum_{a,b}$是满足上列(1)～(5)的最小元对集.

命题 14 (i) (a,b)是方程③的解，$(a_1,b_1)\varepsilon\sum_{a,b}$，那么$(a_1,b_1)$是方程③的解.

(ii) (a_l,b_l)是(a,b)经过用命题 11 的办法求出来的短解，那么

$$\sum_{a_l,b_l}=\sum_{a,b}.$$

命题 15 如果方程③有解，那么存在有限个方程③的短解(a_1,b_1)，(a_2,b_2)，…，(a_k,b_k)，使得

方程③的解集合$=\sum_{a_1,b_1}\cup\cdots\cup\sum_{a_k,b_k}$.

证 由命题 12 知，方程③的短解个数有限. 设其为(a_1,b_1)，(a_2,b_2)，…，(a_k,b_k). 如果(x_0,y_0)是方程③的解，由命题 14 知有方程③的短解(a_i,b_i)，使得$(x_0,y_0)\varepsilon\sum_{a_i,b_i}$.

参考文献

[1] Lyndon R C. Equation in free groups. Trans. Amer. Math. Soc. ,96 1960,96: 445－457.

[2] Lorenc A A. The solution of systems of equations in one unknown in free groups. Dokl. Akad. Nauk SSSR,1963,148:1 253－1 256.

[3] Lorenc A A. , Equations without coefficients in free groups. Dokl. Akad. Nauk SSSR,1965,160:538－540.

[4] Appel K I. One variable equations in free groups. Proc. Amer. Math. Soc. , 1968,19:912－918.

[5] Appel K I. On two-variable equations in free groups. Proc. Amer. Math. Soc. , 1969,21:179－185.

[6] Lyndon R C & Sehupp P E. Combinatorial group theory, Springer-Verlag,1977.

数学学报,1981,24(3):472～480

可换群中无限生成元直和项消去条件的探讨①

Research on the Conditions of Cancellation of Direct Summands of Abelian Groups

Kaplansky 在文[1]中,提出了 3 个关于可换群同构的问题,其中第 3 个问题是:

"如果 F 是一个有限生成元的可换群,G 和 H 是可换群,使得 $F\oplus G\cong F\oplus H$,G 和 H 同构吗?"

Cohn[2]和 Walker[3]分别解决了上述问题. Walker 并在[3]中举例说明了原问题中为什么要限制 F 是有限生成元的. 本文证明了虽然无限多个循环群的直和一般地不能从一个直和式中消去,但是如果 F 是无限个循环群的直和,其中无限循环群作为直和项的个数 r_0,以及细分之后对每一个素数 p_i,F 中幂 p_i 循环群直和项的个数 r_i 具有共同的上限 $M\geqslant r_i$ $(i=0,1,\cdots)$,那么 F 是可以从一个直和式中消去的.

文中对可换群一律简称群. 有限个或无限个群的直和记作 $\sum\oplus A_i$. x 生成的子群记作$\{x\}$. n 元循环群记作 $C(n)$.

定义 设 $G=\sum\oplus A_i$,$\overline{A}_i=\sum\limits_{j\neq i}\oplus A_i$ 叫作 A_i 在 G 内的补项.

预理 1 设 G 是群,x 是 G 的 p^n 阶元,H 是 G 的子群. 如果有某一 $kx\in H$,$0<k<p^n$,那么 $p^{n-1}x\in H$.

证 设 p^i 是满足 $p^i\mid k$ 的极大者,$o\leqslant i<n$,只需证 $p^ix\in H$. 设 $k=qp^i$,$p\nmid q$,$(p^{n-i},q)=1$. 有整数 u,v 满足 $up^{n-i}+vq=1$.

---

① 收稿日期:1979-10-16.

$$p^i = p^i(up^{n-i}+vq)$$
$$=up^n+vqp^i=up^n+vk.$$
$$p^ix=up^nx+vkx=v(kx)\in H.$$

预理 2 G是群,H是G的子群. 如果 a 是 p^n 阶元,$p\nmid k$,$ka+lb\in H$. 那么存在整数 w 使得 $a+wb\in H$.

证 选取 $up^n+vk=1$.

$$v(ka+lb)=vka+vlb$$
$$=(1-up^n)a+vlb=a+wb\in H.$$

预理 3 设 $G=A\oplus\overline{A}=C\oplus\overline{C}$. a,c 各是 A,C 的生成元,$A\cong C\cong C(p^n)$. $E=\overline{A}\cap\overline{C}$. 那么这组对应的直和分解具有下列性质(Ⅰ)或(Ⅱ):

(Ⅰ)存在 $b\in\overline{A}$,$c=a+b$,$p^{n-1}a\notin\overline{C}$.

(Ⅱ)又分为两种可能.

(Ⅱ·1)存在 p^n 阶元 $b\in\overline{A}$,$c=a+b$,$p^{n-1}a\in\overline{C}$,

或者

(Ⅱ·2)存在 p^n 阶元 $b\in\overline{A}$,$c=p^ma+b$,$0<m<n$.

当满足性质(Ⅰ)时,又可推出存在整数 u 使得 $ua+b\in\overline{C}$. 当满足性质(Ⅱ)时,又可推出存在 $x\in(\overline{A}-\overline{C})$,$y\in(\overline{C}-\overline{A})$(集合差),$x$ 和 y 都是 p^n 阶元,并且

$$\overline{A}=\{x\}\oplus E,\quad \overline{C}=\{y\}\oplus E.$$

证 设 $c=ka+b$,$b\in\overline{A}$. 如果 $p\nmid k$,那么由预理 2 知有 $c_1=a+vb=a+b_1\times(b_1\in\overline{A})$,$C_1$ 也可以作 C 的生成元. 对 b_1 的阶再划分如下:b_1 的阶为 p^m,$0\leqslant m<n$,即下面情形(1). b_1 的阶为 p^n 即下面情形(2). 如果$c=ka+b$,$b\in\overline{A}$,$p|k$,那么 b 的阶是 p^n,c 的生成元可以换为 $c_1=p^ma+b_1$,$0\leqslant m\leqslant n$. b_1 也是 p^n 阶元. 这里 $m=0$ 已由情形(2)讨论,$0<m\leqslant n$ 归入情形(3).

(1)$c=a+b$,$b\in\overline{A}$,b 是 p^m 阶元,$0\leqslant m<n$(包括 $b=0$).

①$p^{n-1}c=p^{n-1}a+p^{n-1}b=p^{n-1}a\in C$. 由直和关系知 $p^{n-1}a\notin\overline{C}$.

②因为 $a\in C\oplus\overline{C}$,所以有整数 k 和 $x+y\in\overline{C}$(其中 $x\in A$,$y\in\overline{A}$)满足

$a=k(a+b)+(x+y)=(ka+x)+(kb+y)$.

$a=ka+x$,$0=kb+y$(直和分解唯一性).

$x+y=(1-k)a-kb\in\overline{C}$,$(k-1)a+kb\in\overline{C}$.

③$p\nmid k$. 否则 $p|k, p\nmid(k-1)$, $(p^n,(k-1))=1$. 有整数 u,v 满足

$$up^n+v\cdot(k-1)=1.$$

$$v((k-1)a+kb)=v(k-1)a+vkb$$

$$=(1-up^n)a+vkb=a+vkb\in\overline{C}.$$

又因 $p|k$,

$$p^{n-1}(a+vkb)=p^{n-1}a+vkp^{n-1}b=p^{n-1}a\in\overline{C},$$

但由(1·1) $p^{n-1}a\notin\overline{C}$,矛盾.

④$(k-1)a+kb\in\overline{C}$, $p\nmid k$,由预理 2 知有 $wa+b\in\overline{C}$,性质(Ⅰ)的条件和结论得到满足.

(2)$c=a+b$, $b\in\overline{A}$, b 是 p^n 阶元.

①设 $p^{n-1}a\notin\overline{C}$,这时仿照(1·2)~(1·4)可知性质(Ⅰ)的条件和结论都已经满足.

②设 $p^{n-1}a\in\overline{C}$. 这时对任意 $0<k<p^n$ 均有 $kb\notin\overline{C}$. 否则由预理 1 知 $p^{n-1}b\in\overline{C}$, $p^{n-1}(a+b)\in\overline{C}$. 但另一方面又有 $p^{n-1}(a+b)\in C$,矛盾于 $C\oplus\overline{C}$ 是直和.

由此得出 $\{b\}\cap\overline{C}=\{0\}$. 所以

$$\{b\}\cap E=\{b\}\cap\{\overline{A}\cap\overline{C}\}=0,$$

$\{b\}+E$ 是直和

(i)因为 $b\in C\oplus\overline{C}$,所以有整数 k 和 $x+y\in\overline{C}$(其中 $x\in A, y\in\overline{A}$),满足

$$b=k(a+b)+(x+y)=(ka+x)+(kb+y).$$

$$ka+x=0, kb+y=b.$$

$$x+y=-ka+(1-k)b\in\overline{C},\quad ka+(k-1)b\in\overline{C}.$$

(ii)$p\nmid k$. 否则 $p|k, p\nmid(k-1)$,有整数 u,v 满足 $up^n+v(k-1)=1$.

$$v(ka+(k-1)b)=vka+v(k-1)b$$

$$=vka+(1-up^n)b=vka+b\in\overline{C}.$$

$$p^{n-1}(vka+b)=vkp^{n-1}a+p^{n-1}b$$

$$=p^{n-1}b\in\overline{C}.$$

矛盾于②中已证出的 $p^{n-1}b\notin\overline{C}$.

(iii)由 $p\nmid k$, $ka+(k-1)b\in\overline{C}$ 及预理 2 知存在整数 w 能使

$$a+wb\in\overline{C}.$$

(iv)因为 $b\in\overline{A}$，$\overline{A}$ 中不可能含有 $k(a+wb)(0<k<p^n)$ 形之元所以

$$\{a+wb\}\cap\overline{A}=\{0\},$$

$$\{a+wb\}\cap E=\{a+wb\}\cap\overline{A}\cap\overline{C}=\{0\},$$

$\{a+wb\}+E$ 是直和.

(v)对任意 $la+b_1\in\overline{C}$（其中 $b_1\in\overline{A}$），$(la+b_1)-l(a+wb)=b_1-lwb\in\overline{C}$. 但 $b_1-lwb\in\overline{A}$，所以

$$b_1-lwb\in(\overline{A}\cap\overline{C})=E.$$

$$la+b_1=l(a+wb)+(b_1-lwb)$$
$$\in(\{a+wb\}\oplus E).$$

反之，由(iii)及 E 的定义显然有

$$\{a+wb\}\oplus E\subseteq\overline{C}.$$

所以
$$\overline{C}=\{a+wb\}\oplus E.$$

由于 a,b 分别是 $A,\overline{A}$ 中的 p^n 阶元，且 $A\cap\overline{A}=\{0\}$，易见 $a+wb$ 是 p^n 阶元.

(vi)对任意 $b_1\in\overline{A}$，因为 $b_1\in C\oplus\overline{C}$，$=C\oplus(a+wb)\oplus E$，所以有整数 k,l 和 $e\in E$ 满足

$$b_1=k(a+b)+l(a+wb)+e$$
$$=(k+l)a+(b+lwb+e).$$
$$(k+l)a=0,\quad b_1=(k+lw)b+e.$$

即有 $b_1\in\{b\}\oplus E$.

反之，显然有$\{b\}\oplus E\in\overline{A}$.

所以
$$\overline{A}=\{b\}\oplus E.$$

以上证明了属于②的情形满足性质(Ⅱ·1)的条件和结论.

(3)$c=p^ma+b$，$0<m<n$，b 是 $\overline{A}$ 的 p^n 阶元.

①$p^{n-1}(p^ma+b)$

$$=p^{n+m-1}a+p^{n-1}b=p^{n-1}b\in C.$$

所以 $p^{n-1}b\notin\overline{C}$. 由预理 1 知对任意 $0<k<p^n$ 均有 $kb\notin\overline{C}$. 由此$\{b\}+E$ 是直和.

②因为 $a\in C\oplus\overline{C}$，所以有 k 和 $x+y\in\overline{C}$（其中 $x\in A$，$y\in\overline{A}$），满足

$$a=k(p^ma+b)+(x+y)$$
$$=(kp^ma+x)+(kb+y).$$

$$a=kp^m a+x,\quad 0=kb+y.$$

$$x+y=(1-kp^m)a-kb\in\overline{C},$$

$$(kp^m-1)+kb\in\overline{C}.$$

$(1-kp^m)$与 p 互素，由预理 2 知存在整数 w 能使 $a+wb\in\overline{C}$.

③$\{a+wb\}\cap E=\{a+wb\}\cap\overline{A}\cap\overline{C}=\{0\}$，其中$\{a+wb\}\cap\overline{A}=\{0\}$仿前(iv)，所以$\{a+wb\}+E$ 是直和.

④对任意 $x+y\in\overline{C}$(其中 $x\in A,y\in\overline{A}$)，设 $x=ka$.

$$(ka+y)-k(a+wb)=y-kwb\in\overline{C}.$$

又易见 $y-kwb\in\overline{A}$. 所以 $y-kwb\in E$.

$$x+y=ka+y=k(a+wb)+(y-kwb)$$

$$\in\{a+wb\}\oplus E.$$

所以 $$\overline{C}=\{a+wb\}\oplus E.$$

仿(v)知 $a+wb$ 是 p^n 阶元.

⑤对任意 $b_1\in\overline{A}$，因为 $b_1\in C\oplus\overline{C}=C\oplus\{a+wb\}\oplus E$. 所以有 k,l 和 $e\in E$ 满足

$$\begin{aligned}b_1&=k(p^m a+b)+l(a+wb)+e\\&=(kp^m+l)a+(k+lw)b+e.\end{aligned}$$

$$(kp^m+l)a=0,\quad (k+lw)b+e=b_1.$$

即有 $b_1\in\{b\}\oplus E$.

所以 $$\overline{A}=\{b\}\oplus E.$$

以上证明了属于(3)的情形满足(Ⅱ·2)的条件和结论.

在证明预理 4,5 之前，我们先介绍一些记号. 设

$$G=\sum_{i=1}^{\infty}\oplus A_i\oplus B=\sum_{i=1}^{\infty}\oplus C_i\oplus D.$$

a_i,c_i 各是 A_i,C_i 的生成元，$\overline{A}_i$ 和 $\overline{C}_i$ 各是 A_i 和 C_i 在 G 内的补项.

$$\overline{A}_i\cap\overline{C}_i=E_i,\quad \bigcap_{i=1}^{\infty}E_i=E.$$

$$A_i\cong C_i\cong C(p_i^{n}i).$$

$p_1,p_2,\cdots,p_i,\cdots$是互不相同的素数. 另外 B 的元素记作 $b,b_1,\cdots,D$ 的元素记作 $d,d_1,\cdots$

预理 4 设 $G=A_i\oplus\overline{A}_i=C_i\oplus\overline{C}_i$ 对所有 i 均满足性质(Ⅰ)，那么知所有 $b\in B$，存在唯一的有限整数列 $k_1,k_2,\cdots,k_r$ 满足 $\sum\limits_{i=1}^{r}k_ia_i+b\in D$，并且

$$f=\sum_{i=1}^{r}k_ia_i+b\rightarrow b$$

是 D 到 B 上的同构对应.

证 (1)由性质(Ⅰ)知有 $c_i=a_i+b_i, b_i\in\overline{A}_i, p^{n_i-1}a\notin\overline{C}_i$,并且存在整数 u_i 满足 $u_ia_i+b_i\in\overline{C}_i$.

(2)易见 b_i 的阶为 $p_i^{n}i$ 的因子.

(3)若 $\sum\limits_{i=1}^{r}k_ia_i\in D$,则必定 $k_i\equiv 0(\mathrm{mod}\ p_i^{n}i)$. 否则例如 $k_1\not\equiv 0(\mathrm{mod}\ p_1^{n_1})$取

$$\pi=p_2^{n_2},p_3^{n_3},\cdots,p_r^{n_r},\quad(\pi,p_1)=1.$$

$$\pi\sum_{i=1}k_ia_i=\pi k_1a_1\in D.$$

$\pi k_1\not\equiv 0(\mathrm{mod}\ p_1^{n_1})$,由预理 1 知 $p_1^{n_1-1}a_1\in D$. 但 $D\subseteq\overline{C}_1$,又可得 $p_1^{n_1}a_1\in\overline{C}_1$,矛盾于(1).

(4) $\sum\limits_{i=1}^{r}k_ia_i+b\in D$, $\sum\limits_{i=1}^{s}l_ia_i+b\in D$(不妨设 $r=s$,否则以 0 补齐)能推出 $k_ia_i=l_ia_i(i=1,2,\cdots,r)$. 否则相减由(3)推出矛盾. 证明了 f 是异元异象的.

(5)对任意 $b\in B$. 因为 $b\in\left(\sum\limits_{i=1}^{\infty}\oplus C_i\oplus D\right)$,所以有 $k_1,k_2,\cdots,k_r$; $l_1,l_2,\cdots,l_s$ 和

$$d=\sum_{i=1}^{s}l_ia_i+b'\in D$$

(不妨认为 $r=s$)满足

$$\begin{aligned}b&=\sum_{i=1}^{r}k_ic_i+d\\&=\sum_{i=1}^{r}k_i(a_i+b_i)+\sum_{i=1}^{r}l_ia_i+b'\\&=\sum_{i=1}^{r}(k_i+l_i)a_i+\left(\sum_{i=1}^{r}k_ib_i+b'\right).\end{aligned}$$

由直和分解的唯一性知

$$l_i\equiv -k_i(\mathrm{mod}\ p_i^{n_i})\quad(i=1,2,\cdots,r),$$

$$b=\sum_{i=1}^{r}k_ib_i+b'.$$

(6)由性质(Ⅰ)知有 $u_ia_i+b_i\in\overline{C}_i$. $u_ia_i+b_i$ 的阶是 p_i 的幂,它不可能 $\in(\overline{C}_i-D)$(集合差),否则 $u_ia_i+b_i=\sum\limits_{j\neq i}^{s}v_jc_j+d$, 其中至少有一个 $v_j\not\equiv 0(\mathrm{mod}\ p_j^{n_j})$得到 $u_ia_i+b_i$ 的倍数是 $p_j(j\neq i)$的倍数,矛盾. 所以 $u_ia_i+b_i\in D$.

$$\begin{aligned}&d+\sum_{i=1}^{r}k_i(u_ia_i+b_i)\\=&\sum_{i=1}^{r}l_ia_i+b'+\sum_{i=1}^{r}k_iu_ia_i+\sum_{i=1}^{r}k_ib_i\\=&\sum_{i=1}^{r}(l_i+k_iu_i)a_i+\sum_{i=1}^{r}k_ib_i+b'\\=&\sum_{i=1}^{r}(l_i+k_iu_i)a_i+b\in D.\end{aligned}$$

$$f(\sum_{i=1}^{r}(l_i+k_iu_i)a_i+b)=b.$$

证明了 f 是 D 到 B 上的映象.

(7)以上证明了 f 是 D 到 B 上的一一对应,其保持运算性也明显,所以
$$B\cong D.$$

预理 5 设 $G=A_i\oplus\overline{A}_i=C_i\oplus\overline{C}_i$ 对所有 i 均满足性质(Ⅱ),那么$B\cong D$.

证 由条件(Ⅱ)知有 $p_i^{n_i}$ 阶元
$$b_i\in(\overline{A}_i-\overline{C}_i),\quad d_i\in(\overline{C}_i-\overline{A}_i)$$
满足 $\overline{A}_i=\{b_i\}\oplus E_i$,$C_i=\{d_i\}\oplus E_i$.

(1)$b_i\in B$,$d_i\in D$. 否则例如 $b_i\notin B$,那么 $b_i\in\overline{A}_i-B$. 即有表达式
$$b_i=\sum_{j\neq i}^{r}k_ja_j+b',$$
其中至少有一个 $k_j\not\equiv 0(\mathrm{mod}\ p_j^{n_j})$,但 b_i 的阶为 $p_i^{n_i}$,由阶数计算推出矛盾.

(2)下证 $\sum\limits_{i=1}^{\infty}\{b_i\}+E$ 是直和.

①假设 $\{b_i\}\cap(\sum\limits_{j\neq i}^{\infty}\{b_j\}+E)\neq\{0\}$, 由预理 1 知 $p_i^{n_i-1}b_i\in\sum\limits_{j\neq i}^{\infty}\{b_j\}+E$. 仿前由阶数计算可得 $p_i^{n_i-1}b_i\in E$,但前面又可知 $p_i^{n_i-1}b_i\notin E_i$,$E_i\supseteq E$,矛盾.

②对每一个正整数 r

$$(E_1\cap E_2\cap\cdots\cap E_r)\cap(\{b_1\}+\{b_2\}+\cdots+\{b_r\})=\{0\}.$$

否则设 $\sum_{i=1}^{r}k_ib_i\in(\bigcap_{i=1}^{r}E_i)$，其中某 $k_i\not\equiv 0(\mathrm{mod}\ p_i^{n_i})$. 取 $\pi=\prod_{j\neq i}^{r}p_j^{n_j}$.

$$\pi\sum_{j=1}^{r}k_jb_j=\pi k_ib_i\in\bigcap_{j=1}^{r}E_j\subseteq E_i.$$

由预理 1，$p_i^{n_i-1}b_i\in E_i$，矛盾于 $\{b_i\}\oplus E_i$ 是直和.

③ $E\cap(\sum_{i=1}^{\infty}\{b_i\})=\{0\}$. 否则有 $\sum_{i=1}^{r}k_ib_i\in E$（至少有一个 $k_i\not\equiv 0(\mathrm{mod}\ p_i^{n_i})$），$E\subseteq(\bigcap_{i=1}^{r}E_i)$ 得 $\sum_{i=1}^{r}\{b_i\}\cap(\bigcap_{i=1}^{r}E_i)\neq\{0\}$，与(2・2)矛盾.

(3) 下证 $B=\sum_{i=1}^{\infty}\oplus\{b_i\}\oplus E$.

①对任意 $b\in B$，因为

$$B\subseteq\overline{A}_i=\{b_i\}\oplus E_i,$$

所以有 $b=k_ib_i+e_i(e_i\in E_i)$.

② $k_1,k_2,\cdots,k_i,\cdots$ 中只有有限个 $\not\equiv 0(\mathrm{mod}\ p_i^{n_i})$. 否则有无限多个 $k_i\not\equiv 0(\mathrm{mod}\ p_i^{n_i})$. 对这样的 k_i，$b=k_ib_i+e_i\notin E_i=(\overline{A}_i\cap\overline{C}_i)$，$(e_i\in E_i)$. 而 $b\in C_i\oplus\overline{C}_i$ 中又有 $b=l_ic_i+y_i(y_i\in\overline{C}_i)$. 这里 $l_i\not\equiv 0(\mathrm{mod}\ p_i^{n_i})$.（否则 $b=y_i\in\overline{C}_i$，$b\in E_i$ 矛盾于前面已设 $b\notin E_i$.）所以有无限多个 $l_i\not\equiv 0(\mathrm{mod}\ p_i^{n_i})$. 但另一方面 $b\in\sum_{i=1}^{\infty}\oplus C_i\oplus D$，如果 $b=\sum_{i=1}^{r}l'_ic_i+d$，容易看出应该

$$l'_i\equiv l_i(\mathrm{mod}\ p_i^{n_i}),\quad i=1,2,\cdots,r.$$

$$0\equiv l_i(\mathrm{mod}\ p_i^{n_i}),\quad i>r.$$

这样只有有限个 $l_i\not\equiv 0(\mathrm{mod}\ p_i^{n_i})$，矛盾.

③设 $k_{r+1}=k_{r+2}=\cdots=0$. 考虑

$$b'=b-\sum_{i=1}^{r}k_ib_i.$$

$b'\in B\subseteq\overline{A}_i$，$b'$ 在 $\overline{A}_i=\{b_i\}\oplus E_i$ 的分解中设为 $b'=k'_ib_i+e'_i$，则

$$\begin{aligned}k'_ib_i+e'_i=b'&=b-\sum_{i=1}^{r}k_ib_i\\&=(k_ib_i+e_i)-(k_ib_i+\sum_{\substack{j=1\\j\neq i}}^{r}k_jb_j)\\&=\sum_{\substack{j=1\\j\neq i}}^{r}-k_jb_j+e_i\in E_i.\end{aligned}$$

由直和分解的唯一性知 $k'_i b_i=0(i=1,2,\cdots)$. 由此得出 $b'\in E_i(i=1,2,\cdots)$,

$$b'\in E(=\bigcap_{i=1}^{\infty}E_i).$$

$$b=\sum_{i=1}^{r}k_ib_i+b'\in(\sum_{i=1}^{\infty}\oplus\{b_i\}\oplus E).$$

所以 $$B=\sum_{i=1}^{\infty}\oplus\{b_i\}\oplus E.$$

(4)类似地可以证明

$$D=\sum_{i=1}^{\infty}\oplus\{d_i\}\oplus E.$$

所以 $$B\cong D.$$

在进行预理 6 的证明之前,我们需要对前面的记号作一些调整,设

$$G=\sum_{i=1}^{\infty}\oplus A_i\oplus B=\sum_{i=1}^{\infty}\oplus C_i\oplus D,$$

按照 $G=A_i\oplus\overline{A}_i=C_i\oplus\overline{C}_i$, $A_i\cong C_i\cong C(p_i^{n_i})$这组对应的直和分解满足性质(Ⅰ)或(Ⅱ)进行分类. 凡是满足性质(Ⅰ)的仍记作 A_i, C_i,除满足性质(Ⅰ)的之外,由预理 3 知应满足性质(Ⅱ),这种直和项记作 F_i, H_i,我们得到 $G=\sum_{i=1}^{\infty}\oplus A_i\oplus\sum_{i=1}^{\infty}\oplus F_i\oplus B=\sum_{i=1}^{\infty}\oplus C_i\oplus\sum_{i=1}^{\infty}\oplus H_i\oplus D$.(不妨设满足性质(Ⅰ)的及性质(Ⅱ)的都有无限多个对应的直和项,否则可以利用[2]的定理将其消去而归入前面两个预理的情形.)其中 $A_i\cong C_i\cong C(p_i^{m_i})$, $F_i\cong H_i\cong C(q_i^{n_i})$. $p_1,p_2,\cdots,p_i,\cdots,q_1,q_2,\cdots,q_i,\cdots$是互不相同的素数. A_i, C_i, F_i 和 H_i 的生成元各是 a_i, c_i, f_i 和 h_i,它们各自在 G 内的补项分别记作 $\overline{A}_i$, $\overline{C}_i$, $\overline{F}_i$ 和 $\overline{H}_i$.

预理 6 $B\cong D$.

证 (1)设 $U=\sum_{i=1}^{\infty}\oplus F_i\oplus B$, $V=\sum_{i=1}^{\infty}\oplus H_i\oplus D$. 由预理 4 知有同构对应 $g:V\to U$,对任意 $v\in V$(由第一个直和式)

$$\begin{aligned}v&=\sum_{i=1}^{r}k_ia_i+\sum_{j=1}^{s}l_jf_j+b\\&=\sum_{i=1}^{r}k_ia_i+u.\quad(u=\sum_{j=1}^{s}l_jf_j+b).\end{aligned}$$

$$g(v)=u,u\in U.$$

(2)设 $g(h_i)=h'_i$，$g(H_i)=H'_i$ 和 $g(D)=D'$. 因为同构对应保持直和，所以

$$U=\sum_{i=1}^{\infty}\oplus F_i\oplus B=\sum_{i=1}^{\infty}\oplus H'_i\oplus D',$$

$$H_i\cong H'_i\cong C(q_i^{n_i}).$$

(3)设 F_i 在 U 内的补项是 $\overline{F}_i$，H'_i，在 U 内的补项是 $\overline{H}_i$.

(4)由设 $G=F_i\oplus\overline{F}_i=H_i\oplus\overline{H}_i$ 满足性质(Ⅱ)，下面我们证明 $U=F_i\oplus\overline{\overline{F}}_i=H'_i\oplus\overline{\overline{H}}'_i$ 也满足性质(Ⅱ).

①由性质(Ⅱ)知又可分为两种可能：

(i) $h_i=f_i+b_i$，$b_i\in\overline{F}_i$，b_i 是 $q_i^{n_i}$ 阶元，$q_i^{n_i-1}f_i\in\overline{H}_i$，或者

(ii) $h_i=q_i^{t_i}f_i+b_i$，$b_i\in\overline{F}_i$，b_i 是 $q_i^{n_i}$ 阶元，$0<t_i<n_i$.

②对(4·1·1)的情形，利用阶数计算知 $b_i\in B\subseteq\overline{\overline{F}}_i$.

$h'_i=g(h_i)=f(f_i+b_i)=f_i+b_i=h_i$.

$q_i^{n_i-1}f_i\in\overline{H}_i$，由阶数计算知 $q_i^{n_i-1}f_i\in D$.

$q_i^{n_i-1}f_i=g(q_i^{n_i-1}f_i)\in g(D)=D'\subseteq\overline{\overline{H}}'_i$.

以上证明了存在 $q_i^{n_i}$ 阶元 $b_i\in\overline{\overline{F}}_i$，$h'_i=f_i+b_i$，$q_i^{n_i-1}f_i\in\overline{\overline{H}}'_i$，即 $U=F_i\oplus\overline{\overline{F}}_i=H_i\oplus\overline{\overline{H}}'_i$ 满足性质(Ⅱ).

③对(4·1·2)的情形，利用阶数计算可以得到 $b_i\in B\subseteq\overline{\overline{F}}_i$.

$$\begin{aligned}h'_i&=g(h_i)=g(q_i^{t_i}f_i+b)\\&=q_i^{t_i}f_i+b_i=h_i.\end{aligned}$$

即存在 $g_i^{n_i}$ 阶元 $b_i\in\overline{\overline{F}}_i$，$h'_i=q_i^{t_i}f_i+b_i$，$0<t_i<n_i$，所以也有 $U=F\oplus\overline{\overline{F}}_i=H'_i\oplus\overline{\overline{H}}'_i$ 满足性质(Ⅱ).

(5)考虑 $U=\sum_{i=1}^{\infty}\oplus F_i\oplus B=\sum_{i=1}^{\infty}H'_i\oplus D'$，对每一个 i 我们证明了 $U=F\oplus\overline{\overline{F}}_i=H'_i\oplus\overline{\overline{H}}'_i$ 都满足性质(Ⅱ)，所由预理 5 得到 $B\cong D'$.

所以 $$B\cong D'\cong D.$$

定理 设

$$G=A_0\oplus\sum_{i=1}^{\infty}\oplus A_i\oplus B=C_0\oplus\sum_{i=1}^{\infty}\oplus C_i\oplus D,$$

$A_0\cong C_0\cong$(有限生成元的可换自由群)，

$A_i\cong C_i\cong$(不超过 M 个其阶数为 p_i 的幂的循环群的直和). p_i，p_2，…，p_i，…是互不相同的素数，那么 $B\cong D$.

证 由[2]知可消去 A_0,C_0. 故不妨设 $H=\sum_{i=1}^{\infty}\oplus A_i\oplus B=\sum_{i=1}^{\infty}\oplus C_i\oplus D$.

设 $A_i=\sum_{j=1}^{r_i}\oplus A_{ij}$, $C_i=\sum_{j=1}^{r_i}\oplus C_{ij}$, 其中

$$A_{ij}\cong C_{ij}\cong C(p_i^{n_{ij}}).$$

$$H=\sum_{i=1}^{\infty}\oplus A_{i1}\oplus\sum_{i=1}^{\infty}\oplus\sum_{j=2}^{r_i}\oplus A_{ij}\oplus B$$

$$=\sum_{i=1}^{\infty}\oplus C_{i1}\oplus\sum_{i=1}^{\infty}\oplus\sum_{j=2}^{r_i}\oplus C_{ij}\oplus D.$$

由预理 6 我们得到

$$H_1=\sum_{i=1}^{\infty}\oplus\sum_{j=2}^{r_i}\oplus A_{ij}\oplus B\cong\sum_{i=1}^{\infty}\oplus\sum_{j=2}^{r_i}\oplus C_{ij}\oplus D.$$

这样反复运用不超过 M 次之后我们得到

$$B\cong D.$$

参考文献

[1] Kaplansky I. Infinite Abelian groups. Ann Arbor, 1954.

[2] Cohn P M. The complement of a finitely generated direct summand of an Abelian group. Proc. Amer. Math. Soc., 1956, 7: 520—521.

[3] Walker E A. Cancellation in direct sums of groups. Proc. Amer. Math. Soe., 1956, 7: 898—902.

数学通报,1996,(9):24～27

计算机科学发展漫谈

A Ramble Talk on the Developments of Computer Science

§1. 历史的回顾(手摇计算机和山本五十六事件的关系引起)

很多计算机科学家把计算机的历史追溯到古希腊的亚里士多德,但我并不认为是这样的. 直到 1940 年人们还只有极其简单的手摇计算机,这种计算机算 3 715×2 864 就要摇 2+8+6+4=20 下. 第二次世界大战快结束时,德国希特勒制成了速度较高的密码机,它的名字叫作 Enigma. 日本人大概也是用这种机器来处理日常的通讯的. 他们认为这样既方便又可靠,而且协约国方面是不可能破译他们的密码的. 可是由于情报机关的努力,英国搞到了这种密码机. 于是英、美方面展开了研破德、日的密码的工作. 当时英国的机构设在 Bletchley Park,有名的计算机科学家兼数学家 Turing 参加了这个机构. 美国的机构设在 Washington. 有名的数学家兼计算机科学家 Von Neumann 主持这里的研究工作. 这个机构的名字叫作密码分析局,直接隶属于海军部. 据当时的另一个负责人和其中的一位手摇计算机手的回忆说他们对外称为一个公司,里面的计算机手有一百多人,全部都是女的,她们对外说是公司里面的会计. 每天上班就人手一台手摇计算机进行工作. 据这位会计说她们实际是按国家工作人员待遇,也知道所从事的工作非常重要,但却不知道自己在做什么工作. 这个密码局最大的成绩就是破译了关于山本五十六的行踪的密码. 于是他所乘的飞机被打了下来. 这种能把对方的最高军事指挥官打掉是战争史上极成功的范例. 而手摇计算机小姐们的一份功劳是不可磨灭的.

从这个密码局的分工来看,我们大致得到了近代电子计算机的一个框架.

(1)每一个操作员相当于一个运算器

(2)有几个人扮演执行程序的 Obj,exe 或 a. out 的任务

(3)核心的几个人相当于输入和输出的 Device

(4)有平行计算的雏形

后来 Von Neumann 向国家建议拨款研制电子计算机是和他这一段时间的工作经验分不开的. 第 1 个电子计算机叫作 ENIAC(1945 年研制成,1946 年投入使用). 它有 18 000 个电子管,其中每周要坏掉 3 个. Von Neumann 参加了第 2 个电子计算机的研制工作. 它的名字叫作 ED—VAC. 1945 年 Von Neumann 给出了关于这个机器的第 1 个报告. 据说这是一个非常重要的报告,但从它一提出便被立即划为国家的机密文件,只能传阅不能发表. 直到 1981 年它才被公之于世,这时 Von Neumann 已经去世十几年了. 由于这个报告受到如此的重视,总工程师 Eckert Jr. 出来争功. 他说那是很多人的集体成果而 Von Neumann 只完成其中很小的一部分,到底应如何记功那就不是我们的事了. 和 Von Neumann 同时 Turing 也参加了英国的第 1 个电子计算机 ACE 的研制工作. 他也写过一个关于 ACE 的报告,但是没有受到如此的重视. 他是 Von Neumann 的博士生,但却很遗憾地过早去世. 我们直到现在还纪念他,以上这一段就是介绍电子计算机的缘起.

§ 2.　Main Frame(主框架)时期

IBM 向国家买到了制造计算机的专利. 最初的商业用机都是 IBM 系列产品. 它的操作系统叫作 CMS. 当时只有大公司才买得起. 例如 Ford 汽车公司,Douglas 飞机公司等. 要运行这种机器要三班倒,每班至少要 6 个人,加上星期六、日轮休共需约 30 人. 由于 Basic,Pascal 和 Fortran 等算法语言的出现使 CMS 能做很多计算工作. 但是到了 20 世纪 70 年代后期这些机器纷纷退役. 其原因是太费人工. 于是工厂便把这批机器作为礼物送给学校. 几乎每所大学都得到一台这种礼物,当然对人工的耗费也转给了学校. CMS 在学校中得了一个雅号叫作 Main Frame. Main Frame

很有生命力，他们在各个大学中扎下了根. 整个 20 世纪 80 年代各个学校的教学、科研都是以 Main Frame 为主体来实现的. 到了 20 世纪 90 年代各种型号的机器相继出现. Main Frame 举足轻重的时代已经过去. 但它们仍然可以作为机器中的一员而继续服务，日本某些大学也用 Main Frame，但不像在美国那么普遍.

我 1981 年到美国，由于我的一位导师是计算机教授，于是开始接触计算机. 我先从打论文入手，逐渐熟悉机器. 当时 TEX 软件还在研制之中，Michigan 用的是自己编的文字处理程序叫作 STAT. 这时计算使用机器的费用非常严密. 我总共从系里要到 200 美元. 有时一运行就要用 3 美元，还要不进到一个反复不停的 loop. 如果进到了 loop，一下手就要十几美元. 机器每走一步就告诉你花了多少钱还剩多少，有时一运行结束，不但钱没有了反而欠了机器的钱. 为此我不得不改为晚上上机. 晚上的费用是白天费用的十分之一. 我在美国、日本工作过的学校都有 Main Frame. 一直到我离开美国，这些机器仍然在运行着.

§3. 全面发展时期

第一个向 IBM 挑战的是 Apple，它的 PC 机以图形来操作别具一格，后来 Vax 和 Digital 的工作站出现，Sun 又更进一步抢去了很多市场，我在日本的学校与 Apple 签了合同，他们使用的 Power Book 中的商业软件对学生很有用. 当时学校规定是人手一机，经验证明学校的决策是对的，该校的毕业生每一个人都有三个工作可以挑选，而其他学校的学生则有些只有半个位置. 在 Mississippi 我利用了半年时间来熟悉 Ceber 超级计算机，在国际会议上我和与会者讨论过平行计算的问题，据报告 Yale 有一个超级魔团平行机，它有 2 000 000 个处理器.

在这一段时间里我还访问过台湾，我和台大、台湾中研院的计算研究所以及资讯研究所都讨论过计算机的发展的问题.

到 1993 年我再去美国，他们已经有了非常高级的照片处理软件，网络系统也已大体完成，这时全国发生了几起打入国防部数据库和大银行现金管理网的事件，这大概是由于联网工作做得太好了的缘故.

大家看《天方夜谭》知道有一个小孩走到一个山洞门口喊一声“芝麻开门”，于是洞门大开，他跑进去装了一大堆财宝回家，现在你可以坐在家里打开你的计算机，当你和银行的网接上头时你只要打一个适当的口令（例如“芝麻开门”），你便可以看到一张大纸，这张大纸横向是有 12 张 A4 大小的样子，而竖向则是无限的. 在大纸上找到你的户头，在上面自己写上存入 1 百万美元，于是下个月你便可以随便花钱了.

§4. 人机象棋大战

大家都在报上看到许多关于人机象棋大战的报道，机器深蓝（Deep Blue）是由两个处理器所组成的平行机，它之所以敢于向人类象棋冠军挑战是因为它有很高的速度，极大的硬盘储存和完善的软件包，可以想象这些硬盘里已经录入了历代有名的比赛和各种研究的棋谱，但即使是这样计算机也不能保证稳赢. 其原因是象棋的复杂程度超过了机器所能控制的范围，象棋的马最多可以有 8 个点作为下一步的选择，而后的选择就更多了，它最多可以有 27 种不同的选择作为下一步，由于这种特性，我们把它叫作非确定性自动机，而如果要制定一个程序来检查胜负就要将非确定性自动机转化为确定性自动机. 后者粗算约有 $2^{2^{128}}$ 个状态，所以人和机器都不可能完全把握胜负.

下棋在最初阶段棋盘上有 32 个棋子，所以变化极多，这时想要判断哪一步棋最好是不可能的. 有些下法好像是不行，但却可能是对你有利. 有些下法好像是稳妥，但偏偏你要吃亏，这时除了计算三、五步之后的变化就要凭经验或者根据棋谱来下，机器呢，它一般不记经验，只好检查棋谱了，随着时间的前进，到了中局、残局，棋的变化就少了，如果机器和人都各剩下了两个棋子，机器是不会头昏的，它会每步都挑选最好的下法，因为它已经能算出每步棋所产生的最后结果. 而人呢，却还有犯错误的可能，因此对人来说最好的策略是在开局的阶段下一步“离谱”的棋. “这步棋太离谱了”这就对了. 例如中国象棋，什么“中炮局”“飞马局”“仙人指路”等都已被人们研究得非常透彻，而“过河炮”“挺边兵”等奇怪的开局就是可以用来对付机器的办法. 在 Kasparov 和 Deep Blue 的对局中 Kasparof 不知走了一步什么样的“离谱”的棋，机器算了半天无法对付只好白白地送一匹马给 Kasparof 吃了. 从这一点就说明他成功地运用了自己的策略，如果他一开始就照着棋谱走，那就非输不可了.

§5. 《未来世界》中人和计算机的战争

人们说“小说是现实生活的前驱”.《未来世界》是一部电影. 假想在未来的什么时候,电子计算机都已经具有人的形状和能力,他们拥有制造电子计算机的工厂和流水线,每一台机器出厂之后要给它取一个名字. 怎么取法? 这台新机器跑去找一个人,找到之后把他的名字和记忆据为己有,然后把这个人杀掉,虽然有了人类的记忆,但他仍然听从计算机王国的命令. 这样就和人类之间发生了战争. 在战争中一台机器的手被打坏了,他可以在地上随便找一个“死”机器,从它身上拔下一只手来往自己的身上一插,便可再战,当然换头也不成什么问题. 这不是很可怕吗? 会不会这样呢? 人类费了九牛二虎之力研制计算机最后反过来和人类自己作对. 其实人们的确在计算机人形化上做了很多努力. 据说最成功的机器已经有了一个很漂亮的脸了,她是一位小姐,至于机器为什么要和人类发生战争我找不到理由. 自古至今人类和人类之间的战争都和重新瓜分生存资料有关,而机器人自己为什么要生存都还不明确,为什么要去瓜分生存资料呢? 如果真像电影中那样一个机器人“走火入魔”,它得到了一个希特勒式的人的脑子,于是立誓要和人类作战,那就很难说了. 至少目前一二十年之内我们还不必担心机器人会联合起来造反.

§6. 计算机发展的一些方向

上面谈的《未来世界》有点远,目前看来计算机已经进到能处理光声的所谓多媒体阶段,因而以下几方面的进展是可以设想的.

§6.1 一体多用

现代化的城市空间很贵,在一个家庭中摆一大堆电器最初几天很神气,久而久之便觉得占地太多而产生不方便了. 因此一体多用成为必要,如果你买来机器既可以当电视机、录音机,又可以做计算机和电话机,那便可以节省一大块地方,用起来也方便. 原来 Commodore 计算机就能连在电视机上用的,后来不知为什么停止了生产,我觉得总有一天人们会继续在这个方向做下去.

§6.2 网络图书馆

念书人爱买书、存书,但是一个家无论如何装不下一个中型图书馆的书,书又重又占地方,一遇到搬家,书是最头疼的东西,印书花掉很多宝贵的木材,对保护地球环境很不好,还有如果你想借一本 Washington 图书馆才能有的书,你就要买飞机票和办签证.如果有了网络图书馆,你就可以在自己的家中读到世界每一个角落的书,所以这真是造福于人类了.

§6.3 电视购物

现在电视购物已经开始,由售货员在电视上展示商品和价格,如果观众满意便打电话去买,这种做法还不甚完善,首先顾客不能自由地挑选反复地看,他们也不能讨价还价.如果联网的工作做得好,能够克服以上的缺点那就更好了.另外送货收款还不牵涉信用问题.我在 1981 年刚到美国时曾经打电话到纽约去买照相机,机器很好,现在还用着,去年有人买了邮购推销的照相机却根本不能用.所以美国现在普遍存在着信用下降的现象,这不是计算机所能解决的了.

§7. 计算机科学和数学的关系

我前面讲过打从一开始制造电子计算机,数学就起着极大的作用,制造机器首先要解决进位制、微程序和操作语言,这些都是利用数学工具来构造实现的.现在计算机能做多方面的事,但是它是通过什么来实现的呢?无论什么事,如果想要利用计算机来做,必须首先将问题数字化,光、声数字化,图像数字化,文字处理、音乐、写作数字化,连思维也要数字化.只有在这些被研究的对象数字化了之后才能在机器中进行运算处理.所以如果不懂数学,计算机将寸步难行.

1988 年在 Mississippi 时计算机系主任邀请我为他们开设与数学有关联的科目,如理论计算机科学数值分析和算法论等.我问他为什么要开这些课,他说这是为了系的升格.看来国家对各地的计算机系还是没有统一的标准的,Ford 等大汽车公司经常招聘既懂数学又懂计算机的人才.这也说明了数学的重要性,所以希望有志者除了熟悉机器之外也要对数学下一番功夫.

至于计算机科学是否也反过来影响数学的发展呢?当然是的,我觉得从计算机的角度出发来认识数的话,数只有有限个.利用这有限个数,计算机可以解决几乎一切从实际中产生出来的问题.它也用到微分、积分等概念,但在它看来这些对象都是有限的.看来建立一个只有有限个数的完整的数学系统已经提到日程上来了.

数学学报,1998,41(3):617～622

多个一元关系上的 Vaught 猜想①

Vaught's Conjecture on Unitary Relations

摘要 我们对多个一元关系理论系统证明了 Vanght 猜想并对其模型作了一个完整的分类.

关键词 Vaught 猜想,一元关系.

Vaught 猜想②已经提出了 40 多年至今未能得到解决. 猜想的内容是:一个完备的可数一阶理论的互不同构的可数模型数是$\leqslant\omega$ 或者$=2^{\omega}$. 这个猜想只有少数几种情形得到了解决. Morley 在[1]证明了这个数总是$\leqslant\omega$ 或者$=\omega_1$ 或者$=2^{\omega}$. Steel 在[2]证明了树的理论的 Vaught 猜想. Shelah,Harrington 和 Makkai 在[3]证明了 ω-稳定理论的 Vaught 猜想. 我们在[4]证明了齐次模型的 Vaught 猜想. Rubin 在[5]证明了一个完备的无限线性理论的互不同构的可数模型数是$=1$ 或者$=2^{\omega}$,这样也就对这类模型肯定了 Vaught 猜想. 本文从研究多个一元关系的模型入手对关于这种结构的理论也肯定了 Vaught 猜想. 我们同时也对这种模型得到了一个完全的分类.

全文中$\mathcal{L}$是含有可数多个一元关系以及等号的语言. T 是$\mathcal{L}$上的一个完备理论. 以 L_0 表示$\mathcal{L}$上的含有变数$\{x_0,x_1,\cdots,x_n,\cdots\}$的全体一阶公式

① 收稿日期:1997-09-05;接收日期:1997-12-08.
国家自然科学基金资助项目.

② 本书出版时,此问题仍未解决——编者注.

所成集. 利用 L_0 的公式我们可以按下列规则定义 L_1 的公式.

(i)如果 φ 是一个 L_0 公式,那么它也是一个 L_1 的公式;(ii)如果 σ 是一个 L_0 的型,那么 $\wedge\sigma$ 是一个 L_1 的公式;(iii)如果 φ,ψ 是 L_1 的公式,那么 $\neg\varphi$、$\varphi\vee\psi$, $\varphi\wedge\psi$, $(\exists x)\varphi$ 和 $(\forall x)\varphi$, 是 L_1 的公式;(iv) L_1 的公式只能由以上三规则反复运用有限次而组成.

我们首先定义一些 L_0 的公式的简记号.

(1)$(\exists 0x)\varphi(x)$的含义是不存在 x 满足 $\varphi(x)$;

(2)$(\exists nx)\varphi(x)$的含义是至少存在 n 个元素满足 $\varphi(x)$,其中 $1\leqslant n<\omega$;

(3)$(\exists !\ nx)\varphi(x)$的含义是恰好存在 n 个元素满足 $\varphi(x)$,其中 $1\leqslant n<\omega$;

(4)$(\exists\infty x)\varphi(x)$的含义是存在无限多个元素满足 $\varphi(x)$;

(5)$(+1)R(x)$的含义是 $R(x)$;

(6)$(-1)R(x)$的含义是 $\neg R(x)$.

其中情形(4)要用无限多个一阶公式:$(\exists 1x)\varphi(x),\cdots,(\exists nx)\varphi(x),\cdots$ 来写出.

定理 1 对每一个完备的理论 T 存在一个量词消去过程使得任意公式 $\varphi(x_1,x_2,\cdots,x_n)$ 是 T 等价于一个不含量词的公式 $\psi(x_1,x_2,\cdots,x_n)$.

证 设 $\varepsilon_0,\varepsilon_1,\cdots,\varepsilon_k$ 为一个 $\{0,1\}$ 的序列而 $\varphi(x)=\varepsilon_0P_0(x)\wedge\varepsilon_1P_1(x)\wedge\cdots\wedge\varepsilon_kR_k(x)$. 因为 T 是一个完备的理论下面三条之一成立:

(1)$(\exists 0x)\varphi(x)$;

(2)$(\exists !nx)\varphi(x)$,其中 $1\leqslant n<\infty$,或者

(3)$(\exists\infty x)\varphi(x)$.

我们用 $\Phi_{\varepsilon_0,\varepsilon_1,\cdots,\varepsilon_k}$ 来表示相应的成立的句子(或句子集). 这样一来

$$T\models\Phi=\{\Phi_{\varepsilon_0,\varepsilon_1,\cdots,\varepsilon_k}\mid\varepsilon_0,\varepsilon_1,\cdots,\varepsilon_k\in\{0,1\}\}.$$

由[6],Lemma 1.5.1 并设原子式子为基本式子. 我们只需考虑以下的合取式子,其中 x 是待消去的变元而 $y_1,y_2,\cdots,y_p$ 是自由变元.

$$\varphi(x,y_1,y_2,\cdots,y_p)\leftrightarrow(\varepsilon_{i_1}R_{i_1}(x)\wedge\cdots\wedge\varepsilon_{i_l}R_{i_l}(x)\wedge x=y_{j_1}\wedge x=y_{j_2}\wedge,\cdots,\wedge x=y_{j_m}$$
$$\wedge x\neq y_{k_1}\wedge x\neq y_{k_2}\wedge,\cdots,\wedge x\neq y_{k_n}\wedge\psi(y_1,y_2,\cdots,y_p)),$$

其中 ψ 是一个不含有 x 的(±1)基本式子的合取式. 要消去$(\exists x)\varphi(x,y_1,y_2,\cdots,y_p)$,我们考虑以下情形:

(4)我们不必考虑 $\psi(y_1,y_2,\cdots,y_p)$,因为它不含有 x;

(5)如果 $m\neq 0$,在合取式中至少有一个等式 $x=y_{j_1}$. 我们可以在式子中将所有 x 换成 y_{j_1} 以消去 x;

(6)如果 $m=0$,那么在合取式中没有等式,我们考虑式子

$(\exists x)(\varepsilon_{i_1}R_{i_1}(x)\wedge\varepsilon_{i_2}R_{i_2}(x)\wedge\cdots\wedge\varepsilon_{i_l}R_{i_l}(x)\wedge x\neq y_{k_1}\wedge x\neq y_{k_2}\wedge\cdots\wedge x\neq y_{k_n})$.

对 $(\varepsilon_{i_1},\varepsilon_{i_2},\cdots,\varepsilon_{i_l})$ 我们检查 Φ 的句子 $\Phi_{\varepsilon_{i_1},\varepsilon_{i_2},\cdots,\varepsilon_{i_l}}$ 并按下列情形来讨论;

(6.1)如果 $\Phi_{\varepsilon_{i_1},\varepsilon_{i_2},\cdots,\varepsilon_{i_l}}$ 是 $(\exists 0x)(\varepsilon_{i_1}R_{i_1}(x)\wedge\varepsilon_{i_2}R_{i_2}(x)\wedge\cdots\wedge\varepsilon_{i_l}R_{i_l}(x))$,那么整个式子等价于一个伪值句子. 我们可以把它全部消去;

(6.2)如果 $\Phi_{\varepsilon_{i_1},\varepsilon_{i_2},\cdots,\varepsilon_{i_l}}$ 是 $(\exists!\ n_1x)(\varepsilon_{i_1}R_{i_1}(x)\wedge\varepsilon_{i_2}R_{i_2}(x)\wedge\cdots\wedge\varepsilon_{i_l}R_{i_l}(x))$ 和 $1\leqslant n_1\leqslant\omega$,那么我们比较 n 和 n_1:

(6.2.1)如果 $n_1>n$,那么

$(\exists x)(\varepsilon_{i_1}R_{i_1}(x)\wedge\varepsilon_{i_2}R_{i_2}(x)\wedge\cdots\wedge\varepsilon_{i_l}R_{i_l}(x)\wedge x\neq y_{k_1}\wedge x\neq y_{k_2}\wedge\cdots\wedge x\neq y_{k_n})$

是等价于一个真值的句子,因此整个式子就等价于 $\psi(y_1,y_2,\cdots,y_p)$;

(6.2.2)如果 $n_1\leqslant n$,那么式子

$(\exists x)(\varepsilon_{i_1}R_{i_1}(x)\wedge\varepsilon_{i_2}R_{i_2}(x)\wedge\cdots\wedge\varepsilon_{i_l}R_{i_l}(x)\wedge x\neq y_{k_1}\wedge x\neq y_{k_2}\wedge\cdots\wedge x\neq y_{k_n})$

等价于:

"在变元 $y_{k_1},y_{k_2},\cdots,y_{k_n}$ 中至多有 n_1-1 个互不相等并且满足式子 $\varepsilon_{i_1}R_{i_1}(x)\wedge\varepsilon_{i_2}R_{i_2}(x)\wedge\cdots\wedge\varepsilon_{i_l}R_{i_l}(x)$."

上述式子可以用基本式子的布氏组合写出. 量词消去法完成. 这样一来每一个式子都 T 等价于一个不含量词的式子.

定理 2 在定理 1 中所定义的句子集 Φ 等价于理论 T.

证 由定理 1 所提供的量词消去法知 Φ 是完备的,再由 $T\models\Phi$ 和 Φ 的完备性知 $\Phi\models T$.

定理 3 一个 T 型 $\sigma(x)$ 被符号 $\{\varepsilon_0,\varepsilon_1,\cdots,\varepsilon_n,\cdots\}$ 所唯一决定,其中 $\varepsilon_nR_n(x)\varepsilon\sigma$ 对 $0\leqslant n<\omega$.

证 (略,详见本刊英文版 14 卷 2 期,下同.)

我们证明一个关于 T 的 L_0 型个数的定理.

定理 4 一个完备理论 T 的含有一个变数的 L_0 型的个数是 $\leqslant\omega$ 或者 $=2^\omega$.

证 假设 T 的型的个数是不可数. 我们列出所有一元关系

$$\{R_0,R_1,\cdots,R_n,\cdots\}.$$

$R_0(x)$或者$\neg R_0(x)$至少有一个是被包含在不可数多个型之内. 我们把它写成$\varepsilon_0 R_0(x)$. 现在证明一个一般的论断.

(1) 对任意被包含在不可数多个型之内的序列$\{\varepsilon_{i_1} R_{i_1}(x) \wedge \varepsilon_{i_2} R_{i_2}(x) \wedge, \cdots, \wedge \varepsilon_{i_k} R_{i_k}(x)\}$我们可以找到一个$R_{i_{k+1}}$使得下面两者

$$\{\varepsilon_{i_1} R_{i_1}(x), \varepsilon_{i_2} R_{i_2}(x), \cdots, \varepsilon_{i_k} R_{i_k}(x), R_{i_{k+1}}\}$$

和

$$\{\varepsilon_{i_1} R_{i_1}(x), \varepsilon_{i_2} R_{i_2}(x), \cdots, \varepsilon_{i_k} R_{i_k}(x), \neg R_{i_{k+1}}\}$$

都被包含在不可数多个型之内.

因为如果(1)不成立, 那么我们可以在$R_{i_k}(x)$之后列出所有其他关系形成类型树的一个分支

$$\begin{array}{ccccccc} \varepsilon_{i_1} R_{i_1}(x) & \Rightarrow & \cdots \varepsilon_{i_k} R_{i_k}(x) & \Rightarrow & \varepsilon_{i_{k+1}} R_{i_{k+1}} & \Rightarrow & \cdots \\ & & \downarrow & & \downarrow & & \\ & & (-\varepsilon_{i_{k+1}}) R_{i_{k+1}} & & (-\varepsilon_{i_{k+2}}) R_{i_{k+2}} & \cdots & \end{array}$$

其中双箭头有不可数多个型通过而单箭头则只有可能多个型通过. 但这样一来最上面的一行的结点至多只有一个型通过. 而通过$\{\varepsilon_{i_1} R_{i_1}(x), \varepsilon_{i_2} R_{i_2}(x), \cdots, \varepsilon_{i_k} R_{i_k}(x)\}$的型的总数成为可数, 矛盾. 所以(1)成立.

反复利用(1)我们可以构造一个无穷的完全二分树使得树上的每一个结点都有不可数多个型通过. 这棵树有2^ω条路径. 由定理3每一条路径确定一个T的型. 不同的路径确定不同的型. 所以T的型的总数是$=2^\omega$.

容易看出如果理论T有2^ω个型, 那么它有2^ω个互不同构的模型.

现在我们用语言L_1来给出两个T模型同构的条件.

定理5 两个T模型同构当且仅当它们是L_1等价的.

证 不难证明两个T模型同构当且仅当下面两个条件得到满足.

(i)它们实现同样多的含有一个变元的L_0的型;

(ii)每一个型在两个模型中被实现的元素个数完全相同($\leqslant$或$=\omega$).

考虑到(i)和(ii)可以用L_1的句子表示出来, 定理得到证明.

下面我们假设T是一个完备的具有无限模型的$\mathcal{L}$理论. 从根开始我们构造如下的一棵无限树:

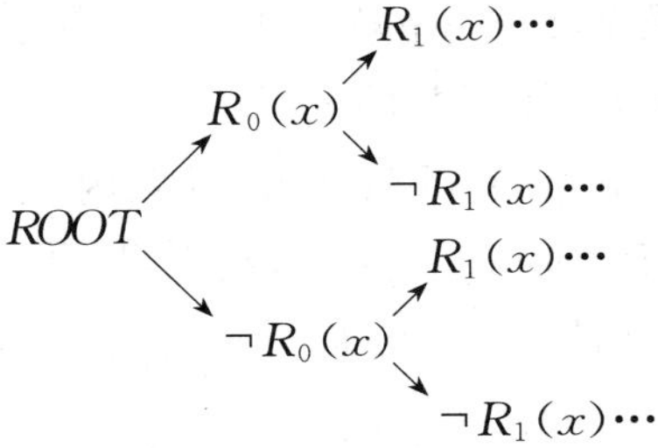

它就叫作型的树.我们从树上取出任意的合取式

(1)$\varepsilon_0 R_0(x)\wedge\varepsilon_1 R_1(x)\wedge\cdots\wedge\varepsilon_k R_k(x)$.

完备的理论 T 必须回答这个问题:有多少个元素满足(1).答案是下列之一:

(iii)$(\exists 0x)(\varepsilon_0 R_0(x)\wedge\varepsilon_1 R_1(x)\wedge\cdots\wedge\varepsilon_k R_k(x))$;

(iv)$(\exists !\ nx)(\varepsilon_0 R_0(x)\wedge\varepsilon_1 R_1(x)\wedge\cdots\wedge\varepsilon_k R_k(x))$,对所有 $1\leqslant n\leqslant\infty$;

(v)$(\exists\infty x)(\varepsilon_0 R_0(x)\wedge\varepsilon_1 R_1(x)\wedge\cdots\wedge\varepsilon_k R_k(x))$.

设 $N_{\varepsilon_0,\varepsilon_1,\cdots,\varepsilon_k}$ 是对应的数.我们可以在上面的树中以对应的数来替换每一个结点的公式而得到另一棵树:

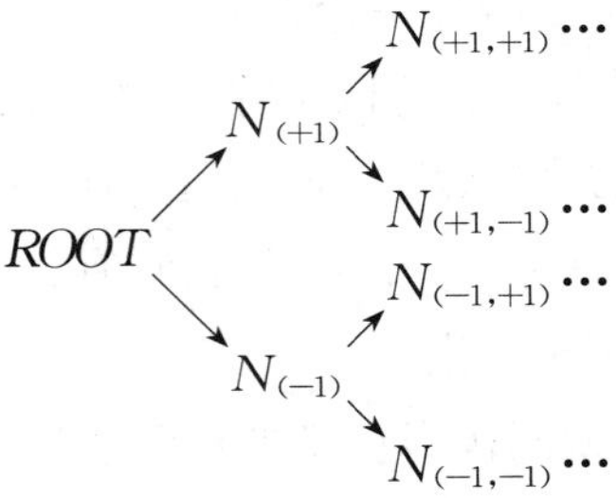

我们把这第二棵树叫作数的树也可以把第二棵树的数看成是第一棵树上的结点的标记.容易看出每一个结点上的数是等于它的两个子结点上的数的和,也等于它的所有同一代的后裔结点上的数的和.我们同样考虑树的无限分支并且试图确定在这条分支上的数.由于一个型是被第一棵树的一条路径所唯一确定,对应于这条分支的数实际上就是实现这个型的数.在第二棵树上的某些数也可以是无限并且如果我们使用无限多个数的加法,那么每一个结点上的数也等于它的所有 ω 代的后裔结点上的数的和.

由定理1,如果 T 有多于 ω 个型,那么它有 2^{ω} 个型,因此也有 2^{ω} 个

互不同构的可数模型. 下面我们假设 T 的型的数是 $\leqslant\omega$. 另外我们需要一个关于极限型的定义.

定义 一个型 σ 叫作型的集 $\Sigma=\{\sigma_0,\sigma_1,\cdots,\sigma_i,\cdots\}$ 的一个极限型,如果 σ 的任何结点有一个 $\sigma_i\in\Sigma$,使得 σ_i 也通过这个结点. 下面的一些预备定理讨论极限型的性质.

预理 1 标记在一个 T 型 σ 的结点的数是一个不升链. 它们有以下三种可能:

(i) $n_0\geqslant n_1\geqslant\cdots\geqslant n_k=n_{k+1}=\cdots$,其中 $n_k=0$,

(ii) $n_0\geqslant n_1\geqslant\cdots\geqslant n_k=n_{k+1}=\cdots$,其中 $1\leqslant n_k<\infty$,或者

(iii) $\infty=n_0=n_1=\cdots=n_k=n_{k+1}=\cdots$

情形(i)不允许 σ 成为 T 的型. 所以 T 的型只可能是(ii)或者(iii).

证 这些是仅有的 $0\leqslant n_k\leqslant\infty$ 中数的非升链的可能情形.

预理 2 如果标记在 T 的型 $\sigma=\varepsilon_0R_0(x),\varepsilon_1R_1(x),\cdots,\varepsilon_kR_k(x),\cdots$ 的结点上的数是 $n_0\geqslant n_1\geqslant\cdots\geqslant n_k=n_{k+1}=\cdots$,其中 $0<n_k<\infty$,那么每一个 T 模型含有恰好 n_k 个元素实现型 σ.

证 有 n_k 个元素满足

$$\varepsilon_kR_0(x)\wedge\varepsilon_kR_1(x)\wedge\cdots\wedge\varepsilon_kR_k(x)$$

因为我们有

$$T\models(\exists\ !\ n_kx)(\varepsilon_0R_0(x)\wedge\varepsilon_1R_1(x)\wedge\cdots\wedge\varepsilon_kR_k(x))$$

$$T\models(\exists\ !\ n_kx)(\varepsilon_0R_0(x)\wedge\varepsilon_1R_1(x)\wedge\cdots\wedge\varepsilon_kR_k(x)\wedge\varepsilon_{k+1}R_{k+1}(x))$$

实现型 $\sigma(x)$ 的元素个数不可能是 $>n_k$. 因为标记在全体同代后裔结点(包含第 ω 代)的数的和是 n_k. 这个元素个数也不可能是 $<n_k$. 因为由 $n_k=n_{k+1}=\cdots$ 我们知道所有在结点 $\varepsilon_kR_k(x)$ 之后离开型 σ 的其他型的后裔结点都带有标记数 0. 因此 T 不可能实现在 $\varepsilon_kR_k(x)$ 之后的其他型.

预理 3 如果 σ 是 T 型集 $\Sigma=\{\sigma_0,\sigma_1,\cdots,\sigma_i,\cdots\}$ 的极限型,那么

(i) σ 自己是一个 T 的型;

(ii) 所有标记在 σ 的结点上的数都是 ∞;

(iii) 存在不同的 T 模型在实现其他型的元素个数不变的情况下而实现 σ 的元素个数可以从 0 直到 ∞.

证 (i) 可以用紧致性定理证明. 由预理 1 标记在一个 T 型的有限结点上的数至少是 1. 由于有无限多个不同的 Σ 型通过每一个 σ 的结点而

后又分开，所以标记在一个结点上的数是∞. 这证明了(ii). 为了证明(iii)，假设有一个模型$\mathfrak{A}$实现所有$\sum$的型(由紧致性定理这种模型存在). 由$\mathfrak{A}$中增减任意多个实现型σ的元素而得到另一个模型$\mathfrak{B}$. 由(ii)我们知道，标记在σ的每一个结点上的数总是∞，与实现型σ的元素个数无关. 模型$\mathfrak{B}$的数的树与模型$\mathfrak{A}$完全相同. 所以由定理 2 知道$\mathfrak{B}$是一个T模型.

预理 4 如果T有无限多个型，那么它有一个极限型.

证 有无限多个T型通过$\varepsilon_0 R_0(x)$，其中又有无限多个通过$\varepsilon_1 R_1(x)$，…这样一来我们找到型的树上的一系列结点

$$\sum=\{\varepsilon_0 R_0(x),\varepsilon_1 R_1(x),\cdots,\varepsilon_n R_n(x),\cdots\}.$$

这些结点唯一地决定一个T的型σ，而它就是T的极限型.

预理 5 如果T有无限多个极限型，那么我们可以找到两个T型的集$\sum=\{\sigma_0,\sigma_1,\cdots,\sigma_i,\cdots\}$和$\prod=\{\pi_0,\pi_1,\cdots,\pi_i,\cdots\}$使得

(i)$\sum\cap\prod=\varnothing$;

(ii)$\sum$的型不是自己的型的极限；

(iii)存在无限多个型的集$(\prod_0,\prod_1,\cdots,\prod_i,\cdots)$使得$\prod_i\subset\prod$对$i=0,1,\cdots$并且$\prod_i\cap\prod_j=\varnothing$对$i\neq j$并且$\sigma_i$是一个$\prod_i$的型的极限型；

(iv)对任意模型$\mathfrak{A}\models T$保持实现$\notin\sum$的型的元素个数不变，而任意地改变实现型$\sum$的元素个数仍然得到T模型.

证 设$\Delta=(\delta_0,\delta_1,\cdots,\delta_i,\cdots)$是$T$的极限型的集，由预理 4 的证明我们可以找到一个$\Delta$的型的极限型$\sigma$. 设$\sigma_0\in\Delta$使它由第$k_0$个结点起离开$\sigma$. 再选$\sigma_1\in\Delta$，使它从第$k_1(>k_0)$个结点起离开$\sigma$. 如此进行下去……我们就找到了$\sum=\{\sigma_0,\sigma_1,\cdots,\sigma_i,\cdots\}$. 类似地对每一个$\sigma_i\in\sum$我们可以找到$T$型的集$\pi_{i0},\pi_{i1},\cdots,\pi_{ij},\cdots$在第$k_i$个结点之后离开$\sigma_i$并且它们的极限是$\sigma_i$. 这样我们得到$T$的型的集$\prod_i=\{\pi_{i0},\pi_{i1},\cdots,\pi_{ij},\cdots\}$. 再定义$\prod=\pi_0\cup\pi_1\cup\cdots\cup\pi_i\cdots$我们得到下面的树：

$$\begin{array}{cccccccccc}
k_0 & \rightarrow & \cdots & k_{00} & \rightarrow & \cdots & k_{01} & \rightarrow & \cdots & \sigma_0(m_0) \\
\downarrow & & & \downarrow & & \cdots & \downarrow & & \cdots & \\
\downarrow & & & \pi_{00} & & \cdots & \pi_{01} & & \cdots & \\
\downarrow & & & & & & & & & \\
k_1 & \rightarrow & \cdots & k_{10} & \rightarrow & \cdots & k_{11} & \rightarrow & \cdots & \sigma_1(m_1) \\
\downarrow & & & \downarrow & & \cdots & \downarrow & & \cdots & \\
\vdots & & & \pi_{10} & & \cdots & \pi_{11} & & \cdots & \\
\sigma & & & & & & & & &
\end{array}$$

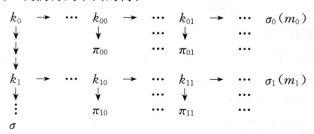

容易看出Σ，Π和Π_i，$i=0,1,\cdots$满足条件(i)～(iii)．而条件(iv)也可以仿前证明．

综合上面结果我们得到下面的定理．

定理 6 假设T是一个完备的多个一元关系理论并且具有无限模型，下列各条将T模型作了一个完全的分类．

(i)如果T只有有限多个型，那么T只有一个可数模型；

(ii)如果T的型的个数是ω，那么我们将它分为两个情形：

(ii.1)如果T只有有限多个极限型，那么T有ω个互不同构的可数模型；

(ii.2)如果T有无限多个极限型，那么T有2^ω个互不同构的可数模型．

(iii)如果T有2^ω个型，那么T有2^ω个互不同构的可数模型．

Abstract We prove the Vaught's conjecture and obtain a complete classification for unitary relations.

Keywords **Vaught's conjecture, Unitary relation.**

参考文献

[1] Morley M D. The number of countable models. J Symbolic Logic, 1970, 35: 14—18.

[2] Steel J. On Vaught's conjecture. In Cabal Seminar 76—77, ed. Kechris A S, Moschovakis Y N. Lecture Notes in Math 689, Berlin Springer, 1984, 193—208.

[3] Shelah S, Harrington L A, Makkai M. A proof of Vaught's conjecture for ω-stable theories. Israel J Math, 1984, 49: 259—280.

[4] Lo L. On the number of homogeneous models. J Symbolic Logic, 1983, 48: 539—541.

[5] Rubin M. Theories of linear orderings. Israel J Math, 1974, 17: 392—443.

[6] Chang C C, Keisler H J. Model Theory. North-Holland. 3rd ed, 1990.

数学研究,2004,37(2):144～154

无原子布氏代数理论的计算复杂性①

The Computational Complexity of the Theory of Boolean Algebras without Atomic Elements

摘要 研究无原子布氏代数的计算复杂性.得到了下面的新定理:

定理1 无原子布氏代数理论 Δ 具有完全的量词消去法,也就是说每一个式子都 Δ 等价于一个开式子.

定理2 无原子布氏代数的初等型 $\Gamma(x_1,x_2,\cdots,x_n)$ 是由型内的不含量词的全体开式子所唯一决定.

定理3 无原子布氏代数的一个长度为 n 的语句的判断过程所消耗的 Turing 时间和空间都是属于 $2^{2^{cn}}$ 指数级.

关键词 无原子布氏代数,量词消去法,模型数,计算复杂性.

§1. 问题的来源

在计算机理论的研究中,计算复杂性是关于一个公理系统的最重要的问题之一.要解决一个公理系统的计算复杂性首先要解决它的判定问题.布氏代数的判定问题是 Tarski 在[1]已经证明了的.但是这个证明只有不到一页的篇幅.有很多关于布氏代数的问题在这里找不到答案.由于判定问题的解决我们自然地要问关于它的命题(或者其中的一部分)是否可以用机器来计算?如果能够计算,那么所消耗的时间和空间是多少,这就是计算复杂性.与之有关联的问题是:它的初等类的形式是什么?由于

---

① 收稿日期:2003-09-23.

基金项目:广东女子职业技术学院研究基金.

布氏代数理论不是完备的，那么增加什么样的条件能够得到完备的系统呢？这些问题对布氏代数来说都没有得到明晰的回答. 利用量词消去法解决计算复杂性是一条途径. 量词消去法有两种：一种是在消去量词的过程中将带有量词的式子变成较为简单、便于处理的带有量词的式子，这种情况一般是不能完全消去量词的. 另一种是在消去量词的过程中将带有量词的式子变成完全没有量词的式子，最后能彻底消去量词. 本文通过量词消去法和直接的计算研究了一类布氏代数，也就是无原子布氏代数. 我们证明了这类布氏代数的理论具有完全的量词消去法，因此它也就是完备的理论. 由于无原子布氏代数的理论是 ω 范畴的，它的势为 ω 的模型数等于 1. 也就是说关于它的 Vaught 猜想已经解决，但是在计算其他基数的模型数时完全的量词消去法也是很有用的. 在众多的理论系统中只有为数不多的理论系统具有这种完全的量词消去法的，首先无端稠密有序集是一个这样的系统. 在某种意义上来说无原子布氏代数是无端稠密有序集的一个多维化推广. 但从量词消去法的角度来看无原子布氏代数比无端稠密有序集的计算要复杂得多. 这一点可以从我们下面的计算中看出. 应该说一般布氏代数比无原子布氏代数又要更复杂一些，对无原子布氏代数的研究为对一般布氏代数的量词消去法的进一步研究做了准备. 有了完全的量词消去法，无原子布氏代数的计算复杂性也就有了一个解决方法. 在表达计算复杂性的度量中也有各种各样的方式，例如交替 Turing 机等. 我们使用普通 Turing 机时间和空间来表达计算复杂性的度量. 这样对将来研究机器证明较为方便. 近年来作者和他的研究生用量词消去法研究了一些公理系统，[2]是其中的一篇研究报告，其他的结果也将陆续发表.

§2.　无原子布氏代数的公理系统

布氏代数的语言是由下列记号来组成：$\mathscr{L}=\langle \cup, \cap, \bar{x}, 0, 1\rangle$，其中 $\cup$ 是并运算，$\cap$ 是交运算，$\bar{x}$ 是余运算，0 代表最小元，1 代表最大元，它们满足下列公理(共 20 条).

幂等律(1)$x\cup x=x$，　　(2)$x\cap x=x$，

交换律(3)$x\cup y=y\cup x$，　　(4)$x\cap y=y\cap x$，

结合律(5)$x\cup(y\cup z)=(x\cup y)\cup z$，(6)$x\cap(y\cap z)=(x\cap y)\cap z$，

分配律(7)$x\cup(y\cap z)=(x\cup y)\cap(x\cup z)$,

(8)$x\cap(y\cup z)=(x\cap y)\cup(x\cap z)$,

吸收律(9)$x\cup(x\cap y)=x$, (10)$x\cap(x\cup y)=x$

De Morgan 律(11)$\overline{x\cup y}=\bar{x}\cap\bar{y}$, (12)$\overline{x\cap y}=\bar{x}\cup\bar{y}$,

0,1 运算律(13)$x\cup 0=x$, (14)$x\cap 0=0$,

(15)$x\cup 1=1$, (16)$x\cap 1=x$,

(17)$x\cup\bar{x}=1$, (18)$x\cap\bar{x}=0$,

(19)$0\neq 1$, (20)$\bar{\bar{x}}=x$.

除以上的 20 条公理之外无原子性质写成的公理是:

(21)$(\exists y)(0\neq y.\wedge.\forall z(z\cup y=y.\rightarrow.z=0.\wedge.z=y))$

以上的 21 条公理所组成的集合记做 Δ.

关于布氏代数有一个有名的定理,也就是布氏代数的表现定理(Stone[3]):每一个布氏代数可以同构于某一个集合的一些子集所作成的集合代数,设 $\mathscr{B}$ 是一个原子布氏代数,对任意元素 $b\in\mathscr{B}$,找出所有比 b 小的原子做成集合 S_b,这样一来全体 $\mathscr{U}=\{S_B\mid b\in\mathscr{B}\}$ 可以看成是 B 的全体原子所成集的一些子集(不一定是全体子集)所做成的集合代数. 容易证明 $\mathscr{U}\cong\mathscr{B}$,在证明定理时有时可利用这个集合表现定理将关于布氏代数的定理转化为关于集合的定理来证明.

§3. 量词消去法的作用与过程

一个理论有了量词消去法,对这个理论就了解得比较清楚. 关于它的逻辑问题就比较容易解决(例如理论的完全性,式子的计算……),量词消去法用得很多,主要解决判定问题. 量词消去法的最好结果是能将所有量词都消去,无原子布氏代数就是这样的一个系统. 在消去量词的过程中有时要用到带有量词的式子,这种情况一般是不能完全消去量词的. 我们这里的消去量词的中间过程只用到不带有量词的式子,以确保能彻底消去量词.

我们所用到的记号和常见的定理在[4]中都可以找到. 下面这个记号我们要特别说明一下. Σ 是语句集,φ 是一个公式,(含有自变量 x_0, $x_1,\cdots,x_n$) $\Sigma\models\varphi$ 称 φ 是 Σ 的推论,当且仅当在每一个 Σ 的模型 $\mathscr{U}$ 中任取 $a_0,a_1,\cdots,a_n$ 都有 $\mathscr{U}\models\varphi(a_0,a_1,\cdots,a_n)$,上述条件也可以写成 $\mathscr{U}\models\forall x_0x_1\cdots x_n\varphi(a_0,a_1,\cdots,a_n)$. 两个式子 φ,ψ 称为互相等价,如果它们满足 $\Sigma\models\varphi\leftrightarrow\psi$. 也就是 $\forall(\varphi\leftrightarrow\psi)$ 而不是 $\forall\varphi\leftrightarrow\forall\psi$.

设 T 是 $\mathscr{L}$ 的一个理论，为了做量词消去法，先选择一个式子集，Γ(开式子或闭式子都有，一般不能单用闭式子)叫作基本式子集，想要证明每一个式子都是 T 等价于 Γ 的一些式子的布氏组合，关键的一步就是要消去量词.

引理 1 T 是一个理论，Γ 是一个公式集，叫作基本式子集，想要证明任意一个公式都 T-等价于一个基本式子的布氏组合. 下面两条就能达到结果；

(i)原子式子是等价于基本式子的布氏组合.

(ii)如果 θ 是一个基本式子的布氏组合，那么 $\exists v\theta$ 也等价于一个基本式子的布氏组合.

引理的证明我们这里就不录了.

§4. 无原子布氏代数的性质

下面我们就用量词消去法来研究无原子布氏代数，公理是 Δ.

基本式子集 Γ：利用个体变量 $v_0, v_1, \cdots, v_n, \cdots$($n$ 为自然数)，$\cap$(交运算)，$\cup$(并运算)，和 $\overline{x}$(余运算)所构造出来的全体原子式子.

基本式子的布氏组合：全体开式子.

引理 2 设 $t(x_1, x_2, \cdots, x_n)$是一个布氏项，那么 t 等价于如下形状的一个布氏项

$$\bigcup_{i=1}^{k} \bigcap_{j=1}^{l} \varepsilon_{ij} x_{ij}.$$

其中 $\varepsilon_{ij}=1$ 或者 -1 是一个简记号，如果 $\varepsilon_{ij}=1$，那么 $\varepsilon_{ij}x_{ij}=x_{ij}$. 如果$\varepsilon_{ij}=-1$，那么 $\varepsilon_{ij}x_{ij}=\overline{x_{ij}}$.

证 利用布氏算律.

引理 3 设 $t(x_1, x_2, \cdots, x_n)$是一个不为零的布氏项，那么 t 等价于如下形状的一个布氏项

$$\bigcup_{i=1}^{k} \bigcap_{j=1}^{n} \varepsilon_{ij} x_{j}.$$

其中每一个连交项都含有正的或负的 $x_1, x_2, \cdots, x_n$，所以每一个连交项的长度都是 n(不计正负号).

证 由布氏算律知同一个项中不会同时含有两个 x_i，或两个 $\overline{x}_i$ 或一个 x_i 一个 $\overline{x}_i$，有这些情况时都可以另外加以处理. 如果一个项 t 不含有 x_i 也不含有 $\overline{x}_i$，就将这个项变成 $t\cap(x_i\cup\overline{x}_i)$，展开后所得的两项的并与原来的项互相等价.

引理 4 设 $t_1(x_1,x_2,\cdots,x_n)$，$t_2(x_1,x_2\cdots,x_n)$是两个布氏项，t_1，t_2 分别等价于如下形状的式子

$$t_1(x_1,x_2,\cdots,x_n)=\bigcup_{i=1}^{k}\bigcap_{j=1}^{n}\varepsilon_{ij}x_j,$$

$$t_x(x_1,x_2,\cdots,x_n)=\bigcup_{i=1}^{l}\bigcap_{j=1}^{n}\delta_{ij}x_j.$$

其中$\bigcap_{j=1}^{n}\varepsilon_j x_j$在项$t_1$ 中出现而不在项 t_2 中出现，$\bigcap_{j=1}^{n}\delta_j x_j$在项$t_2$ 中出现而不在项 t_1 中出现，那么

$$t_1=t_2\Rightarrow\bigcap_{j=1}^{n}\varepsilon_j x_j=0,\bigcap_{j=1}^{n}\delta_j x_j=0.$$

证 分别化简

$$t_1\cap\bigcap_{j=1}^{n}\varepsilon_j x_j=t_2\cap\bigcap_{j=1}^{n}\varepsilon_j x_j,\quad t_1\cap\bigcap_{j=1}^{n}\varepsilon_j x_j=t_2\cap\bigcap_{j=1}^{n}\delta_j x_j.$$

可以得到

$$\bigcap_{j=1}^{n}\varepsilon_j x_j=0,\bigcap_{j=1}^{n}\delta_j x_j=0.$$

引理 5 设 $t_1(x_1,x_2,\cdots,x_n)$，$t_2(x_1,x_2,\cdots,x_n)$是两个布氏项，t_1，t_2 分别等价于如下形状的式子

$$t_1(x_1,x_2,\cdots,x_n)=\bigcup_{i=1}^{k}\bigcap_{j=1}^{n}\varepsilon_{ij}x_j\cup\bigcup_{i=1}^{m}\bigcap_{j=1}^{n}\eta_{ij}x_j,$$

$$t_2(x_1,x_2,\cdots,x_n)=\bigcup_{i=1}^{l}\bigcap_{j=1}^{n}\delta_{ij}x_j\cup\bigcup_{i=1}^{m}\bigcap_{j=1}^{n}\eta_{ij}x_j.$$

其中$\bigcup_{i=1}^{k}\bigcap_{j=1}^{n}\varepsilon_{ij}x_j$是所有在项 t_1 中出现而不在项 t_2 中出现的连交项的并，$\bigcup_{i=1}^{l}\bigcap_{j=1}^{n}\delta_{ij}x_j$是所有在项 t_2 中出现而不在项 t_1 中出现的连交项的并，那么

$$t_1=t_2\Leftrightarrow\bigcap_{j=1}^{n}\varepsilon_{ij}x_j=0,i=1,2,\cdots,k,\bigcap_{j=1}^{n}\delta_{ij}x_j=0,i=1,2,\cdots,l$$

(反之)

$t_1\neq t_2$ 成立的充分必要条件是下列的不等式中有一个成立：

$$\bigcap_{j=1}^{n}\varepsilon_{ij}x_j\neq 0,i=1,2,\cdots,k,\bigcap_{j=1}^{n}\delta_{ij}x_j\neq 0,i=1,2,\cdots,l$$

证 $\Rightarrow$：由引理 4$\Leftarrow$：由直接计算可得.

引理 6 设 $t_1(x_1,x_2,\cdots,x_n)$，$t_2(x_1,x_2,\cdots,x_n)$是两个如引理 5 的布

氏项,那么其结论又可以改写为

$$t_1=t_2\Leftrightarrow\left(\bigwedge_{i=1}^{k}\bigcap_{j=1}^{n}\varepsilon_{ij}x_j=0\cdot\wedge\cdot\bigwedge_{i=1}^{l}\bigcap_{j=1}^{n}\delta_{ij}x_j=0\right).$$

$$t_1\neq t_2\Leftrightarrow\left(\bigvee_{i=1}^{k}\bigcap_{j=1}^{n}\varepsilon_{ij}x_j\neq 0\cdot\vee\cdot\bigvee_{i=1}^{l}\bigcap_{j=1}^{n}\delta_{ij}x_j\neq 0\right).$$

引理 6 使我们可以将原子式子改写为只含一端为 0 的式子.

现在我们来定义记号≤并证明与它有关的一些性质.

定义 1 记号 $x\leqslant y$ 的含义是 $x\cup y=y$.

引理 7 记号 $x\leqslant y$ 满足下列性质:

(i)$x\leqslant y\Leftrightarrow x\cup y=y\Leftrightarrow x\cap y=x\Leftrightarrow x\cap\bar{y}=0\Leftrightarrow\bar{x}\cup y=1$,

(ii)在 $x\leqslant y$ 的作用之下布氏代数做成偏序集,

(iii)$x\cup y=z\Leftrightarrow x$ 和 y 的最小上界是 z,

(iv)$x\cap y=z\Leftrightarrow x$ 和 y 的最大下界是 z.

除了记号 $x\leqslant y$ 之外我们还可以定义记号<.

定义 2 记号 $x<y$ 的含义是 $x\leqslant y\wedge x\neq y$.

利用定义 2 可以将无原子性质(21)改写为:

(21′)$(\forall y)(0\neq y.\rightarrow.\exists z(0<z<y))$.

这是一种稠密性,我们还可以将这种稠密性推广到任意两个元之间.

引理 8 对无原子布氏代数下式成立:

(22)$(\forall xy)(x<y.\rightarrow.\exists z(x<z<y))$.

证 $x<y\Rightarrow\bar{y}<\bar{x}$,

考虑 $y\cap\bar{x}\neq 0$,否则

$$y\cap\bar{x}=0,y\cup\bar{x}\geqslant y\cup\bar{y}=1$$

$$\Rightarrow\bar{y}=\bar{y}\cap(y\cup\bar{x})=(\bar{y}\cap y)\cup(\bar{y}\cap\bar{x})=0\cup\bar{x}=\bar{x}\text{ 矛盾!}$$

由性质(21)知存在 z 满足 $0<z<y\cap\bar{x}$.

下证 $x<x\cup z<x\cup(y\cap\bar{x})=y$ 否则

$$x\cup z=y\Rightarrow(x\cup z)\cap\bar{x}=(y\cap\bar{x})\Rightarrow z=y\cap\bar{x}\text{ 矛盾!}$$

引理 9 要证明关于无原子布氏代数的定理只需在可数模型中证明即可.

证 因为无原子布氏代数的理论是可数的,它的任意模型都有可数初等子模型,所以只需在可数模型中验证即可.(请参看[4] Theorem 3.1.6.)

引理 10 在无原子布氏代数的可数模型中从任何一个元 a 开始存在无限递降序列

$$a_0=a, a_1, \cdots, a_n, \cdots$$

使它满足

$$x<a_0, x<a_1, \cdots, x<a_n, \cdots \Rightarrow x=0.$$

证 由于在无原子布氏代数的可数模型中任意一个元素与 0 之间的序型是可数稠密的，由集合论中的可数序列的共尾性知，可找到序列

$$a_0, a_1, \cdots, a_n, \cdots$$

使其满足

$$x<a_0, x<a_1, \cdots, x<a_n, \cdots \Rightarrow x=0.$$

引理 11 设 $a_0>a_1>\cdots>a_n>\cdots$ 和 $b_0>b_1>\cdots>b_n>\cdots$ 是两个序列，它们分别满足 $x<a_0, x<a_1, \cdots, x<a_n, \cdots \Rightarrow x=0$，和 $x<b_0, x<b_1, \cdots, x<b_n, \cdots \Rightarrow x=0$.

令序列 $c_0>c_1>\cdots>c_n>\cdots$ 为 $c_i=a_i\cup b_i, i=0,1,\cdots,n,\cdots$

那么也有 $x<c_0, x<c_1, \cdots, x<c_n, \cdots \Rightarrow x=0$.

证 用反证法，假设存在 y 满足 $y<c_0, y<c_1, \cdots, y<c_n, \cdots$，而$y\neq 0$.

考虑 $y\cap a_0\geqslant y\cap a_1\geqslant\cdots\geqslant y\cap a_n\geqslant\cdots$ 是一个序列，它满足

$$x<y\cap a_0, x<y\cap a_1, \cdots, x<y\cap a_n, \cdots \Rightarrow x=0, \quad ①$$

类似地

$$x<y\cap b_0, x<y\cap b_1, \cdots, x<y\cap b_n, \cdots \Rightarrow x=0. \quad ②$$

这样我们推出

$$c_0=a_0\cup b_0\geqslant y\Rightarrow (y\cap a_0)\cup(y\cap b_0)=y\cap(a_0\cup b_0)=y,$$

$$c_1=a_1\cup b_1\geqslant y\Rightarrow (y\cap a_1)\cup(y\cap b_1)=y\cap(a_1\cup b_1)=y, \cdots$$

①②两个下极限为 0 的序列，它们的对应项的和都是 $y\neq 0$，我们要从这一点推出矛盾. 为了化简记号我们将①②两个序列的记号改写如下：

令 $d_i=y\cap a_i, e_i=y\cap b_i$，我们有

$$x<d_0, x<d_1, \cdots, x<d_n, \cdots \Rightarrow x=0, \quad ③$$

$$x<e_0, x<e_1, \cdots, x<e_n, \cdots \Rightarrow x=0, \quad ④$$

和 $\qquad y=d_0\cup e_0=d_1\cup e_1=\cdots=d_n\cup e_n=\cdots$

将每一个 $d_i\cup e_i$ 分为三部分并对相邻的两组进行比较：

$$d_i \cup e_i = (d_i \cap \bar{e}_i) \cup (d_i \cap e_i) \cup (\bar{d}_i \cap e_i) = y.$$

由于 $d_i \cup e_i$ 所分成的三部分是互不相交的，而前两部分的和是 d_i，后两部分的和是 e_i.

不可能对于所有 i 都有 $d_i \cap \bar{e}_i = \bar{d}_i \cap e_i = 0$，因为这与序列③④的结构性质不符. 由对称性和任意设开始讨论点我们不妨设 $d_0 \cap \bar{e}_0 \neq 0$.

$\cap$	$\bar{e}_i$	$e_i \cap \overline{e_i+1}$	e_{i+1}
$\bar{d}_i$	$\bar{d}_i \cap \bar{e}_i$	$\bar{d}_i \cap (e_i \cap \overline{e_{i+1}})$	$\bar{d}_i \cap e_{i+1}$
$d_i \cap \overline{d_{i+1}}$	$(d_i \cap \overline{d_{i+1}}) \cap \bar{e}_i$	$(d_i \cap \overline{d_{i+1}}) \cap (e_i \cap \overline{e_{i+1}})$	$(d_i \cap \overline{d_{i+1}}) \cap e_{i+1}$
d_{i+1}	$d_{i+1} \cap \bar{e}_i$	$d_{i+1} \cap (e_i \cap \overline{e_{i+1}})$	$d_{i+1} \cap e_{i+1}$

观看上面列表，表中第一行的后三个元是将 1 分为三个互不相交的部分，其中后两者的和为 e_i，类似地表中第一列的后三个元是将 1 分为三个互不相交的部分，其中后两者的和为 d_i. 右下方的 9 个元是它们的左边和上面的对应元的交. 由此得出

$d_i \cup e_i = (\bar{d}_i \cap (e_i \cap \overline{e_{i+1}})) \cup (\bar{d}_i \cap e_{i+1}) \cup ((d_i \cap \overline{d_{i+1}}) \cap \bar{e}_i) \cup ((d_i \cap \overline{d_{i+1}}) \cap (e_i \cap \overline{e_{i+1}})) \cup ((d_i \cap \overline{d_{i+1}}) \cap e_{i+1}) \cup (d_{i+1} \cap \bar{e}_i) \cup (d_{i+1} \cap (e_i \cap \overline{e_{i+1}})) \cup (d_{i+1} \cap e_{i+1})$,

$d_{i+1} \cup e_{i+1} = (\bar{d}_i \cap e_{i+1}) \cup ((d_i \cap \overline{d_{i+1}}) \cap e_{i+1}) \cup (d_{i+1} \cap \bar{e}_i) \cup (d_{i+1} \cap (e_i \cap \overline{e_{i+1}})) \cup (d_{i+1} \cap e_{i+1})$.

由于上面的两个和式中的项都是互不相交的，并且它们的和都等于 y，所以

$$\bar{d}_i \cap (e_i \cap \overline{e_{i+1}}) = (d_i \cap \overline{d_{i+1}}) \cap \bar{e}_i = (d_i \cap \overline{d_{i+1}}) \cap e_{i+1} = 0.$$

这样一来我们推出

$$\bar{d}_i \cap \bar{e}_i = ((d_i \cap \overline{d_{i+1}}) \cap \bar{e}_i) \cup (d_{i+1} \cap \bar{e}_i) = d_{i+1} \cap \bar{e}_i. \qquad ⑤$$

由 $e_{i-1} \geqslant e_i$ 反过来 $\overline{e_{i-1}} \leqslant \bar{e}_i$，再推出

$$d_i \cap \overline{e_{i-1}} \leqslant d_i \cap \bar{e}_i. \qquad ⑥$$

反复运用⑤⑥我们得到

$$d_i \cap \bar{e}_i \geqslant d_i \cap \overline{e_{i-1}} = d_{i-1} \cap \overline{e_{i-1}}.$$

再反复运用这个式子我们推出了

$$d_i \cap \bar{e}_i \geqslant d_{i-1} \cap \overline{e_{i-1}} \geqslant \cdots \geqslant d_0 \cap \bar{e}_0 \neq 0.$$

序列 $d_i \cap \bar{e}_i$ 也是下界为 0 的序列，这样得出了矛盾！引理 11 证完.

引理 12 设 $b_{i0} > b_{i1} > \cdots > b_{ij} > \cdots, j=0,1,\cdots, i=1,2,\cdots,n$ 是 n 个序列，它们都满足

$x < b_{i0}, x < b_{i1}, \cdots, x < b_{ij}, \cdots \Rightarrow x=0$，对 $i=1,2,\cdots,n$.

令序列 $c_0 > c_1 > \cdots > c_n > \cdots$ 为 $c_j = b_{1j} \cup \cdots \cup b_{nj}, j=0,1,\cdots$，

那么也有 $x < c_0, x < c_1, \cdots, x < c_n, \cdots \Rightarrow x=0$.

证 反复运用引理 11 即可.

引理 13 设 $\mathscr{B}$ 是无原子布氏代数的可数模型. $a_1, a_2, \cdots, a_n \neq 1, 0$ 是 $\mathscr{B}$ 的有限个元素，那么存在 $\mathscr{B}$ 的元素 x 使得

$$x \not\leqslant a_1, \cdots, x \not\leqslant a_n; x \not\geqslant a_1, \cdots, x \not\geqslant a_n.$$

证 令 $b_1 = \bar{a}_1, b_2 = \bar{a}_2, \cdots, b_n = \bar{a}_n \neq 1, 0$.

对每一个 $b_i, i=1,2,\cdots,n$ 如引理 10 找一个序列

$b_{i0} = b_i > b_{i1} > \cdots > b_{ij} > \cdots$ 可以使它满足

$$x < b_{i0}, x < b_{i1}, \cdots, x < b_{ij}, \cdots \Rightarrow x=0.$$

再令 $c_j = b_{1j} \cup b_{2j} \cup \cdots \cup b_{nj}, j=0,1,\cdots$ 由引理 12

$$x < c_0, x < c_1, \cdots, x < c_j, \cdots \Rightarrow x=0.$$

因为对每一个 i 比 a_i 大的 c_j 只有有限个，所以可以找到一个 c_m，使得它不比任何一个 a_i 大. 下面我们要证明它也不比任何一个 a_i 小.

假设 $c_m < a_i$ 我们得到

$$c_m = b_{1m} \cup \cdots \cup b_{nm} < a_i, \Rightarrow b_{im} < a_i \Rightarrow b_{im} \cap a_i = b_{im} \neq 0.$$

但是由于

$$b_i = \bar{a}_i, b_{im} < b_i \Rightarrow b_{im} \cap a_i \leqslant b_i \cap a_i = 0$$

与上面矛盾. 这样一来 c_m 就是我们所要找的既不大于任何一个 a_i 也不小于任何一个 a_i 的元素.

引理 14 设 $\mathscr{B}$ 是无原子布氏代数的可数模型. $S < G$ 是 $\mathscr{B}$ 的两个元素. $S < a_1, a_2, \cdots, a_n < G$ 是 S, G 之间的有限个元素，那么存在 S, G 之间的元素 x 使得

$$x \not\leqslant a_1, x \not\leqslant a_2, \cdots, x \not\leqslant a_n; x \not\geqslant a_1, x \not\geqslant a_2, \cdots, x \not\geqslant a_n.$$

证 与引理 13 类似，这里就不再重复.

以上 14 个引理为我们证明定理做了充分的准备.

§5. 无原子布氏代数的量词消去法

定理 1 无原子布氏代数理论具有完全的量词消去法，也就是说每一个式子都 Δ 等价于一个开式子.（一般布氏代数理论本定理不成立.）

证 (1)将式子变为 $Q_1x_1Q_2x_2\cdots Q_nx_n\varphi$ 形式，其中 $\varphi(x_1,x_2,\cdots,x_n)$ 是含有 $x_1x_2\cdots x_n$ 开式子，量词 Q_i 是∃记号或者∀记号 $1\leqslant i\leqslant n$，Q_n 只是∃记号.

(2)利用引理 6 将原子式子改写为只含一端为 0 的式子再利用分配律展开得到

$$\bigvee_{i=1}^{l}((\bigwedge_{j=1}^{\varepsilon}\bigcap_{k=1}^{n-1}\alpha_{ijk}x_k\cap x_n=0)\wedge(\bigwedge_{j=1}^{t}\bigcap_{k=1}^{n-1}\beta_{ijk}x_k\cap\overline{x}_n=0)$$
$$\wedge(\bigwedge_{j=1}^{u}\bigcap_{k=1}^{n-1}\gamma_{ijk}x_k\cap x_n\neq0)\wedge(\bigwedge_{j=1}^{v}\bigcap_{k=1}^{n-1}\delta_{ijk}x_k\cap\overline{x}_n\neq0)). \quad ⑦$$

要求每一项中每一个 x_i 都出现，这是可以做到的. 其中 $\alpha_{ijk},\beta_{ijk},\gamma_{ijk},\delta_{ijk}=\pm1$.

(3)为了把式子写得简单一些，我们定义一些记号：

$$A_{ij}=\bigcap_{k=1}^{n-1}\alpha_{ijk}x_k,\quad B_{ij}=\bigcap_{k=1}^{n-1}\beta_{ijk}x_k,$$
$$C_{ij}=\bigcap_{k=1}^{n-1}\gamma_{ijk}x_k,\quad D_{ij}=\bigcap_{k=1}^{n-1}\delta_{ijk}x_k.$$

将 φ 变为

$$\bigvee_{i=1}^{l}((\bigwedge_{j=1}^{s}A_{ij}\cap x_n=0)\wedge(\bigwedge_{j=1}^{t}B_{ij}\cap\overline{x}_n=0)$$
$$\wedge(\bigwedge_{j=1}^{u}C_{ij}\cap x_n\neq0)\wedge(\bigwedge_{j=1}^{v}D_{ij}\cap\overline{x}_n\neq0)). \quad ⑧$$

(4)将 $\exists x\varphi$ 的∃记号移入 φ 需消去的式子变成 $\exists x_n\wedge\vee$ 式子(8)变为 $\vee(\exists x_n)$. 在只讨论一个连与项的 $x=x_n$ 的消去时下标(i)可以不必写出，我们得到

$$(\exists x)((\bigwedge_{j=1}^{s}A_j\cap x=0)\wedge(\bigwedge_{j=1}^{t}B_j\cap\overline{x}=0)\wedge(\bigwedge_{j=1}^{u}C_j\cap x\neq0)\wedge(\bigwedge_{j=1}^{v}D_j\cap\overline{x}\neq0)) \quad ⑨$$

(5)下面讨论消去量词的步骤.

(i) x 满足 $A_j\cap x=0$ 的充分必要条件是 $A_j\leqslant\overline{x}$. 所以 x 满足 $\bigwedge_{j=1}^{\varepsilon}A_j\cap$

$x=0$ 的充分必要条件是 $A_j\leqslant\bar{x},j=1,2,\cdots,s\Leftrightarrow\bar{A}_j\geqslant x,j=1,2,\cdots,s,\Leftrightarrow(\bigcap_{j=1}^{s}\bar{A}_j)\geqslant x.\Leftrightarrow(\overline{\bigcup_{j=1}^{s}A_j})\geqslant x.$

令 $A=\bigcup_{j=1}^{i}A_j$，上面条件又可以写成 $\bar{A}\geqslant x.$

(ii)类似地 x 满足 $B_j\cap\bar{x}=0$ 的充分必要条件是 $B_j\leqslant x$. 所以 x 满足 $\bigwedge_{j=1}^{i}B_j\cap\bar{x}=0$ 的充分必要条件是 $B_j\leqslant x,j=1,2,\cdots,s\Leftrightarrow(\bigcup_{j=1}^{i}B_j)\leqslant x.$

令 $B=\bigcup_{j=1}^{s}B_j$，上面条件又可以写成 $B\leqslant x.$

(iii)x 满足 $C_j\cap x\neq 0$ 的充分必要条件是 $C_j\nleqslant\bar{x}$. 所以 x 满足 $\bigwedge_{j=1}^{u}C_j\cap x\neq 0$ 的充分必要条件是 $C_j\nleqslant\bar{x},j=1,2,\cdots,u(\Leftrightarrow\bar{C}_j\ngeqslant x,j=1,2,\cdots,u).$

(iv)x 满足 $D_j\cap\bar{x}\neq 0$ 的充分必要条件是 $D_j\nleqslant x$. 所以 x 满足 $\bigwedge_{j=1}^{v}D_j\cap\bar{x}\neq 0$ 的充分必要条件是 $D_j\nleqslant x,j=1,2,\cdots,v.$

(v)使上述条件(i)～(iv)得到满足的 x 存在充分必要条件是

$$(\bar{A}\geqslant B)\wedge(\bigwedge_{j=1}^{u}\bar{C}_j\ngeqslant\bar{A})\wedge(\bigwedge_{j=1}^{v}D_j\nleqslant B).\qquad ⑩$$

条件(10)的必要性易见. 下证它的充分性. 如果 $\bar{A}=B$，那么 $x=B$ 可以满足(i)～(iv). 在 $\bar{A}>B$ 令

$$\bar{C}'_j=\bar{C}_j\cap\bar{A},\quad D'_j=D_j\cup B.$$

我们要在 $\bar{A},B$ 之间找一个 x 使它满足①②③④.

如果 $\bar{C}'_j\ngeqslant B$，只要 x 在 $\bar{A},B$ 之间，③便自然满足.

如果 $D'_j\nless\bar{A}$，只要 x 在 $\bar{A},B$ 之间，④便自然满足.

最后剩下的情形是

$$B<\bar{C}'_1,\cdots,\bar{C}'_u,D'_1,\cdots,D'_v<\bar{A}.$$

由引理 14 可以找到在 $\bar{A},B$ 之间的元素 x 满足

$$x\nleqslant\bar{C}'_1,x\nleqslant\bar{C}'_2,\cdots,x\nleqslant\bar{C}'_u;x\ngeqslant D'_1,x\ngeqslant D'_2,\cdots,x\ngeqslant D'_v.$$

也就是满足了(i)～(iv). 条件⑩的充分性得到了证明.

条件⑩是一个不含有 x 的条件，这样变量 x 就消去了.

(6)反复运用上述过程直到量词完全消去为止.

这样每一个含有量词的式子都 Δ 等价于一个不含有量词的开式子.

推论 每一个闭式子(语句)都 Δ 等价于真值式子 $\mathscr{T}$ 或假值式子 $\mathscr{F}$. 因此无原子布氏代数的理论 Δ 是一个完备的理论.

证 由量词消去法直接可得.

定义 3 一个理论的含有 n 个变量的初等型 $\Gamma(x_1,x_2,\cdots,x_n)$ 是含有这个变量的极大和谐式子集.

定理 2 无原子布氏代数的初等型 $\Gamma(x_1,x_2,\cdots,x_n)$ 是由型内的不含量词的全体开式子所唯一决定.

证 由定理 1 知凡是含有变量的式子经过消去量词都等价于一个开式子,所以它们在型中都是不起作用的.

利用定理 3 可以证明关于无原子布氏代数的很多有用性质,例如 ω 范畴型定理等. 我们就不在这里列出了.

§6. 无原子布氏代数的计算复杂性

利用量词消去法可以判断式子的真假,但是由于反复运用分配律上述判定闭式子真假的过程所用的时间(最坏的情况)是 $2^{2^{\cdot^{\cdot^{2^n}}}}$ ($n-1$ 层),这里 n 是量词的个数. 因此我们需要找出一个不用分配律的办法来得到展开式,这就是赋值计算法.

引理 15(赋值计算法) 在一个含有 p 个命题变量的命题演算式子 $\varphi(A_1,A_2,\cdots,A_p)$ 中,以一组真假值赋值 $(\varepsilon_1,\varepsilon_2,\cdots,\varepsilon_p)$ 代入得到 $\varphi(\varepsilon_1,\varepsilon_2,\cdots,\varepsilon_p)=\mathscr{T},\mathscr{F}$. 那么 $\varepsilon_1A_1,\varepsilon_2A_2,\cdots,\varepsilon_pA_p$ 是 $\varphi(A_1,A_2,\cdots,A_p)$ 的展开式中的一个连加项的充分必要条件是 $\varphi(\varepsilon_1,\varepsilon_2,\cdots,\varepsilon_n)=\mathscr{T}$.

如果要将所有连与项全部找出至多需要计算 2^p 个赋值式.

计算一个布氏项的展开项也有类似结果.

引理 16 引理 15 中计算展开式所需的时间和空间都属于 2^{cp} 指数级.

引理 15,引理 16 纯属命题演算定理,这里就不再证明了.

定理 3 无原子布氏代数的一个长度为 n 的语句的判断过程所消耗的 Turing 时间和空间都是属于 $2^{2^{cn}}$ 指数级.

证 由于语句的长度为 n,它至多含有 n 个量词. 我们逐步计算定理的量词消去过程所用的时间.

(1)将式子变为 $Q_1x_1Q_2x_2\cdots Q_nx_n\varphi$ 形式所用的时间是 c_1n. 因为我们只需将量词移到式子的最前面. 这个时间和后面的时间来比较是可以忽略不计的.

(2)利用引理 6 将原子式子改写为只含一端为 0 的式子,再利用分配律展开得到

$$\bigvee_{i=1}^{l}((\bigwedge_{j=1}^{s}\bigcap_{k=1}^{n-1}\alpha_{ijk}x_k\cap x_n=0)\wedge(\bigwedge_{j=1}^{t}\bigcap_{k=1}^{n-1}\beta_{ijk}x_k\cap\overline{x}_n=0)$$

$$\wedge\left(\bigwedge_{j=1}^{u}\bigcap_{k=1}^{n-1}\gamma_{ijk}x_k\cap x_n\neq 0\right)\wedge\left(\bigwedge_{j=1}^{v}\bigcap_{k=1}^{n-1}\delta_{ijk}x_k\cap\overline{x}_n\neq 0\right)\Big), \qquad ⑪$$

要求每一项中每一个 x_i 都出现. 其中 $\alpha_{ijk},\beta_{ijk},\gamma_{ijk},\delta_{ijk}=\pm 1$.

这里要分下列步骤来讨论:

(i)把每一个原子式子中的项对 $\cap,\cup,\overline{x}$ 展开,为了节省时间,我们也可以改用引理 15,引理 16 的办法,这样得到

$$\left(\bigcup_{j=1}^{s}\bigcap_{k=1}^{n}\alpha_{ijk}x_k\right)=\left(\bigcup_{j=1}^{t}\bigcap_{k=1}^{n}\beta_{jk}x_k\right).$$

(ii)上式中形如$\left(\bigcap_{k=1}^{n}\alpha_{jk}x_k\right)$的项有 2^n 个. 因此形如$\left(\bigcup_{j=1}^{s}\bigcap_{k=1}^{n}\alpha_{jk}x_k\right)$的项中的 s 满足 $0\leqslant s\leqslant 2^n$.

(iii)利用引理 4 将如下形状的式子

$$t_1(x_1,x_2,\cdots,x_n)=\bigcup_{i=1}^{k}\bigcap_{j=1}^{n}\varepsilon_{ij}x_j=t_2(x_1,x_2,\cdots,x_n)=\bigcup_{i=1}^{l}\bigcap_{j=1}^{n}\delta_{ij}x_j.$$

改写为如下形状的式子

$$t_1=t_2\Rightarrow\bigcap_{k=1}^{n}\varepsilon_k x_j=0,\bigcap_{k=1}^{n}\delta_k x_k=0.$$

这种式子至多有 2^n 个.

(iv)所以原式中每一个原子式子改写后有 2^n 个=号或≠号. 上述过程所用到的时间是 2^{c2^n}.

(v)按引理 5 将上式改写为上面⑪的形式. 其中用到 $\wedge,\vee,{}^{-}$ 分配律展开时,改用引理 15,引理 16,由于原式的长度为 n,至多含有 n 个原子式子,每一个原子式子含有至多 2^n 个=式或≠式. 用引理 15,引理 16 求展形式所用到的时间是 $2^{2^{c3^n}}$,空间是 $2^{2^{c4^n}}$. 其计算式如下:

$$2^{2^{cn}}\times 2^{2^{cn}}=2^{2^{cn}+2^{cn}}=2^{2^{cn+1}}$$

(3)移动量词改写式子并不增加时间和空间的指数级.

(4)将 $\exists x_n$ 的 $\exists$ 记号移入 φ,需消去的式子由 $\exists x_n(\vee\wedge)$变为 $\vee(\exists x_n)\wedge$ 并不增加时间和空间的指数级.

(5)消去量词 $\exists x_n$ 的步骤中,只是调整式子,也并不增加时间和空间的指数级.

(6)反复运用上述过程直到量词完全消去为止. 其中如果有两个相邻的量词是不同的,需要再求分配律的展开式,我们还是用引理 15,引理 16 的办法来做,时间和空间的指数级还能保持. 总共有 n 个量词要消去,所用的时间和空间的指数级仍然不变. 其计算式如下:

$$(2^{2^{cn}})^n = 2^{2^{cn}\times n} = 2^{2^{cn}\times 2^{\log_2 n}} = 2^{2^{cn+\log_2 n}} \leqslant 2^{2^{(cn+1)n}}.$$

定理 3 证毕.

§7. 结论

在一般的布氏代数的研究工作中其证明方式都是通过用文字描述来证明定理. 我们则是用形式化的方法先证明量词消去法,然后再通过它来证明计算复杂性和其他定理. 我们办法的优点是可计算性强. 用来做研究工作有其实用价值.

参考文献

[1] Tarski A. Arithmetical classes and types of Boolean algebras. Bull. Amer. Math. Soc. ,1949,55:1 192.

[2] 刘吉强,廖东升,罗里波. 完全二叉树的量词消去. 数学学报,2003,46(1):95—102.

[3] Stone M H. The representation theorem of Boolean algebra. Trans. Amer. Math. Soc. ,1936,40:37—111.

[4] Chang C C,Keisler H J. Model Theory. North-Holland,3rd ed. ,1990.

Abstract In this paper we study the computational complexity of the theory of Boolean algebra without atomic elements. The following new theorems are obtained:

Theorem 1 The Theory of Boolean algebra without atomic elements has a complete quantifier elimination i. e. under the theory every formula is equivalent to an open formula.

Theorem 2 The elementary type $\Gamma(x_1, x_2, \cdots, x_n)$ of Boolean algebra without atomic elements is uniquely decided by its open formulas.

Theorem 3 In the theory of Boolean algebra without atomic elements to decide a sentence with length n the Turing time and space spent in the procedure is in the exponential degree of $2^{2^{cn}}$.

Keywords Boolean algebra without atomic elements, complete quantifier elimination, number of models, computational complexity.

数学通报,2004,(8):40～43

利用计算机计算古典数论问题[①]

Computing Problems in Classical Number Theory Using Microcomputers

§1. 问题的来源

判断一个大的整数是否素数,如果它不是素数的话又如何将它分解为若干个因数的乘积是古典数论的一个重要问题. 由于计算机科学和密码学的发展,上述的古典问题又焕发出了新的光亮. 因为有一种很简单的密码是用素数模乘法变换来构造的. 如果你不知道这个素数,你就无法解开这个密码. 有人甚至将这个密码的钥匙半公开:一般是把两个素数乘起来产生一个合数. 因为这个数很大,如果你不会分解它,你是无法解开这个密码的. 而制造这个密码的人就需要记住合数分解的方法,密码随时可以解开. 长整数分解是一个很难的问题. 形如 $F_n=2^{2^n}+1$ 的数称为费马数. 如果它也是素数,就叫作费马素数. 费马验证了 $F_0=3, F_1=5, F_2=17, F_3=257, F_4=65\ 537$ 五个数都是素数. 于是它大胆猜测对任意 n,F_n 都是素数(1660 年). 但是过了几十年到了 1732 年欧拉分解了 $F_5=2^{2^5}+1=641\times 6\ 700\ 417$,从而否定了费马的猜想. 现在如果我们利用计算机去分解这个费马数,只需要不到一秒钟.

人们对于如何利用计算机判断素数和分解合数的研究做的很多,但是对在长整数计算和编程的问题上却研究得很少. 原因是以前的研究工作者和编程人员的分工太清. 研究工作者负责设计算法而不管程序,编程人员负责设计程序而不管算法,编程人员甚至可以不知道研究人员在研

① 本文与龚成清、蒋桂梅和陈永遥合作.

究什么课题. 这种研究人员和编程人员的脱节现象也是历史所造成的. 以前资深的研究人员都是一些大学问家. 他们肚子里头的学问对于一般人来说是可望而不可即的. 于是编程人员也就只管编程. 反过来,由于以往计算这些难题必须用到大实验室里的超级计算机,而控制这些机器也是一门高深的学问,这对于研究工作者来说也是摸不着头脑,更不用说其他的人了. 这种研究人员和编程人员的脱节现象使对素数判别和整数分解的研究落到了极少数的研究人员和编程人员的手中. 一般的没有大实验室支持的研究人员无法从事这项研究工作. 现在情况大大的不同了. 现在的一台普通计算机的计算能力就比在 20 世纪 70 年代的高级实验室中用来分解 $F_8=2^{2^8}+1$ 的计算机要强. 吸引更多人来参加这样的研究工作成为可能. 我们从 1996 年就开始研究计算机上的长整数计算. 近年来又吸收了一些年轻的同志来参加我们的研究工作. 本文就长整数计算的计算机编程中所遇到的一些问题作一些介绍,并对如何解决这些问题做出探讨,我们也想将对这方面的研究工作进行推广普及,从而吸引更多人来一起进行研究,以使研究工作能取得更大的进展.

§2. 整数计算方法的选择

我们所需要考虑的第一个问题是编程语言的选择. 我们要从长整数的计算过程的特征来考虑. 长整数的计算过程往往是用扫描的方式在大量的整数中挑出所要的整数. 假定所要分解的数是长度为 80 位的十进位整数,而我们采用一个一个地试验的办法去找,就要试验所有的 40 位的整数作为一个可能的因子,这些数至少有 1 亿亿亿亿个. 用最高效率的计算机假定每秒钟能试验 1 亿个,一年的秒数还不到 1 亿,所以至少要用 1 亿亿年才能试验完毕. 无论是人还是机器都是不可能耐受这么长的时间. 由于到目前为止人们还没有发现消耗时间为整数长度的多项式函数的程序来判定素数或分解合数,我们所用的程序具有极大的偶然性,一般来说程序中必然出现大量循环和搜索,为了减少程序语言本身的循环,用汇编语言是比较好的. 但是汇编语言很繁杂,又不容易控制,用起来很不方便. C 语言在位运算上有它的长处,但是在定义输出输入方面又很不方便. FORTRAN 语言在超级计算机上用的较多,但是我们的目标程序是尽量利用计算机或者较为小型的机器,面向对象的语言在循环控制上有它的不便之处,因此也不是我们的首选. 一般常识是 PASCAL 语言最适于科学计算程序,这也是我们的选择.

§3. 普通整数计算的数据

在程序设计语言中整数分为三类，它们分别叫作短整数，整数(为了更好地区分它们，我们把这一类整数叫作普通整数)和长整数. 它们的性质和范围在下表列出：

类型	取值范围	占字节数	bit 位数
shortint	−128..129	1	8
integer	−32768..32767	2	16
longint	−2147483648..2147483647	4	32

以上规定是为了节省程序运算所使用的资源，问题可以用小整数解决的就尽量不用普通整数，问题可以用普通整数解决的就尽量不用长整数. 在所有的编程语言中都有最大的整数. PASCAL 的普通整数的最大数是 32 767，它有一个特殊的记号就是 maxint，长整数的最大数是 2 147 483 647，它也有一个特殊的记号 maxlongint. 超过了这些相应的最大整数，程序就不能运转或者出错. 在这一节中我们先就普通整数的计算问题进行讨论. 下面是一个构造并打印出32 767以下的全体素数的程序.

```
program comprime(output);
(* compute and print out the prime numbers between 1 and 32767. *)
var i,j,k,l,m,r:integer;
outfl:text;
begin
assign(outfl,'C:prime. out');
rewrite outfl
write(outfl,2:6);
l:=1;r:=0;
for i:=3 to 32767 do
begin
m:=i;j:=1;k:=0;
repeat j:=j+1;
if(m mod j=0)then k:=1;
until(k=1)or(j>=m- 1);
if(k=0)then
```

```
begin write(outfl,m:6);l:=(l+1)mod 10;
if l=0 then begin writeln (outfl);r:=r+1;writeln(r:6) end;
end;
end;
close(outfl);
end.
```

首先介绍一下我们的机器,我们用的是一种家庭版的计算机,它的内存有 56 GB,CPU 的型号是奔 3-667,硬盘储存量有 15 GB.机器是再普通不过了,我们的程序就是在这样的一个机器上运行的.上面列出的程序叫作程序 A.整个程序的运行时间是 3.08 s.在上列的程序中我们省略了源程序中计算时间的部分.由于运行的时间很短我们只求程序简单而暂不考虑优化程序的问题.程序的优化问题我们放在后面的程序中来讨论.程序 A 运行的结果是打印出了 352 行每行 10 个字共3 512个素数,应该说虽然这个程序不是构造素数表的最好程序,3 s 的等待对于一个人来说是没有什么问题的.它的缺点是素数只能算到32 767,再大的素数就不能用它来算了.如果将计算的上限改大,程序就不再能运行.

§4. 长整数计算方法的比较

现在考虑长整数计算的程序设计.如果将上面程序中的整数类型 integer 改成长整数类型 longint 上面的程序就可以对较大的整数上限运行了.计算的速度问题是长整数计算中的重要问题,某些问题由于受到时间的限制本来是不能计算的,由于计算速度的提高变成了是可以计算的了.下面就举一个例子来说明如何提高计算的速度.比如说现在要构造一个 6 位的素数表,我们用试除的办法来做.可以设计以下的方案:任意给定一个长整数 x,我们可以用另一个数 y 来试除它,如果能够整除,那么 x 就不是素数,如果不能整除,就找下一个 y 来进行试除.对于 x 的选择可以有两种.第一种 x 是任意整数;第二种是除了 2 之外,我们可以只选奇数,第二种算法看起来应该是优于 2 之外我们可以只选奇数.第二种算法看起来应该是优于第一种,因为它可以少试一半的数.但是第一种算法也有它的优点,因为它可以免去判断 x 是否奇数的过程.如果进而设想我们在选择 x 时是否可以去掉 3 的倍数,5 的倍数等,从而节省时间呢?当然这样就增加了判断 x 是否这些倍数的过程,而这也是消耗时间的.

对于 y 的选择我们考虑以下几个不同的方案:

①按照第一种办法选择 x，也就是把整数从 2 开始一个一个地选下去，不做任何事先的区分. 对 y 也是将 $2\leqslant y<x$ 的范围内的全体整数一个一个地选下去. 素数的判断是在选取之后由程序的其他部分来完成. 这种选取方式写出来的程序叫作程序 A.

②按照第一种办法选择 x，也就是把整数从 2 开始一个一个地选下去，不做任何事先的区分. 对 y 则是选取在 $2\leqslant y<\sqrt{x}$ 的范围内的全体整数. 然后再作素数的判断，这种选取方式本文不做更多的讨论. 它是前面程序 A 和后面程序 B 的一种中间状态.

③按照第二种办法选择 x，也就是把整数从 2 开始以后只选奇数整数. 对 y 则只选取在 $2\leqslant y<\sqrt{x}$ 的范围内的全体奇数整数. 素数的判断也是在选取之后由程序的其他部分来完成. 这种选取方式写出来的程序叫作程序 B.

④按照第二种办法选择 x，也是把整数从 2 开始以后只选奇数整数. 对 y 则只选取在 $2\leqslant y<\sqrt{x}$ 的范围内的全体素数整数. 素数的判断也是在选取之后由程序的其他部分来完成. 这种选取方式写出来的程序叫作程序 C.

我们先来看程序 A 的运行. 在程序 A 中被除数 x 是逐个地选取的整数，除数 y 也是逐个地选取 $2\leqslant y<x$ 中的所有整数，我们将程序 A 中的上限分别按 1000，10000，100000 和 1000000 四种情形进行运行并将运行所消耗的时间记录如下：

上限	1000	10000	100000	1000000	10000000
A 程序	0.06 秒	1.20 秒	146.89 秒	约两小时	约一周

这四种情形所产生的数据如下：

上限	行数	数字数	页数	储存量	耗时
1000	17	168	1	2KB	0.06 秒
10000	123	1229	3	8KB	1.20 秒
100000	960	9592	22	59KB	146.89 秒
1000000	7850	78498	179	476KB	6879.75 秒

下面再来看程序 B 的运行. 在程序 B 中被除数 x 只选取奇数整数，除数 y 也是只选取 $2\leqslant y<\sqrt{x}$ 中的奇数整数，这样就大大地加快了运算速度. 由于程序 B 的速度较快，我们除了运行 10^3，10^4，10^5 和 10^6 四种情形之外还运行了 10^7 和 10^8 两种情形. 所消耗的时间记录如下：

上限	10^3	10^4	10^5	10^6	10^7	10^8
程序 B	0.01 s	0.11 s	1.10 s	15.38 s	299.23 s	7 130.42 s

最后一种情形的运行所得出的数据是：上限 100000000，运行时间 7 130.42 s，打印出来的数字有5 761 455个，用 44 行一页来分页共有13 095 页，储存量是57 390 KB. 用 PASCAL 程序和写字板已经打不开这些数据了. 用 word 来打开这些数据需要十几分钟. 这个程序的运行结果就是构造了一个含有全体≤8 位的素数表. 在后面的程序中我们还用到了这个表.

在程序 C 中被除数 x 只选取奇数整数，除数 y 进一步只选取 $2\leqslant y<\sqrt{x}$中的素数整数. 这种选取方式所选取的除数是达到了最低的限度. 这样是不是就可以大大地加快了运算速度呢？其实情况并不那么好. 首先由于没有可以有效地计算素数的公式，我们不可能像程序 A 和程序 B 那样，由一个 y 能够很快地计算下一个 y. 我们必须将已经计算好了的素数保存在某一个地方，在计算下一个素数时再把它们调出来用做除数. 在 PASCAL 程序中保存大量数据的最好存储工具是 Array（数组），但是 PASCAL 程序对 Array 有很多限制. 一维 Array 的下标至多是 maxint. 用 10000×10000 的二维 Array 又因为结构太大而不被 PASCAL 所接受. 这样一来程序运行中的动态存储单元是放不下这么多素数的了. 还剩一个办法是直接利用硬盘存储单元. 而反复将数据送到硬盘存储单元，再把它提取出来进行计算又增加了程序的运行时间. 因此只有直接用机器来运行才可以判断出哪一个方案是最好的了.

对程序 C 我们运行了 10^3，10^4，10^5，10^6 和 10^7 五种情形. 其结果如下：

上限	10^3	10^4	10^5	10^6	10^7
程序 C	0.39 s	2.96 s	28.18 s	251.51 s	3 233.4 s

从实际运行的结果看来程序 C 不如程序 B 的运行速度快. 程序 C 中对每一个被试除的数来说，试除的次数肯定要比程序 B 的相应次数要少，而它的运行速度反而比程序 B 要慢. 唯一的解释就是上面所说的程序与磁盘之间的数据交换比与缓冲区之间的数据交换要费时间. 下表将各个不同程序和不同上限所运行的时间集中起来进行一个比较. 其中有些数据是我们估计的.

上限	10000	100000	1000000	10000000	100000000
程序 A	1.20 s	146.89 s	约 2 h	约一周	约半年
程序 B	0.11 s	1.10 s	15.38 s	299.23 s	7 130.42 s
程序 C	2.96 s	28.18 s	251.51 s	3 233.4 s	约一天

这里我们对用时太长的情形没有进行试运行，因为运行过了的情形已经可以充分说明问题了. 上面表中我们对这些情形只是给出了参考性的约略估计时间. 另外把程序放在不同的机器内进行运行所使用的时间是不同的，同一个机器同一个程序两次不同的运行所花的时间也稍有不同. 经过比较我们可以看出程序 B 的效果最好. 所以在上机运行时很多因素都要加以考虑. 观察表中的数据我们可以得出以下初步结论：

①被试除检验的范围扩大十倍，时间的消耗量就增加十倍以上并且越往后增加的幅度越大. 这个增加的幅度在数学中是可以写出渐近函数的. 我们这里只是大体上说是在十倍到一百倍之间.

②被试除检验的范围扩大十倍，计算出来的素数并没有增加十倍，它也是有渐近公式可以算出来的，我们这里暂时不涉及这些计算公式.

③由以上的数据可以推算出在计算机上用这种硬的试除法来构造素数表，再算下去至多能构造十位的素数表. 计算时间至少要一个月. 计算出来的结果要用 5 GB 来储存.

§5. 超长整数计算的数据

整数超过了 maxlongint 就不能用 PASCAL 的整数来进行计算了. 幸好我们还有 array，可以用来定义更长的整数. 作为研究对象的整数选择多少位为最合适，一般地说应该是根据问题来决定. 作为一个适应性较强的程序我们使用 80 位的十进位整数. 因为 80 位适于在屏幕上打印出来成为一行. 由行推算所用的二进位整数应当在 270 位左右. 我们用{0，1}的 array 来代表超长的二进位整数，{0—9}的 array 来代表超长的十进位整数. 由于我们现在所用的 PASCAL 的版本在程序中不能使用函数值为数组的函数，我们使用了过程并且利用全局变量将函数值从过程中带出来. 在运算上由于二进制整数的模运算比较好定义，我们将整数化为二进制以进行模运算和带余数的除法运算. 二进制的模运算用减法来定义是非常方便的. 十进制的整数只定义加法和乘法运算.

下面是程序的定义部分：

```
program factors(output);
const bnumln=270;d=10;dnumln=79;twpwmun=250;
label 10,20;
```

```
type bits=0..1;
blnum=array[0..bnumln]of bits;
dgts=0..9;
dlnum=array[0..dnumln]of dgts;
store=array[0..twpwmun]of dlnum;
var h,i,j,k,l,m,n,p,q,r,s,mc,md:integer;
infl1,infl2,outfl: text;
pw,pw1,pw2,pm,pm1,rm,rm1,bpls,btim,bmod,bdiv:blnum;
bs,tp,dpls,dtim: dlnum;
twpwls: store;
```

这里常数 bnumln，dnumln，各是二进位数的长度和十进位数的长度；twpwmun 是存放 250 个 2 的幂的类型；blnum，dlnum 各是二进位长整数和十进位长整数的类型；infl1，infl2，outfl 是文件变量；bpls，btim，bmod，bdiv 用来定义二进位的加法、乘法、模运算和除运算，dpls，dtim 用来定义十进位的加法和乘法运算. 由于我们已经定义了超长的整数，在计算中我们不再使用长整数，这样可以节省一些时间.

我们设计了一个对超长整数进行分解的程序. 由于前面我们已经构造出了一个 8 位素数表，在程序中我们就利用这个表直接调用素数这样可以节省判断除数是否素数的时间. 用我们的程序来分解费马数 $F_5=2^{2^5}+1=641\times 6\ 700\ 417$，所用时间不到一秒钟. 用我们的程序来分解费马数 $F_6=2^{2^6}+1$，其运行结果及所得数据如下：

$F_6=274177\cdot 67280421310721$，所用时间是 391.40 s.

§6. 今后研究的方向

下面简单地提出几条今后研究的方向以供读者参考：(1)找出更快速的计算方法.(2)扩大计算能力.(3)分解费马数 F_9（直到 2010 年，已经知道当 $5\leqslant n\leqslant 32$ 时 F_n 是复合数，知道分解方式的是 $0\leqslant n\leqslant 11$. 对 $n=20$ 和 $n=24$，还不知道它们的任何因子. 已知的最大可分解费马数是 $F_{2747497}$，它的一个素因子 $57\times 2^{2747499}+1$ 是被 M. Bishop 在 2013—05—13 发现的).

参考文献

孙琦，旷京华. 素数判定与大数分解. 辽宁教育出版社，1987.

数学研究,2008,41(1):72~78

康托尔实数的局限性[①]

The Limitation of Cantor's Real Numbers

摘要 康托尔为我们建立了集合论,并且证明了实数的不可数性,但是其中留下了很多疑点.

1. 一个实数能在每一个集合论模型中出现的充分必要条件是它是可以被集合论来定义的.那些在集合论模型中不出现的实数,我们可以把它们叫作看不见的实数.

2. 在实数的十进位无穷小数表示法中有些是我们能确切地知道它的第几位是什么,但是对另外的一些实数我们对它们就只能有模糊的认识,也就是说它的第几位是什么我们不可能全部知道.我们可以把它们叫作写不出的实数.

3. 由于 Cantor 关于实数是不可数的证明不是构造性的证明,而是用所谓的归谬证法.它们中有很多是看不见写不出的实数.因此说它们是虚拟的实数.

4. 虚拟实数就像银行中的虚拟货币,你可用它来买东西,它可从一个户头转拨到另一个户头,但是钱的实体是不存在的,这个现象也让我们对某些数学工具的合法性提出质疑.我们用对角线法来证明实数的基数比自然数的基数大,但是我们并没有真正有效地构造出那么多的实数.因此我们没有办法来确切地定义它们.也可以说它们中的绝大多数是不可以定义的.在一般的情况下虚拟实数是不可以个别地使用的.

关键词 非标准集合论模型,实数的相对性,虚拟实数.

---

① 收稿日期:2007-05-08.

§1. 引言

在20世纪80年代回国访问期间王浩曾经多次说过:“什么是连续统是没有解决的.”最初我们都觉得很奇怪,连续统就是实数的基数难道还有什么疑问吗?直到学了模型论中的Löwenheim-Skolem-Tarski定理我们才大吃一惊,LST定理说任意可数理论都有可数模型,集合论是可数理论,因此它也就有可数模型.实数是一个集合,在集合论模型中,它也就应该是可数的.那么到底实数集合是可数的还是不可数的呢?为此我们请教了王世强先生,他做出了如下的解释:“如果在一个集合论模型中没有足够的一一对应来比较集合,那么一个集合论模型可以是可数的.”问题似乎是解决了,但是留下了很多疑点:**1.** 实数集合真的是不可数吗?**2.** 一个可数集合论模型的结构是怎么样?其中哪些实数被漏掉了?**3.** 我们是否可以使用那些在某些集合论模型中不出现的实数?

§2. 集合论的可数模型

集合论是如此重要,人们对它进行了很多研究.人们总是着眼于大的集合论模型,而几乎没有人研究小的集合论模型.我们现在就主要研究可数集合论模型,另外我们给它们起了一个名字叫作非标准集合论模型.集合论是一切数学的基础,所有数学对象都可以用集合论来定义,因此我们必须用集合论来研究集合论.在进入非标准集合论之前我们需要一个普通集合论来作为讨论的环境.我们承认所有普通集合论中的定理.例如:所有在大学课本中所常见的集合论定理(其中也包括实数集合是不可数的,等).

首先我们要说明集合论是可以用一阶语言来描述.这在很多形式集合论的书中都可以找到.因为这一点很重要,我们在这里节引了Gödel的形式集合论系统[1].

集合论的语言$\mathcal{L}$含有两个一元关系$\mathfrak{Cls}$.$\mathfrak{M}$和一个二元关系$X\in Y$.大写英文字母$X,Y,Z,\cdots$表示类,小写英文字母$x,y,z,\cdots$表示集合,整个系统叫作$\mathcal{ZFG}$.它由以下5组公理来组成:

A组.

1. $\mathfrak{Cls}(x)$,

2. $(X\in Y)\supset\mathfrak{M}(X)$,

3. $(\forall u)[(u\in X)\leftrightarrow(u\in Y)]\rightarrow(x\equiv y)$,

4. $(\forall xy)(\exists z)[(u\in z)\leftrightarrow((u\equiv x)\vee(u\equiv y))]$,

定义 A1. $\mathfrak{Pr}(X)\leftrightarrow\neg\mathfrak{M}(X)$,

定义 A2. $(u\in\{xy\})\leftrightarrow((u\equiv x)\vee(u\equiv y))$,

定义 A3 $\{x\}=\{xx\}$,

定义 A4 $<xy>=\{\{x\}\{xy\}\}$.

类似地我们可以定义 $<x>=x$, $<xyz>=<x<yz>>$. 和 $<x_1x_2\cdots x_n>=<x_1<x_2\cdots x_n>>$.

定义 A5 $(X\subseteq Y)\leftrightarrow(\forall u)[(u\in X)\rightarrow(u\in Y)]$,

定义 A6 $\mathfrak{Em}(X)\leftrightarrow(\forall u)\neg(u\in X)$,

定义 A7 $\mathfrak{Er}(X,Y)\leftrightarrow(\forall u)\neg[(u\in X)\wedge(u\in Y)]$,

定义 A8 $\mathfrak{Un}(X)\leftrightarrow(\forall uvw)[((<v,u>\in X)\wedge(<w,u>\in X))\rightarrow(v\equiv w)]$.

B 组.

1. $(\exists A)(\forall xy)[(<xy>\in A)\leftrightarrow(x\in y)]$,

2. $(\forall AB)(\exists C)(\forall u)[(u\in C)\leftrightarrow((u\in A)\wedge(u\in B))]$,

3. $(\forall A)(\exists B)(\forall u)[(u\in B)\leftrightarrow\neg(u\in A)]$,

4. $(\forall A)(\exists B)(\forall x)[(x\in B)\leftrightarrow(\exists y)(<yx>\in A)]$,

5. $(\forall A)(\exists B)(\forall xy)[(<yx>\in B)\leftrightarrow(x\in A)]$,

6. $(\forall A)(\exists B)(\forall xy)[(<xy>\in B)\leftrightarrow(<yx>\in A)]$,

7. $(\forall A)(\exists B)(\forall xyz)[(<xyz>\in B)\leftrightarrow(<yzx>\in A)]$,

8. $(\forall A)(\exists B)(\forall xyz)[(<xyz>\in B)\leftrightarrow(<xzy>\in A)]$.

定义 B1 $(x\in A\cap B)=[(x\in A)\wedge(x\in B)]$,

定义 B2 $(x\in\overline{A})=\neg(x\in A)$,

定义 B3 $(x\in\mathfrak{D}(A))=(\exists y)(<yx>\in A)$.

C 组.

1. $(\exists a)[\neg\mathfrak{Em}(A)\wedge(\forall x)[(x\in a)\rightarrow(\exists y)((y\in a)\wedge(x\subset y))]]$,

2. $(\forall x)(\exists y)(\forall uv)[((u\in v)\wedge(v\in x))\rightarrow(u\in y)]$,

3. $(\forall x)(\exists y)[(u\subseteq x)\rightarrow(u\in y)]$,

4. $(\forall xA)[\mathfrak{Un}(A)\rightarrow(\exists y)(\forall u)((u\in y)\leftrightarrow(\exists v)((v\in x)\wedge(<uv>\in A)))]$.

D组. $\neg\mathfrak{Em}(A)\to(\exists u)[(u\in A)\wedge(\mathfrak{Er}(u,A))]$.

E组. $(\exists A)[\mathfrak{Un}(A)\wedge(\forall x)(\neg\mathfrak{Em}(x)\to(\exists y)((y\in x)\wedge(<yx>\in A)))]$.

由于 Gödel 已经把他的定理演绎得很完整,我们就不在这里展开了,这里我们只把用到的一些记号和定理列出来:全体自然数所成的集合记作 ω,全体有理数所成的集合记作 $\mathfrak{Q}$. 全体实数所成的集合记作 $\mathfrak{R}$,它是一个不可数集合,基数记作 $\mathfrak{C}$,ω,$\mathfrak{Q}$ 和 $\mathfrak{R}$ 都具有常见的函数等. 现在我们可以讨论集合论的可数模型了.

定理 2.1 如果集合论系统$\mathcal{ZFG}$是和谐的,那么它有一个可数模型.

证 因为$\mathcal{ZFG}$已经假设是和谐的,而语言$\mathcal{L}=\{\mathfrak{Cls},\mathfrak{M},\in\}$是可数的,由下降的 Löwenheim-Skolem-Tarski 定理,$\mathcal{ZFG}$有一个可数模型.

这里我们不去讨论集合论系统$\mathcal{ZFG}$是不是和谐的问题. 现在我们知道,如果$\mathcal{ZFG}$是和谐的话,那么它至少有一个可数模型. 进一步我们要问$\mathcal{ZFG}$有多少个互不同构的可数模型呢? 它们的构造又是怎样呢? 假设 $\mathfrak{S}$ 是一个可数的$\mathcal{ZFG}$模型,因为每一个元素 $a\in\mathfrak{Q}$ 可以用一个式子$f_a(x)='m*x\equiv n'$来定义,其中 m,n 是整数,在 $\mathfrak{S}$ 中必定有一个元素 b 实现式子 $f_a(x)$. $f_a(x)$可以扩充成为一个$\mathcal{L}$的型 $\delta_a(x)$,所有 $\Delta(x)=\{\delta(x)\mid a\in\mathfrak{Q}\}$ 中的型都可以被 $\mathfrak{R}$ 中的有理数所实现. 所以我们得到下面的定理:

定理 2.2 系统$\mathcal{ZFG}$恰好有 $2^{\aleph_0}$ 个互不同构的可数模型.

证 每一个 $r\in\mathfrak{R}$ 可以被一个式子的集合 $\Pi_r(x)=\{x\in\mathfrak{R}\wedge x>a\mid r>a,a\in\mathfrak{Q}\}\cup\{x\in\mathfrak{R}\wedge x\leqslant a\mid r\leqslant a,a\in\mathfrak{Q}\}$所定义,$\Pi_r(x)$可以扩充成为一个$\mathcal{L}$的型$\sum_r(x)$. 设 $\Gamma=\{\sum_r(x)\mid r\in\mathfrak{R}\}$是所有这种单变数的型所成的集合,$\Gamma$ 是不可数的. $\mathfrak{S}$ 作为是可数模型只能实现其中的可数个,所以系统$\mathcal{ZFG}$至少具有 $2^{\aleph_0}$ 个互不同构的可数模型. 另一方面由模型论的其他定理可以知道$\mathcal{ZFG}$作为一个可数语言的理论,它至多可以有 $2^{\aleph_0}$ 个可数模型,所以$\mathcal{ZFG}$恰好有 $2^{\aleph_0}$ 个互不同构的可数模型.

一个可以被一个式子来定义的元素是重要的,例如代数数就是这种元素. 我们给出一个关于可定义性的定义.

定义 2.1 一个$\mathcal{ZFG}$模型的元素 A 叫作是可以定义的,如果存在一个$\mathcal{L}$式子 $\varphi(X)$使得$\mathcal{ZFG}\vdash(\exists!\ X)\varphi(X)$和 $\mathfrak{S}\models\varphi(A)$.

下面的定理告诉我们关于实数出现在$\mathcal{ZFG}$模型的条件.

定理 2.3 对集合论系统$\mathcal{ZFG}$下面 3 条性质成立：

(i)如果 r 是一个代数数，那么它的定义式子在所有$\mathcal{ZFG}$模型中实现，也就是说它不可能被$\mathcal{ZFG}$省略.

(ii)如果 r 是一个超越数并且能被一个集合论式子所定义，那么它的定义式子在所有$\mathcal{ZFG}$模型中实现，也就是说它不可能被$\mathcal{ZFG}$省略.

(iii)如果 r 是一个超越数并且不能被任何一个集合论式子所定义，那么它被$\mathcal{ZFG}$省略，也就是说它至少在一个$\mathcal{ZFG}$模型中不出现.

证 首先整数是可以定义的，因此有理数也是可以被集合论式子所定义. 对每一个 $r\in\mathfrak{R}$ 式子集$\Pi_r(x)=\{x\in\mathfrak{R}\wedge x>a\mid r>a,a\in\mathfrak{Q}\}\cup\{x\in\mathfrak{R}\wedge x\leqslant a\mid r\leqslant a,a\in\mathfrak{Q}\}$分以下 3 种情形来讨论.

(i)如果 r 是一个代数数. $\Pi_r(x)$可以换成式子集 $x\in\mathfrak{R}\wedge f(x)\wedge a<x<b$，其中 $f(x)$是 r 所满足的整系数多项式，而 a,b 是与 r 很接近的有理数，使得 r 是 a,b 区间内唯一的 $f(x)$的根，因为$(\exists!\ x)(x\in\mathfrak{R}\wedge f(x)\wedge a<x<b)$是$\mathcal{ZFG}$的一个定理，所以$\Pi_r(x)$的定义式子 $f(x)$在所有$\mathcal{ZFG}$模型中实现也就是说它不可能被$\mathcal{ZFG}$省略.

(ii)如果 r 是一个超越数并且可以被一个$\mathcal{L}$式子 $\varphi(x)$所定义，那么 r 所满足的型$\Pi_r(x)$可以替换成为一个式子$\{x\in\mathfrak{R}\wedge\varphi(x)\}$，由于$\mathcal{ZFG}\models$，$(\exists!\ x)(x\in\mathfrak{R}\wedge\varphi(x))$，所以$\Pi_r(x)$在所有$\mathcal{ZFG}$模型中实现，也就是说它不可能被$\mathcal{ZFG}$省略.

(iii)如果 r 是一个超越数并且不可以被任何一个$\mathcal{L}$式子所定义，那么它可以被$\mathcal{ZFG}$省略.

因为我们可以建立一个式子集$\Pi_r(x)=\{x\in\mathfrak{R}\wedge x>a\mid r>a,a\in\mathfrak{Q}\}\cup\{x\in\mathfrak{R}\wedge x\leqslant a\mid r\leqslant a,a\in\mathfrak{Q}\}$.

$\Pi_r(x)$是与$\mathcal{ZFG}$和谐的，因为在原来的集合论环境中 r 是存在的，$\Pi_r(x)$不可能被局部实现，否则存在一个式子 $\varphi(x)$满足下列性质：

(iii.1)$\varphi(x)$与$\mathcal{ZFG}$和谐.

(iii.2)对所有 $\sigma\in\Pi_r(x)$，$\mathcal{ZFG}\models(\forall x)(\varphi(x)\rightarrow\sigma(x))$.

从这两条性质我们可以推导出

$$\mathcal{ZFG}\models(\forall xy)(\varphi(x)\wedge\varphi(y)\rightarrow x=y).$$

因为如果不是这样，那么$\mathcal{ZFG}$与$\neg(\forall xy)(\varphi(x)\wedge\varphi(y)\rightarrow x=y)$和谐. 因此我们可以找到一个集合论模型 $\mathfrak{S}$，使得其中存在两个不同的实数

$x_0 < y_0 \in \mathfrak{R}$ 满足 $\mathfrak{S} \models \varphi(x_0) \wedge \varphi(y_0)$. 设 a 是这两个实数间的一个有理数，它满足 $x_0 < a < y_0$. 设 σ_1, σ_2 分别代表下面两个式子 $x \in \mathfrak{R} \wedge x > a$, $x \in \mathfrak{R} \wedge x < a$. 由 $\varphi(x_0)$以及 $x_0 < a$ 我们知道 $\sigma_2 \in \Pi_r(x)$，再由 $\varphi(y_0)$和 $a < y_0$ 我们知道 $\sigma_1 \in \Pi_r(x)$. 这就与$\Pi_r(x)$的和谐性产生了矛盾. 矛盾的来源是前面假设了$\mathcal{ZFG}$能局部实现$\Pi_r(x)$. 由[2]定理 2.2.9 $\mathcal{ZFG}$应该局部省略$\Pi_r(x)$，也就是说 r 不在$\mathcal{ZFG}$模型 $\mathfrak{S}$ 中出现.

前面已经提到代数数是可以定义的，我们不禁要问有没有可以定义的超越数. 下面就给这样的数.

定理 2.4 设 $\psi(x)$是一个从自然数集 ω 到实数集 $\mathfrak{R}$ 的一个可定义函数，并且极限$\lim\limits_{n\to\infty}\psi(n)=r$ 存在，那么 r 是可以定义的.

证 设 $\varphi(x)=(x\in\mathfrak{R}) \wedge (\forall \varepsilon \in \mathfrak{R})(\exists n \in \omega)(\forall m > n, m \in \omega)(x-\varepsilon < \psi(m) < x+\varepsilon)$. r 满足$(\exists!\ x)\psi(x)$，因此 r 是可以定义的.

定理 2.5 实数 π 和 e 是可以定义的.

证 设 $\psi(x)$是由下式所定义的函数：

$$\psi \mathfrak{F} \mathfrak{n} \omega \wedge \psi(0)=1 \wedge (\forall n \geqslant 0, n \in \omega)\psi(n+1)=\psi(n)+\frac{1}{n!},$$

$\psi(x)$是由 ω 到 $\mathfrak{R}$ 的可定义函数并且$\lim\limits_{n\to\infty}\psi(n)=\mathrm{e}$. 因此由定理 2.4，e 是可以定义的.

类似地设 $\zeta(x)$是由下式所定义的函数：

$$\zeta \mathfrak{F} \mathfrak{n} \omega \wedge \zeta(0)=1 \wedge (\forall n \geqslant 0, n \in \omega)\zeta(n+1)=\zeta(n)+(-1)^n\frac{1}{2n+1},$$

$\zeta(x)$是由 ω 到 $\mathfrak{R}$ 的可定义函数并且$\lim\limits_{n\to\infty}\zeta(n)=\frac{\pi}{4}$. 因此由定理 2.4，$\pi$ 是可以定义的.

现在我们回过头来解释关于一个只含有可数个实数的集合论模型的悖论. 模型论者说：实数的个数与你所站的角度有关. 如果你站在模型的内部，那么实数的个数是不可数的，但是如果你站在模型的外部，那么实数的个数就是可数的了. 因此我们说实数的个数并不是真正不可数的，也就是说它具有相对性. 从上面的定理我们知道可以定义的超越数是可数的. 如果一个超越数是不可以定义的，那么它是很难被拿来使用的，因为我们对它的信息所知甚少. 类似地所有不可以定义的集合也有这个问题.

§3. $\mathcal{ZFG}$模型中元素的不可区分群组

上一节我们讨论了集合论的可定义元素，这一节里我们要讨论对元素的一个较弱的控制概念. 我们给出关于两个集合论模型中的元素的可区分性质，

定义 3.1 集合论模型 $\mathfrak{S}$ 中的两个元素 A,B 是可区分的，如果存在一个集合论式子 $\varphi(X)$，使得 $\mathfrak{S}\models\varphi(A)$ 和 $\mathfrak{S}\models\neg\varphi(B)$ 成立. 如果两个元素不是可区分的我们就说它们是不可区分的.

集合论中的元素可以被划分为不可区分的族类. 集合论模型中（尤其是在一些很大的集合论模型中）存在着很大的集合，例如 $\aleph_0$，$2^{\aleph_0}$，$2^{2^{\aleph_0}}\cdots$ Gödel[1]定理 8.56 说明了全体基数所成的集合 N 与全体序数所成的集合 ON 是同构的，也就是说基数的数目比 $2^{\aleph_0}$，$2^{2^{\aleph_0}}$，$2^{2^{2^{\aleph_0}}}$ 等都多. 不相同的基数应该是可以区分的. 因此在我们的想象中不可区分的族类的总数应该是很大的. 但是下面的定理告诉我们不可区分的族类的总数没有那么多.

定理 3.1 在一个集合论$\mathcal{ZFG}$模型 $\mathfrak{S}$ 中不可区分的族类的总数至多为 $2^{\aleph_0}$.

证 一个式子 $\varphi(X)$ 可以将 $\mathfrak{S}$ 的元素分为两类，$\mathcal{L}$的式子个数是 $\aleph_0$，所以它们至多能将 $\mathfrak{S}$ 的元素划分为 $2^{\aleph_0}$ 个不可区分的族类.

§4. 无穷小数的不确定性

我们虽然有 $2^{\aleph_0}$ 个实数，但是我们只能定义其中的 $\aleph_0$ 个，我们虽然有很多很多的基数，但是我们只能区分其中的 $2^{\aleph_0}$ 个. 为什么会有这样的限制呢？其基本原因是无限性在作祟. 在现实生活中我们所能用到的自然数是有限的，但是在我们的想象中重复“加 1”的过程我们可以把它们一个一个地产生出来并且可把它们放在一起组成一个集合. 实数的情况也与此类似，考虑在[0,1]区间的实数，我们可以把它们表示成十进位的无穷小数形式：$0.\varepsilon_1\varepsilon_2\varepsilon_3\cdots$其中每一个 ε_i 是 0 到 9 之间的任意一个数. 在我们的脑海的想象中有如下的一个图：

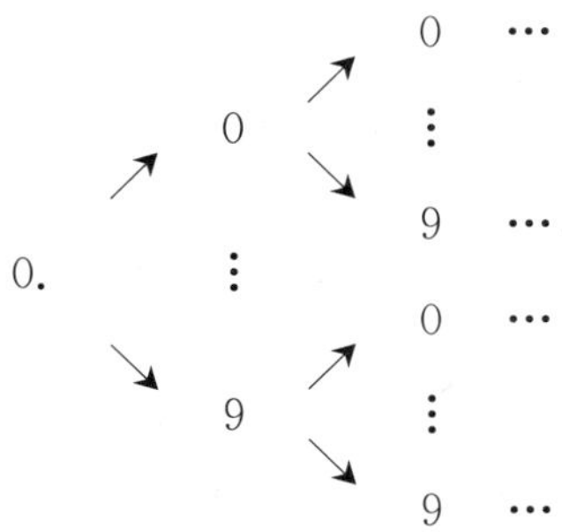

这些数其实是很模糊的. 因为有些数我们是能确切地知道它们的第几位是什么，例如有限小数、无限循环小数和某些无限不循环小数包括 π 和 e. 而对另外一些无限不循环小数我们则只能大概地知道它们是什么样子. 我们是不可能把它们一位一位地写出来的. 因为要写出一个实数我们还必须用一个过程，而以这种过程也只能最后归结到集合论的一个式子(我们认为一切数学对象都可以用集合论来生成). 前面我们已经指出集合论只有 $\aleph_0$ 个式子，因此只有 $\aleph_0$ 个实数是可以被写出来的. 对那些写不出来的实数我们只能大概地知道它们在哪里. 它们也只能被集体地使用. 例如[1,2]区间的实数乘以[2,3]区间的实数，其结果在[2,6]区间之内.

§5. 结论

康托尔为我们建立了集合论，并且证明了实数的不可数性，但是其中留下了很多疑点.

1. 前面我们已经证明一个实数能在每一个集合论模型中出现的充分必要条件是它是可以被集合论来定义的. 这些在集合论模型中不出现的实数，我们可以把它们叫作看不见的实数.

2. 在实数的十进位无穷表示法中有些是我们能确切地知道它的第几位是什么，但是对另外的一些实数，我们对它们就只能有模糊的认识，也就是说它的第几位是什么，我们不可能全部知道. 我们可以把它们叫作写不出的实数.

3. 由于 Cantor 关于实数是不可数的证明不是构造性的证明，而是用所谓的归谬证法. 它们中有很多是看不见写不出的实数. 因此说它们是虚拟的也就不足为怪了.

4. 实数就像银行中的虚拟货币，你可用它来买东西，它可从一个户头转拨到另一个户头，但是钱的实体是不存在的. 这个现象也让我们对某些数学工具的合法性提出质疑. 我们用对角线法来证明实数的基数比自然数的基数大，但是我们并没有真正有效地构造出那么多的实数. 因此我们没有办法来确切地定义它们. 也可以说它们中的绝大多数是不可以定义的.

参考文献

[1] Gödel K. The Consistency of the Axiom of Choice and of the Generalized Continuum Hypothesis with the Axioms of Set Theory. Ann. Math. Studies 3 (Princeton Univ. Press, Princeton, N. J.), 2nd edit, 1951.

[2] Chang C C, Keisler H J. Model Theory. Amsterdam: North-Holland, 3rd ed., 1990.

Abstract Cantor proved that the number of all real numbers is uncountable but from the LST theorem of model theory a countable theory should have a countable model. Set theory is a countable theory. It should have a countable model. Real number is a set. It should be countable in the model. This is so call Skolem paradox. In this paper we further explain this phenomena. Our conclusions are:

1. A necessary and sufficient condition for a real number to appear in every set theory model is that it is definable by a set theory formula. Therefore some of the real numbers can be absent from a set theory model.

2. In the expressions of infinite decimal of real numbers some of them we can know their digits one by one but for the others we can only have fussy knowledge. It is to say we can not know all of their digits.

3. Since the uncountablity proof of all real numbers given by Cantor is not constructible, the major part of the real numbers are fictitious. It is like the fictitious money in the bank. We can use it to do shopping. It goes from one account to another but the money body never exists. We use diagonallization method to prove that the cardinal number of real numbers is greater than the cardinal number of natural numbers, but we do not really produce such amount of real numbers. So now we have trouble to identify(define) them. As a result most of them are indefinable. Because of the undefinability in general fictitious real numbers can not be used individually.

Keywords **Nonstandard models of set theory, The relativity of real number, Fictitious real numbers.**

北京师范大学学报(自然科学版),2009,45(3):221~225

非良基集合论模型悖论①

A New Paradox on Unwell-Founded Set Theory Models

摘要 给出一个新的集合论悖论.用模型论方法证明了非良基性的集合论模型的存在性.这个模型中存在对∈关系下降的无限元素外序列.还证明可以存在集合论模型,其中ω,ω_1等集合,ON和N等类的内部都存在对∈关系无限递降的元素外序列.这种对∈关系是无限递降的外序列还可以插入于一个没有可数共尾的对∈是上升的序列的后面,插入后成为同名集合的一部分.用这种模型第一次定义并给出了外集合不是内集合的例子.

关键词 非良基性,集合论模型,外集合,内集合.

§1. 集合论的新悖论

很多人认为在一个集合论模型中的序数从0,1,2,…,n,…开始,下面的第一个序数是ω.接下来的序数是:$\omega+1$,…然后是2ω,…,3ω,…,$\omega\omega$,…,直到$\omega_1,\omega_2,\cdots,\omega_\omega,\cdots$它们被认为是最遵守良基性原则的.也就是说在序数中不存在集合序列f,f是ω上的函数,使得$f'(n+1)\in f'n$,$f'n\in\omega$,这里面f是一个序数的序列.实际的情况却恰好相反.我们用模型论的方法证明了非良基的集合论模型的存在性,也就是说存在集合论模型,其中存在对∈关系是下降的无限元素外序列.我们还证明了可以存在集合论模型,其中ω,ω_1等集合,ON和N等类的内部都存在着对∈关系是无限递降的元素外序列.这种对∈关系是无限递降的元素外序列还

① 收稿日期:2008-05-21.

可以插入于一个没有可数共尾的对∈是上升的序列的后面，插入之后成为同名集合的一部分. 这个悖论说明了集合论公理中关于良基性的规定并不能限制集合论模型一定是良基的. 利用这种模型我们还第一次定义并且给出了外集合不是内集合的例子. 下面我就来研究这种模型.

集合论是一切数学的基础，所有数学对象都可以用集合论来定义，因此我们必须用集合论来研究集合论. 在进入形式集合论之前我们需要一个普通集合论模型来作为讨论的环境. 我们承认所有普通集合论中的定理. 例如：所有在大学课本中常见的集合论定理(其中也包括实数集合是不可数的，序数类是良序的，等). 文中经常提到普通自然数就是我们工作环境中的自然数，要将它们与模型中的自然数区分开来.

§2. 良基性定理与非良基的集合论模型

我们知道良基性是集合论的重要基础. 为了要更好地讨论良基性我们在这里节引了 Gödel 的形式集合论系统[1]. 我们使用这个系统是因为它比其他的系统简洁，并且够用.

集合论的语言 $\mathscr{L}$ 含有 2 个一元关系 $\mathfrak{Cls}$，$\mathfrak{M}$ 和一个二元关系 $X\in Y$，大写英文字母 $X,Y,Z,\cdots$ 表示类，小写英文字母 $x,y,z,\cdots$ 表示集合. 整个系统叫作 $\mathscr{ZFG}$. 它由以下 5 组公理来组成：

A 组.

1) $\mathfrak{Cls}(x)$.

2) $(X\in Y)\to\mathfrak{M}(X)$.

3) $(\forall u)[(u\in X)\leftrightarrow(u\in Y)]\to(x\equiv y)$.

4) $(\forall xy)(\exists z)[(u\in z)\leftrightarrow((u\equiv x)\vee(u\equiv y))]$.

定义 A1　$\mathfrak{Pr}(X)\leftrightarrow\neg\mathfrak{M}(X)$.

定义 A2　$(u\in\{xy\})\leftrightarrow((u\equiv x)\vee(u\equiv y))$.

定义 A3　$\{x\}=\{xx\}$.

定义 A4　$\langle xy\rangle=\{\{x\}\{xy\}\}$.

类似地我们可以定义 $\langle x\rangle=x$，$\langle xyz\rangle=\langle x\langle yz\rangle\rangle$ 和 $\langle x_1x_2\cdots x_n\rangle=\langle x_1\langle x_2\cdots x_n\rangle\rangle$.

定义 A5　$(X\subset Y)\leftrightarrow(\forall u)[(u\in X)\to(u\in Y)]$.

定义 A6　$\mathfrak{Em}(X)\leftrightarrow(\forall u)\neg(u\in X)$.

定义 A7　$\mathfrak{Ex}(X,Y)\leftrightarrow(\forall u)\neg[(u\in X)\wedge(u\in Y)]$.

定义 A8 $\mathfrak{Un}(X)\leftrightarrow(\forall uvw)[((\langle v,u\rangle\in X)\wedge(\langle w,u\rangle\in X))\rightarrow(v\equiv w)]$.

B 组.

1) $(\exists A)(\forall xy)[(\langle xy\rangle\in A)\leftrightarrow(x\in y)]$.

2) $(\forall AB)(\exists C)(\forall u)[(u\in C)\leftrightarrow((u\in A)\wedge(u\in B))]$.

3) $(\forall A)(\exists B)(\forall u)[(u\in B)\leftrightarrow\neg(u\in A)]$.

4) $(\forall A)(\exists B)(\forall x)[(x\in B)\leftrightarrow(\exists y)(\langle yx\rangle\in A)]$.

5) $(\forall A)(\exists B)(\forall xy)[(\langle yx\rangle\in B)\leftrightarrow(x\in A)]$.

6) $(\forall A)(\exists B)(\forall xy)[(\langle xy\rangle\in B)\leftrightarrow(\langle yx\rangle\in A)]$.

7) $(\forall A)(\exists B)(\forall xyz)[(\langle xyz\rangle\in B)\leftrightarrow(\langle yzx\rangle\in A)]$.

8) $(\forall A)(\exists B)(\forall xyz)[(\langle xyz\rangle\in B)\leftrightarrow(\langle xzy\rangle\in A)]$.

定义 B1 $(x\in A\cap B)=[(x\in A)\wedge(x\in B)]$.

定义 B2 $(x\in\overline{A})=\neg(x\in A)$.

定义 B3 $(x\in\mathfrak{D}(A))=(\exists y)(\langle yx\rangle\in A)$.

C 组.

1) $(\exists a)[\neg\mathfrak{Em}(A)\wedge(\forall x)[(x\in a)\rightarrow(\exists y)((y\in a)\wedge(x\subset y))]]$.

2) $(\forall x)(\exists y)(\forall uv)[((u\in v)\wedge(v\in x))\rightarrow(u\in y)]$.

3) $(\forall x)(\exists y)[(u\subset x)\rightarrow(u\in y)]$.

4) $(\forall xA)[\mathfrak{Un}(A)\rightarrow(\exists y)(\forall u)((u\in y)\rightarrow(\exists v)((v\in x)\wedge(\langle uv\rangle\in A)))]$.

D 组.

$\neg\mathfrak{Em}(A)\rightarrow(\exists u)[(u\in A)\wedge(\mathfrak{Er}(u,A))]$.

E 组.

$(\exists A)[\mathfrak{Un}(A)\wedge(\forall x)(\neg\mathfrak{Em}(x)\rightarrow(\exists y)((y\in x)\wedge(\langle yx\rangle\in A)))]$.

以上所列出的形式集合论系统记作 $\mathscr{ZFG}$，由于 Gödel 已经把他的定理演绎得很完整，我们就不在这里展开了. 这里我们直接引用他的记号和定理. Gödel 在他的书中并没写出他引进集合记号的过程，我们可用一般逻辑书中的增加常量记号的办法来加以处理. 这样一来语言就膨胀了. 令 $\overline{\mathscr{L}}=\mathscr{L}\cup\{C_0,C_1,\cdots\}$. 将 $\overline{\mathscr{L}}$ 的全部存在形公式列出如下：

$$\exists XA_0(X),\exists XA_1(X),\cdots$$

再增加如下的公理：

$$\exists!\ XA_0(X)\rightarrow A_0(C_0),\exists!\ XA_1(X)\rightarrow A_0(C_1),\cdots$$

加上原来的系统所组成新的形式集合论系统 $\mathscr{ZFG}^+$ 这样一来每个集合论中能证明是存在唯一的集合或类就有了它的记号，其中包括了 ω,ω_1 等基数集合，全体序数所作成的类 ON 和全体基数所作成的类 N 等，对这些集合与类我们仍沿用它们原来的记号.

上面我们介绍了形式集合论系统 $\mathscr{ZFG}^+$. 在下面的讨论中会出现同一个概念的 3 个不同层次的实现的问题. 我们的工作环境就是我们普通所理解的集合论. 我们的工作环境中有它的 $0,1,\cdots,\omega,\omega_1$ 等基数集合，全体序数所作成的类 ON 和全体基数所作成的类 N 等. 我们把它们写成 $\widetilde{0},\widetilde{1},\cdots,\widetilde{\omega}$, $\widetilde{\omega_1}$, $\widetilde{ON}$ 和 $\widetilde{N}$ 等. 这个工作环境当然也可以看成是一个集合论模型，一般地我们认为这是一个常规的集合论模型. 也就是不存在对 $\in$ 关系是无限递降的元素序列并且 ω_1 是由不可数多个元素来组成的模型. 而其他模型就都是在这个工作环境中构造出来的. 第 2 个层次是 $\mathscr{ZFG}^+$ 的记号，我们就直接使用记号 $0,1,\cdots,\omega,\omega_1,ON$ 和 N 等. 在定理中当我们谈到集合论模型时，就是指 $\mathscr{ZFG}^+$ 的模型. 当我们构造这些模型时，在我们的模型中也有与上述记号相应的元素. 我们把模型中的集合与类记成 $\overline{x}$ 和 $\overline{X}$. 所以在模型中我们把与上述记号相应的集合与类写成 $\overline{0},\overline{1},\cdots,\overline{\omega},\overline{\omega_1},\overline{ON}$ 和 $\overline{N}$ 等.

定理 1 （良基性定理）

$$\neg\,\exists A[A\mathfrak{Fn}\omega\wedge\forall n(n\in\omega\rightarrow A'n\in A'(n+1))].$$

以上用的是文献[1]的内定理记号，它的含义是不存在集合序列 A, A 是 ω 上的函数，使得 $A'(n+1)\in A'n,n\in\omega$，这里面 A 是一个内序列.

证 定理的证明可以直接由公理 D 得出，公理 D 说：任意非空集合 $\overline{X}$ 存在一个元素 $\overline{a}$，$\overline{a}$ 与 $\overline{X}$ 不相交. 而使得 $\overline{A}'(\overline{n}+\overline{1})\in\overline{A}'\overline{n},\overline{n}\in\overline{\omega}$ 的集合序列 $\overline{A}$ 如果存在，那么把它们放在一起组成一个集合 $\overline{B}=\overline{A}''\overline{\omega}$.

$$\overline{A}'\overline{1}\in(\overline{B}\cap\overline{A}'\overline{0}),\cdots,\overline{A}'(\overline{n}+\overline{1})\in(\overline{B}\cap\overline{A}'\overline{n}),\cdots$$

在集合 $\overline{B}$ 中不存在与集合 $\overline{B}$ 不相交的元素，与公理 D 发生矛盾，所以定理得证.

良基性定理是集合论的奠基性定理之一，人们希望有了这个定理就能保证集合是建立在健康的基础之上. 也就是说希望集合论模型是像良基性原则所规定的那样不存在无限多个向下的 $\in$ 元素序列 $\overline{a_{\widetilde{0}}}\ni\overline{a_{\widetilde{1}}}\ni\overline{a_{\widetilde{2}}}\ni\cdots$ 但是这个愿望并没有能够实现. 恰恰相反，答案是否定的，下面的定理就来讨论这个问题.

定理 2(非良基集合论模型存在定理)　存在一个集合论模型$\widetilde{\mathfrak{S}}$,其中存在集合外序列:$\overline{a_{\tilde{0}}},\overline{a_{\tilde{1}}},\cdots$,使得$\overline{a_{\widetilde{n+1}}}\in\overline{a_{\tilde{n}}},\tilde{n}=\tilde{0},\tilde{1},\cdots$这里$\tilde{n}\in\tilde{\omega}$是工作环境中的自然数而不是模型中的自然数$\bar{n}\in\bar{\omega}$.而$\in$则是模型中的“属于”关系.

证　以TH^+表示$\mathscr{ZFG}^+$的全部定理所组成的理论,再选取一个全部由新常数所组成的集合$C=\{c_{\tilde{0}},c_{\tilde{1}},\cdots\}$.构造下面的理论

$$T=TH^+\cup\{c_{\widetilde{n+1}}\in c_{\tilde{n}},\tilde{n}=\tilde{0},\tilde{1},\cdots\}.$$

这里$\tilde{n}\in\tilde{\omega}$看成是工作环境中的自然数.容易证明如果$TH^+$是和谐的,那么$T$也是和谐的.因为设$T'$为$T$的任意有限子集.$T'$中所出现的全部新常数为$c_{\tilde{i}_1},c_{\tilde{i}_2},\cdots,c_{\tilde{i}_k}$.可以在$\mathscr{ZFG}^+$的任意一个模型中用其中的有限个自然数来解释$c_{\tilde{i}_1},c_{\tilde{i}_2},\cdots,c_{\tilde{i}_k}$,使$T'$得到满足.所以$T'$有模型$\mathfrak{S}'$.由模型论[2]中的紧致性定理知$T$也是和谐的.于是$T$有模型$\overline{\mathfrak{S}}$.$TH^+\subset T$保证了$\overline{\mathfrak{S}}$是满足$\mathscr{ZFG}^+$的集合论模型.在模型$\overline{\mathfrak{S}}$中用来解释$c_{\tilde{0}},c_{\tilde{1}},\cdots$的元素$\overline{a_{\tilde{0}}},\overline{a_{\tilde{1}}},\cdots$满足$\overline{a_{\widetilde{n+1}}}\in\overline{a_{\tilde{n}}},\tilde{n}\in\tilde{\omega}$.

这样一来一个新的悖论诞生了:它就是存在实际上是非良基的集合论模型,而这个模型的理论中却有良基性定理.对这个新的悖论我们给出如下的解释:

1)和 Skolem 悖论一样,定理 1 是集合论的内定理,而定理 2 是集合论的外定理,定理中的$\overline{a_{\tilde{0}}},\overline{a_{\tilde{1}}},\cdots$不是集合模型的内序列而是一个外序列.所以 2 个定理是可以兼容的.

2)集合论模型不是像我们所想象的那样简单.尤其是其中的结构不是只存在一个一个地向上的$\in$关系,向下只能追溯到空集.却是也可能存在向下的$\in$无限外序列.

3)虽然在模型$\overline{\mathfrak{S}}$中存在着向下的$\in$无限外序列.这只是一个从外部看的序列.它们应该是不能组成一个真正是建立在$\bar{\omega}$之上的内部函数序列,因为集合论具有良基性的内定理.因此这个外部序列$\overline{a_{\tilde{0}}},\overline{a_{\tilde{1}}},\cdots$满足$\overline{a_{\widetilde{n+1}}}\in\overline{a_{\tilde{n}}},\bar{n}\in\tilde{\omega}$在模型内部是如何构成的尚需进一步探讨.

§3.　非良基的集合论模型的精确化

上一节我们构造了一个非良基的集合论模型使其中出现了一个对$\in$是向下的无限外序列,那么这个元素序列能不能组成一个真正是建立在

ω 之上的内部函数序列呢？如果它们组成一个这样的序列，那么就和良基性定理产生了矛盾了. 但是如果我们从一开始就把它们放在一个序列里结果又会怎么样呢？下面我们就来构造这样的模型.

定理 3(局部非良基的集合序列存在定理) 存在一个集合论模型 $\widetilde{\mathfrak{S}}$，其中存在集合序列：$\overline{A}\mathfrak{Fn}\widetilde{\omega}$，满足

$$\{\overline{A}'\overline{n}\in\overline{A}'(\overline{n+1}),\overline{n}=\overline{0},\overline{1},\cdots\},$$

其中 $\overline{n}=\overline{0},\overline{1},\cdots$ 是模型中的自然数.

证 以 TH^{+} 表示 $\mathcal{ZFG}^{+}$ 的理论如前，再选取一个新常数 C. 构造下面的理论

$$T=TH^{+}\cup\{C\mathfrak{Fn}\omega,C'n\in C'(n+1),n=0,1,\cdots\}.$$

这里 $n\in\omega$ 看成是 $\mathcal{ZFG}^{+}$ 中的自然数记号，只有这样才能保证上列 T 是由式子来组成. 仿前证明如果 TH^{+} 是和谐的，那么 T 也是和谐的. 因为设 T' 为 T 的任意有限子集. 不妨设 TH^{+}，$C\mathfrak{Fn}\omega$ 已经在 T' 中，全部其他式子为 $C'(i_1)\in C'(i_1+1),\cdots,C'(i_k)\in C'(i_k+1)$. 可以在 $\mathcal{ZFG}^{+}$ 的一个模型 $\overline{\mathfrak{S}}'$ 中构造一个从 $\overline{\omega}$ 到 $\overline{\omega}$ 的函数，用其中的有限个自然数来解释 $C'(i_1),\cdots,C'(i_k)$，使 T' 得到满足. 所以 T' 有模型 $\overline{\mathfrak{S}}''$. 由模型论[2]中的紧致性定理知 T 也是和谐的. 于是 T 有模型 $\overline{\mathfrak{S}}$. 在模型 $\overline{\mathfrak{S}}$ 中用来解释 C 的元素 $\overline{A}$ 满足 $\overline{A}\mathfrak{Fn}\overline{\omega}$，$\{\overline{A}'\overline{n}\in\overline{A}'(\overline{n+1}),\overline{n}=\overline{0},\overline{1},\cdots\}$.

在模型 $\overline{\mathfrak{S}}$ 中元素 $\overline{A}'\overline{n}$；$\overline{n}=\overline{0},\overline{1},\cdots$ 是不是一个向下的无限 $\in$ 序列呢？如果是的话，那么这会不会与良基性定理产生矛盾呢？其实 $\overline{A}'\overline{n}$；$\overline{n}=\overline{0}$，$\overline{1},\cdots$ 的确是一个向下的无限 $\in$ 序列，但是它仍然是一个外序列. 所以仍然不会与良基性定理产生矛盾. 因为定理中不能排除在 $\overline{A}''\overline{\omega}$ 中除了 $\overline{A}'\overline{n}$；$\overline{n}=\overline{0},\overline{1},\cdots$ 之外另有其他元素 $\overline{b}$，使得 $\overline{b}\cap\overline{0}\overline{A}'\overline{0}=\varnothing$. 也就是说在模型 $\mathfrak{S}$ 中的集合 $\overline{\omega}$ 并不是只由 $\overline{0},\overline{1},\cdots$ 来组成的，而还会有其他的不同元素，关于这一点可以参考非标准数论中的说明.

推论 1 定理 3 中的 $\{\overline{0},\overline{1},\cdots\}$ 在模型 $\overline{\mathfrak{S}}$ 中不组成集合.

证 假设 $\{\overline{0},\overline{1},\cdots\}$ 组成 $\overline{\mathfrak{S}}$ 的集合 $\overline{B}$，那么 $\overline{B}$ 对函数 $\overline{A}$ 的像 $\overline{A}''\overline{B}$ 组成 $\overline{A}'\overline{n}$；$\overline{n}=\overline{0},\overline{1},\cdots$ 是一个向下的无限 $\in$ 序列，但是它是一个内序列. 与定理 1 产生矛盾. 所以 $\{\overline{0},\overline{1},\cdots\}$ 不能组成 $\overline{\mathfrak{S}}$ 的集合.

定义 外集合是指工作环境的集合，内集合是指模型里的集合.

可以给出 $\{\overline{0},\overline{1},\cdots\}$ 作为外集合的证明，推论 1 也就给出了外集合不是内集合的例子. 也就是说一个内集合的某些元素，从工作环境的角度来

看,它们组成一个集合,但是在模型中它们并不组成集合.

定理 4 定理 3 中的$\{\bar{0},\bar{1},\cdots\}$在工作环境中组成集合.

证 我们只能用外定义来进行工作. 首先定义自然数记号集 Na,这是可以用归纳法来完成. 再定义每一个 $x\in Na$ 在模型 $\bar{\mathfrak{S}}$ 中的突现. 它们的总体就组成集合$\{\bar{0},\bar{1},\cdots\}$.

§4. 非良基集合论模型中的良序集与类

在文献[1]的集合论系统 $\mathscr{ZFG}$ 中定义了很多对$\in$关系是良序的集与类,例如 ω,ω_1 等基数,全体序数所作成的类 ON 和全体基数所作成的类 N 等. 这些良序的集与类被认为是最遵守良基性原则的. 当然这是从内定理的角度来说的. 那么在集合论模型中它们会不会也很遵守良基性原则呢. 其答案也是否定的. 为此我们将上一节的定理做一个推广.

定理 5(良基类与非良基类的转换定理) 设 A 是一个可定义类,如果存在一个集合论模型 $\bar{\mathfrak{S}}'$,其中 $\bar{A}$ 满足良基性原则并且存在对$\in$关系上升的无限元素序列. 那么存在一个集合论模型 $\bar{\mathfrak{S}}$,其中 $\bar{A}$ 存在元素序列:$\overline{a_{\tilde{0}}},\overline{a_{\tilde{1}}},\cdots$满足

$$\overline{a_{\tilde{n}}}\in\bar{A},\overline{a_{\widetilde{n+1}}}\in\overline{a_{\tilde{n}}},\tilde{n}=\tilde{0},\tilde{1},\cdots$$

其中 $\tilde{n}=\tilde{0},\tilde{1},\cdots$是工作环境中的自然数而不是模型中的自然数.

证 以 TH^+ 表示 $\mathscr{ZFG}^+$ 的理论如前,再选取一个全部由新常数所组成的集合 $C=\{c_{\tilde{0}},c_{\tilde{1}},\cdots\}$. 构造下面的理论

$$T=TH^+\cup\{c_{\tilde{n}}\in A,c_{\widetilde{n+1}}\in c_{\tilde{n}},\tilde{n}=\tilde{0},\tilde{1},\cdots\}.$$

容易证明如果 TH^+ 是和谐的,那么 T 也是和谐的. 因为设 T' 为 T 的任意有限子集. T'中所出现的全部新常数为 $c_{\tilde{i}_1},c_{\tilde{i}_2},\cdots,c_{\tilde{i}_k}$. 可以在模型 $\bar{\mathfrak{S}}'$中用其中 $\bar{A}$ 的有限个对$\in$关系上升的无限序列中的元素来解释 $c_{\tilde{i}_1}$,$c_{\tilde{i}_2},\cdots,c_{\tilde{i}_k}$,使 T'得到满足. 所以 T'有模型 $\bar{\mathfrak{S}}''$. 由模型论[2]中的紧致性定理知 T 也是和谐的. 于是 T 有模型 $\bar{\mathfrak{S}}$. 在模型 $\bar{\mathfrak{S}}$ 中用来解释 $c_{\tilde{0}},c_{\tilde{1}},\cdots$的元素$\overline{a_{\tilde{0}}},\overline{a_{\tilde{1}}},\cdots$满足

$$\overline{a_{\tilde{n}}}\in\bar{A},\overline{a_{\widetilde{n+1}}}\in\overline{a_{\tilde{n}}},\tilde{n}=\tilde{0},\tilde{1},\cdots$$

以上的插入过程是只针对一个集合进行插入,对多个集合同时进行插入这种序列也是可以做到的,我们这里不再多加讨论. 这样一来我们找到了集合论模型 $\bar{\mathfrak{S}}$ 其中 $\bar{\omega},\bar{\omega}_1$ 等集合,$\overline{ON}$和 $\bar{N}$ 等类的内部都可以存在着

对∈关系是无限递降的元素序列. 原来我们认为集合论模型中的序数是最遵守良基性原则的. 而现在在它们之间插入了一些对∈关系是下降的无限元素序列,顺序就完全被搞乱了. 还有人会说这个对∈关系是无限递降的元素序列如果插入于 ω 的后段也就同时被插入于 ω_1, ON 和 N 等集合与类,因为 ω 是 ω_1, ON 和 N 等集合与类的前段. 是否可以将这种对∈关系是无限递降的元素序列插入于一个不可数的上升序列之后段呢? 我们知道 ω_1 不存在共尾的可数上升序列,那么能不能插入一个对∈关系是无限递降的元素序列于 ω_1 的不可数多个元素之后使插入之后所得到的模型中这个元素仍然是 ω_1? 答案是肯定的,但是办法要稍微复杂一点.

定理 6(下降的∈序列的定点插入定理) 设 $\mathfrak{S}'$ 是常规集合论模型. 那么存在一个集合论模型 $\overline{\mathfrak{S}}$,其中有一个对∈关系下降的无限元素序列 $\overline{a_{\tilde{0}}}$, $\overline{a_{\tilde{1}}}$,…被插入于 $\overline{\omega_1}$ 的不可数多个元素的后面. 它们同属于新模型 $\overline{\mathfrak{S}}$ 中的 $\overline{\omega_1}$.

证 选取一个新常数集合

$$D=\{d_{0'},\cdots d_{\alpha'}\cdots;\alpha'\in\omega_1{}'\}$$

以它们代表 $\mathfrak{S}$ 中 ω_1' 的元素. 将 $\mathfrak{S}'$ 膨胀成为

$$\mathcal{ZFG}^+\{d_{0'},\cdots d_{\alpha'}\cdots;\alpha'\in\omega_1{}'\}$$

的模型 $\mathfrak{S}S'_{\omega_1{}'}$. 以 TH' 表示 $\mathfrak{S}S'_{\omega_1{}'}$ 的理论,再选取一个新常数集合 $C=\{c_{\tilde{0}}, c_{\tilde{1}},\cdots\}$ 构造下面的理论

$$T=TH'\cup\{c_{\tilde{n}}\in\omega_1, c_{\widetilde{n+1}}\in c_{\tilde{n}}, \tilde{n}=\tilde{0},\tilde{1},\cdots\}\cup\{d_{\alpha'}<c_{\tilde{n}};\tilde{n}=\tilde{0},\tilde{1},\cdots,\alpha'<\omega_1{}'\}.$$

可以证明 T 是和谐的. 因为设 T' 为 T 的任意有限子集. T' 中所出现的全部新常数为 $c_{\tilde{i}_1}, c_{\tilde{i}_2},\cdots,c_{\tilde{i}_k}, d_{\alpha_1{}'}, d_{\alpha_2{}'},\cdots,d_{\alpha_m{}'}$. 可以在模型 $\mathfrak{S}'_{\omega_1{}'}$ 中的 $\omega_1{}'$ 里最大的 $d_{\alpha_j{}'}$ 所对应的元素之后找到有限个符合良序性的元素来解释 $c_{\tilde{i}_1}, c_{\tilde{i}_2},\cdots,c_{\tilde{i}_k}$,使 T' 得到满足. 所以 T' 有模型 $\mathfrak{S}_{\omega_1{}'}''$. 由模型论[2]中的紧致性定理知 T 也是和谐的. 于是 T 有模型 $\overline{\mathfrak{S}_{\omega_1{}'}}$. 在模型 $\overline{\mathfrak{S}_{\omega_1{}'}}$ 中用来解释 $c_{\tilde{0}}, c_{\tilde{1}},\cdots$ 的元素 $\overline{a_{\tilde{0}}}, \overline{a_{\tilde{1}}},\cdots$ 满足 $\overline{a_{\widetilde{n+1}}}\in\overline{a_{\tilde{n}}}$, $\tilde{n}=\tilde{0},\tilde{1},\cdots$ 并且被插入于 $\overline{\omega_1}$ 的不可数多个元素的后面. 模型 $\overline{\mathfrak{S}_{\omega_1{}'}}$ 还可以归约成为 $\mathscr{L}$ 的模型 $\overline{\mathfrak{S}}$. 而上述下降序列的性质是不会随模型的归约而改变.

当然除了上述被插入于 $\overline{\omega_1}$ 的不可数元素序列的后面的元素 $\overline{a_{\tilde{0}}}, \overline{a_{\tilde{1}}},\cdots$ 之外也会带来其他的元素,我们在这里不再加以讨论.

§5. 结论

自从 Hilbert 倡导建立形式化的公理系统以来,很多数学分支都建

立了这样的系统. 在所有常见的数学系统之中,形式化的公理和这个数学系统本身一般是没有什么相悖的现象的. 但是形式化的公理系统用来研究集合论系统时就会有些不可思议的问题出现. Skolem 发现可以存在可数的集合论模型,而这个模型中却存在不可数的集合. Skolem 悖论说明了集合论公理中关于不可数集合存在性的规定并不能限制集合论模型一定是由不可数个元素来组成. 我们在文献[3]里研究了这种模型,并提出了虚拟实数的概念. 虚拟实数就像银行中的虚拟货币,你可用它来买东西,它可从一个户头转拨到另一个户头,但是钱的实体是不存在的. 本文给出了一个新的集合论悖论:我们证明了非良基性的集合论模型的存在性,也就是说存在集合论模型,其中存在对∈关系是下降的无限元素外序列. 我们还证明了可以存在集合论模型其中 ω,ω_1 等集合,ON 和 N 等类的内部都存在着对∈关系是无限递降的元素外序列. 这种对∈关系是无限递降的元素外序列还可以插入于一个没有可数共尾的对∈是上升的序列的后面,插入之后成为同名集合的一部分. 用这种模型我们还给出了外集合不是内集合的例子.

参考文献

[1] Cödel K. The consistency of the axiom of choice and of the generalized continuum hypothesis with the axioms of set theory[M]. 2nd ed. Princeton N J: Princeton Univ Press,1951.

[2] Chang C C,Keisler H J. Model theory[M]. 3th ed. Amsterdam: North-Holland,1990.

[3] 罗里波. 康托尔实数的局限性[J]. 数学研究,2008,41(1):72.

Abstract A new set theory paradox is given in this paper. The existence of a set theory model is proved, where exists an infinite descending outer sequence with respect to ∈ relation of the model. This paper also proves that this kind of sequences may be inserted into well-ordered sets and classes like ω,ω_1 and the class of all ordinal numbers ON of the class of all cardinal numbers N of the new model. It is also possible to insert such a sequence into the end section of an uncountable ascending sequence with respect to ∈ relation of the model like ω_1. This paper for the first time defines and gives an example for an outer set that is not an inner set in a set theory model.

Keywords **unwell-founded, set theory model, outer set, inner set.**

数学学报,2003,46(1):95～102

完全二叉树的量词消去①

Quantifier Elimination for Complete Binary Trees

摘　要　量词消去法已经成为计算机科学和代数模型论中最有力的研究工具之一.本文针对完全二叉树理论所独有的特性,给出了它的基本公式集,然后利用分布公式及有限覆盖证明了完全二叉树的理论可以量词消去.

关键词　完全二叉树,量词消去,基本公式,分布公式,有限覆盖.

§0.　引言

有关量词消去的结果已经出现了很多,尤其是对我们都很熟悉的代数结构如:环、域等.20世纪70年代末80年代初,*Rose*,*Berline* 证明了特征为0可量词消去的环是一个代数闭域[1,2].同时 *Berline* 又给出了特征为 p 的可量词消去的非半单环的完全分类[3],1983年又与Cherlin给出了特征为 p^n 可量词消去的环的完全分类[4].1980年,Boffa,Macintyre和Point给出了无幂零元可量词消去的半单环的一个完全分类[5].1985年,Weispfenning证明了正则环的左 R 模以及一类分配格可以量词消去[6,7].完全二叉树理论无论是在数学中还是在计算机科学中的应用都是

① 收稿日期:2000-02-15;接收日期:2001-12-30.

基金项目:国家自然科学基金资助项目(19571009);北方交通大学基金资助项目.

本文与刘吉强和廖东升合作.

很广泛的，但关于该理论的量词消去一直没有相关的结论，本文研究完全二叉树理论的量词消去，得到了较好的结果，为以后相关问题的研究提供了理论基础，其中主要用到文[8]中关于量词消去的基本引理，另外也沿用了文[8～11]中的记号和术语.

基本引理 设 T 是一阶语言$\mathcal{L}$的理论，Σ是基本公式集. 若满足下列两个条件：每个原子公式都 T-等价于基本公式的布尔组合；如果 φ 是基本公式的布尔组合，则 $\exists x\varphi$ 也 T-等价于基本公式的布尔组合. 则$\mathcal{L}$中每个公式都 T-等价于基本公式的布尔组合.

我们将首先选取基本公式，然后按照引理的条件逐步完成量词消去.

§1. 完全二叉树理论

我们将讨论的完全二叉树理论 T，由一阶语言$\mathcal{L}=\{E,R\}$中以下公理组成的，其中 E 是二元关系符号，R 为常量符号.

(1)图的理论：(a)无圈 $\forall x\neg(xEx)$；(b)无向图 $\forall xy(xEy\leftrightarrow yEx)$.

(2)树：(a)存在唯一的根节点，记为 R，$\exists y_1y_2((y_1\neq y_2)\wedge(REy_1\wedge REy_2)\wedge(\forall z(REz\rightarrow(z\equiv y_1\vee z\equiv y_2))))$；

(b)无环，$\neg\sigma_3$，$\neg\sigma_4$，…，$\neg\sigma_n$，…，其中 $\sigma_n(n>3)$表示公式 $\exists y_1y_2\cdots y_n((\bigwedge_{i\neq j}y_i\neq y_j)\wedge(y_1Ey_2)\wedge(y_2Ey_3)\wedge\cdots\wedge(y_nEy_1))$.

(3)完全二叉树：树中除了根节点外，对于其他节点，有且仅有 3 个不同的节点与之相邻，其中 y_1，y_2，y_3 两两不同. $\exists y_1y_2y_3((xEy_1\wedge xEy_2\wedge xEy_3)\wedge(\forall z(zEx\rightarrow(z\equiv y_1\vee z\equiv y_2\vee z\equiv y_3))))$.

语言$\mathcal{L}$的满足理论 T 的模型 $\mathfrak{A}=\langle A,E,R\rangle$，被称为完全二叉树模型. 其中 A 为树的节点集，xEy 表示节点 x，y 有边相连，R 为唯一的一个根节点.

§2. 基本公式

为使基本公式的形式统一，我们将先给出一些新的记号.

定义 1 在树中，任意两个节点之间都有唯一的通路，我们用 $PW(x,y)$表示节点 x 到 y 的路径. 两个节点间通路所含的边的个数表示两点间的距离. 以 $d(x,y)=n$ 记节点 x，y 的距离为 $n(n\geqslant 0)$，其中当 $n=0$ 时，表示 $x\equiv y$；$n=1$ 时，表示 xEy. 事实上 $d(x,y)=n$ 等价于

$$\exists z_1 z_2 \cdots z_{n-1}\left(\left(\bigwedge_{i\neq j} z_i \neq z_j\right) \wedge \left(\bigwedge_{i=1}^{n-1} x \neq z_i\right) \wedge \left(\bigwedge_{i=1}^{n-1} y \neq z_i\right)\right) \wedge$$
$$(xEz_1) \wedge (z_1Ez_2) \wedge \cdots \wedge (z_{n-1}Ey).$$

以 $h(x)=n$ 记 x 到 R 的距离为 n，即 $d(x,R)=n$，有时我们称为 x 的高度为 n.

性质 1 (1) $d(x,y)=n$ T-等价于 $d(y,x)=n$.

(2)给定节点 y，至多有 $3\times 2^{n-1}$ 个节点 x，使得关系 $d(x,y)=n$ 成立.

选择所有形如 $d(x,y)=n, h(x)=m(n,m=0,1,2,\cdots)$ 的公式组成基本公式集，记为Σ.

定义 2 用 $d(x,y)\neq n, h(x)\neq m$ 表示 $\neg(d(x,y)=n)$，$\neg(h(x)=m)$，并称其为负的基本公式. $d(x,y)>n$ 表示$(d(x,y)\neq 0)\wedge\cdots\wedge(d(x,y)\neq n)$. $d(x,y)\leqslant n$ 表示 $\neg(d(x,y)>n)$. 类似有"$<$""$\geqslant$"的定义.

因为原子公式 xEy 和 $x\equiv y$ 属于基本公式集，所以基本引理中的第 1 条显然成立，我们只需证明第 2 条.

设公式 φ 是基本公式的布尔组合，需要考虑公式 $\exists x\varphi$ 的量词消去，而由于存在量词对析取的可分配性，将把公式 φ 化成析取式的形式而分别考虑每一个析取项的量词消去，其中每一个析取项的一般形式由以下 4 部分合取组成(不妨仍用 φ 来表示).

(1)x 和 y 的距离正公式的合取式

$$(d(x,y_1)=m_1)\wedge(d(x,y_2)=m_2)\wedge\cdots\wedge(d(x,y_s)=m_s). \quad ①$$

(2)x 和 R 的距离正公式的合取式

$$(h(x)=p_1)\wedge(h(x)=p_2)\wedge\cdots\wedge(h(x)=p_q). \quad ②$$

(3)x 和 y 的距离负公式的合取式

$$(d(x,y_1)\neq n_1)\wedge d(x,y_2)\neq n_2\wedge\cdots\wedge(d(x,y_t)\neq n_t). \quad ③$$

(4)x 和 R 的距离负公式的合取式

$$(h(x)\neq l_1)\wedge(h(x)\neq l_2)\wedge\cdots\wedge(h(x)\neq l_r), \quad ④$$

其中，$s,t,r,q\geqslant 0$，且当 s,t,q 或 r 为 0 时，即表示无对应项. 为便于进行量词消去，将作进一步的化简. 实际上是在理论 T 下找到与之等价的公式. 在下文中若不加声明，将都认为是在理论 T 下讨论的.

性质 2

$$\bigwedge_{i=1}^{2}(d(x,y_i)=n_i)\rightarrow \bigvee_{i=0}^{\min(n_1,n_2)}(d(y_1,y_2)=n_1+n_2-2i). \quad ⑤$$

证 由完全二叉树的性质，只需考虑路径 $PW(x,y_1)$ 和路径 $PW(x,y_2)$ 的重合点的个数即可得到上述结论.

很希望⑤式本身就是等价式，但并非如此，如果把式子右边加上一些限制可以得到类似的等价式. 假设 $n_1\leqslant n_2$，需要考虑根节点 R 关于点 x 和 y_1 的相对位置. 设从 x 到 y_1 的不同节点位置分别记为 $1,2,\cdots,n_1-1$，则"根节点 R 在位置 $k(1\leqslant k<n_1)$"可以表示为 $(h(y_1)=n_1-k)\wedge(h(x)=k)$. 为方便起见，用 ψ_k 简记之，并用 ψ_0 记 $\neg(\bigvee_{k=1}^{k=n_1-1}\psi_k)$.

性质 3

$$\bigwedge_{i=1}^{2}(d(x,y_i)=n_i)\wedge\psi_k\rightarrow\bigvee_{i=0,i\neq k}^{\min(n_1,n_2)}(d(y_1,y_2)=n_1+n_2-2i). \qquad ⑥$$

证 此性质只是增加了一个限制条件，即根节点在路径 $PW(x,y_1)$ 之上，而因为与根节点有边相连的节点只有两个，所以 y_1,y_2 不可能在此处分叉，从而在右边的蕴涵项中少一个析取项 $d(y_1,y_2)=n_1+n_2-2k$.

用 $V(y_1,y_2;n_1,n_2)$，记 $\bigvee_{i=0}^{\min(n_1,n_2)}(d(y_1,y_2)=n_1+n_2-2i)$，$V(y_1,y_2;n_1,n_2;k)$（此处 $(1\leqslant k<n_1)$）. 记 $\bigvee_{i=0,i\neq k}^{\min(n_1,n_2)}(d(y_1,y_2)=n_1+n_2-2i)$. 为得到形式上的统一，有时也用 $V(y_1,y_2;n_1,n_2;0)$ 表示 $V(y_1,y_2;n_1,n_2)$.

定理 1 设 $d(x,y_1)=n_1$，根节点 R 不在 x 与 y_1 之间的任何位置，则有

$$\bigwedge_{i=1}^{2}(d(x,y_i)=n_i)\leftrightarrow((d(x,y_1)=n_1)\wedge \neg V(x,y_2;2n_1,n_2;n_1)\wedge V(y_1,y_2;n_1,n_2)).$$

证 (1)子公式 $\neg V(x,y_2;2n_1,n_2;n_1)$ 实际上是形如 $d(x,y_2)\neq k$（其中不含 $k=n_2$ 一项）的公式的合取，因此有 $d(x,y_2)=n_2\rightarrow\neg V(x,y_2;2n_1,n_2;n_1)$. 再结合性质 2，有→方向成立；

$$(2)(d(x,y_1)=n_1)\wedge\left(\bigvee_{i=0}^{\min(n_1,n_2)}d(y_1,y_2)=n_1+n_2-2i\right)$$

$$\leftrightarrow\bigvee_{i=0}^{\min(n_1,n_2)}(d(y_1,y_2)=n_1+n_2-2i\wedge d(x,y_1)=n_1).$$

对每一个析取项利用性质 2 得到

$$\bigvee_{i=0}^{\min(n_1,n_2)}(d(y_1,y_2)=n_1+n_2-2i\wedge d(x,y_1)=n_1)$$

$$\rightarrow\bigvee_{i=0}^{\min(n_1,n_2)}\left(\bigvee_{j=0}^{\min(n_1+n_2-2i,n_1)}d(x,y_2)=2n_1+n_2-2i-2j\right). \qquad ⑦$$

注意到由 $i\leqslant\min(n_1,n_2)$ 和 $j\leqslant\min(n_1+n_2-2i,n_1)$，可以得到 $i+j\leqslant 2n_1$. 同时由 $j\leqslant n_1+n_2-2i$，即 $i+j\leqslant n_1+n_2-i$ 和 $i\leqslant n_1$，可以得到 $i+j\leqslant n_2$，所以有 $i+j\leqslant\min(2n_1,n_2)$，可以将所有析取项合并.

$$\bigvee_{i=0}^{\min(n_1,n_2)}\left(\bigvee_{j=0}^{\min(n_1+n_2-2i,n_1)} d(x,y_2)=2n_1+n_2-2i-2j\right)$$

$$\leftrightarrow\bigvee_{(i+j)=0}^{\min(2n_1,n_2)}(d(x,y_2)=2n_1+n_2-2(i+j))$$

$$\leftrightarrow\bigvee_{l=0}^{\min(2n_1,n_2)}(d(x,y_2)=2n_1+n_2-2l),$$

所以有下面的结果

$$(d(x,y_1)=n_1)\wedge V(y_1,y_2;n_1,n_2)$$

$$\rightarrow\bigvee_{l=0}^{\min(2n_1,n_2)}(d(x,y_2)=2n_1+n_2-2l)\left(\bigvee_{l=0}^{\min(2n_1,n_2)}(d(x,y_2)=2n_1+n_2-2l)\right)$$

$$\wedge\left(\bigwedge_{i=0,i\neq n_1}^{\min(2n_1,n_2)}(d(x,y_2)\neq 2n_1+n_2-2i)\right)\rightarrow d(x,y_2)=n_2,$$

所以 $(d(x,y_1)=n_1)\wedge\neg V(x,y_2;2n_1,n_2;n_1)\wedge V(y_1,y_2;n_1,n_2)\rightarrow(d(x,y_1)=n_1)\wedge(d(x,y_2)=n_2)$.

根据以上(1)(2)两部分所证的结果，定理成立.

定理 2 设 $d(x,y_1)=n_1$，根节点 R 在 x 与 y_1 之间的第 k 个位置，则

$$\bigwedge_{i=1}^{2}(d(x,y_i)=n_i)\leftrightarrow((d(x,y_1)=n_1)\wedge\neg V(x,y_2;2n_1,n_2;n_1)\wedge V(y_1,y_2;n_1;n_2;k)).$$

证 此定理与上一个定理的区别在于根节点的出现，利用性质 3 类似上定理的证明可得.

结合定理 1 和定理 2，我们得到

定理 3 $(d(x,y_1)=n_1)\wedge(d(x,y_2)=n_2)\leftrightarrow((d(x,y_1)=n_1)\wedge\neg V(x,y_2;2n_1,n_2;n_1)\wedge\bigvee_{k=0}^{n_1-1}(\Psi_k\wedge V(y_1,y_2;n_1,n_2;k)))$.

由此定理，我们可以把①式化成只含一个 x 到 y_1 的距离的正公式以及②～④式中的多个合取项，而在只考虑可满足式的情况下，②式也可化成只含一项的形式(x 与根节点的距离只能有一个值)，事实上，根节点 R 完全可以被看作是一个特殊的 y 点(其区别只在与之相邻的点的个数上，可在具体问题上区别对待)，即只需要证如下形式的情况即可(仍设为 φ).

$(d(x,z)=m)\wedge(d(x,y_1)\neq n_1)\wedge\cdots\wedge(d(x,y_t)\neq n_t)$，其中 $t,m\geqslant 1$.

§3. 量词消去

设 $\mathfrak{A}$ 是 T 的模型，对于公式 $\exists x\varphi(x)$，我们发现 A 中是否存在 x，使得 $\varphi(x)$ 成立，只取决于 $z, y_1, y_2, \cdots, y_t$ 的相对位置及它们的高度，而且尤为重要的是不存在 x，使得 $\varphi(x)$ 成立的相对位置只有有限多个. 因此可以通过找到公式 $\neg(\exists x\varphi(x))$ 的 T-等价式（不含变元 x）来完成 $\exists x\varphi(x)$ 的量词消去.

首先，看一个简单的例子，φ 中只有一个否定式：$d(x,z)=2\wedge d(x,y_1)\neq 2$，当 $d(z,y_1)\geqslant 2$ 时，y_1 到 z 的路径中必定有一点（而且只有一点）与 z 的距离为 2，而与 z 的距离等于 2 的点至少有 4 个（至多 6 个），因此必定存在 x，使得 φ 成立. 当 $d(z,y_1)=1$ 时，y_1 存在于某个 x 点（也有且只有一点）与 z 之间的路径上，此时也必存在 x，使得 φ 成立. 当 $d(z,y_1)=0$，即 z, y_1 重合时 φ 等价于 $d(x,z)=2\wedge d(x,z)\neq 2$，此时不存在这样的 x. 因此 $\neg(\exists x\varphi(x))$ 等价于 $d(z,y_1)=0$. 此例属于特殊情形，不加高度就可确定. 下面考虑一般形式，具体步骤与此类似，只不过当有多个否定式存在时，其过程略微复杂一些而已. 另外，需加上对 $z, y_1, y_2, \cdots, y_t$ 的高度的讨论.

定义 3 设 $\mathfrak{A}$ 为 T 的一个模型，用 $\theta(z, y_1, \cdots, y_t)$ 表示 $\neg(\exists x\varphi(x))$，其中 $z, y_1, \cdots, y_t$ 为自由变元. 可以用表示节点间相对位置的公式

$$\Big(\bigwedge_i d(z,y_i)=d_i\Big)\wedge\Big(\bigwedge_{i,j} d(y_i,y_j)=L_{i,j}\Big)\wedge\Big(\bigwedge_i h(y_i)=h_i\wedge h(z)=h_0\Big)$$

（基本公式的布尔组合）来刻画模型 $\mathfrak{A}$ 的一类解释，并称自由变元在模型 $\mathfrak{A}$ 中的一类解释为一个分布. 刻画这个分布的公式，称为分布公式，记作 $F(d_i, h_0, h_i, L_{i,j}\mid 1\leqslant i,j\leqslant t)$.

如果模型 $\mathfrak{A}$ 中只有有限个分布满足公式 θ，则对应的分布公式的析取必与公式 θ 在 T 下等价，且不含变元 x，由此便完成了量词消去.

根据定理 1，当 $d(z,y_i)=d_i, d(z,y_j)=d_j$ 时，$d(y_i,y_j)$ 只能取有限个值，所以和谐的 $F(d_i, h_0, h_i, L_{i,j}\mid 1\leqslant i,j\leqslant t)$ 形的公式只有有限个. 公式 θ 的每个解释确定一个这样的分布公式，由此使模型 $\mathfrak{A}$ 满足 θ 的分布公式只有有限个. 下面将分析确定具体有哪些分布公式析取后与 θ 等价.

定义 4 用集合 C 表示满足 $d(x,z)=m$ 的所有 x 点的集合，即 $C=\{x|d(x,z)=m,x\in A\}$.

z 的高度不同时，集合 C 的元素个数也不同，具体范围为 $2^m\leqslant|C|\leqslant 3\times 2^{m-1}$. 为简化叙述，假定 z 的高度大于 m，此时 $|C|=3\times 2^{m-1}$（在其余的情况下只是元素个数上的差异不存在实质性的差别）.

用 $B(y_i)_{d_i}^{u_i}$ 表示集合 C 中与 y_i 距离等于 u_i 的所有 x 点组成的子集 $(0<i\leqslant t)$：$B(y_i)_{d_i}^{u_i}=\{x|d(x,y_i)=u_i,d(z,y_i)=d_i,x\in C\}$.

当 $z,y_1,\cdots,y_t$ 的相对位置固定后，对于这些 y_i，与集合 C 中的 x 点的距离只有有限种. 根据 y_i 与 x 点的不同距离，可以给出集合 C 分块如下

当 $d_i=0$ 时

$$|B(y_i)_0^m|=3\times 2^{m-1},$$
$$C=B(y_i)_0^m.$$

当 $d_i=1$ 时

$$|B(y_i)_1^{m-1}|=2^{m-1},|B(y_i)_1^{m+1}|=2\times 2^{m-1},$$
$$C=B(y_i)_1^{m-1}\cup B(y_i)_1^{m+1}.$$

当 $2\leqslant d_i\leqslant m-1$ 时

$$|B(y_i)_{d_i}^{m-d_i}|=2^{m-d_i},|B(y_i)_{d_i}^{m-d_i+2}|=2^{m-d_i},\quad |B(y_i)_{d_i}^{m-d_i+4}|=2^{m-d_i+1},$$
$$\cdots,\quad |B(y_i)_{d_i}^{m-d_i+2(d_i-1)}|=2^{m-2},\ |B(y_i)_{d_i}^{m+d_i}|=2\times 2^{m-1},$$

$$C=\bigcup_{k=0}^{d_i}B(y_i)_{d_i}^{m-d_i+2k}.$$

当 $d_i\geqslant m$ 时

$$|B(y_i)_{d_i}^{d_i-m}|=1,|B(y_i)_{d_i}^{d_i-m+2}|=1,\quad |B(y_i)_{d_i}^{d_i-m+4}|=2,$$
$$\cdots,\quad |B(y_i)_{d_i}^{d_i+m-2}|=2^{m-2},|B(y_i)_{d_i}^{d_i+m}|=2\times 2^{m-1},$$

$$C=\bigcup_{k=0}^{m}B(y_i)_{d_i}^{d_i-m+2k}.$$

对于所有其余的 $B(y_i)_{d_i}^{u_i}$，有 $B(y_i)_{d_i}^{u_i}=\varnothing$，$|B(y_i)_{d_i}^{u_i}|=0$. 而且对于每一个给定的 d_i，当 $d_i\leqslant m+n_i$ 时，至多有一组 $B(y_i)_{d_i}^{d_i-m+2k}$ 与 y_i 的距离等于 n_i，当 $d_i>m+n_i$ 时，没有与 y_i 的距离等于 n_i 的点. 也就是说，当 $(d(z,y_i)>m+n_i)$ 时，$(d(x,y_i)\neq n_i)$ 一定成立，因此我们可以把 φ 分解为（只以 y_1 为例）$((d(z,y_1)>n_1+m)\wedge\varphi(x,\hat{y}_1,\cdots,y_t,z))\vee((d(z,y_1)\leqslant n_1+m)\wedge\varphi(x,y_1,\cdots,y_t,z))$，其中用记号 $\varphi(x,\hat{y}_1,\cdots,y_t,z)$ 表示公式

$$((d(x,z)=m)\wedge(d(x,y_2)\neq n_2)\wedge\cdots\wedge(d(x,y_t)\neq n_t)).$$

即 $\varphi(x,y_1,\cdots,y_t,z)$中不含 $d(x,y_1)\neq n_1$ 项. 对 $y_2,y_3,\cdots,y_t$ 重复上述过程,得到公式 $\varphi(x,y_1,\cdots,y_t,z)$的 T 等价公式

$$\frac{\varphi}{\varnothing}\vee\left(\bigvee_{i=1}^{t}\frac{\varphi}{i}\right)\vee\left(\bigvee_{i\neq j}^{t}\frac{\varphi}{i,j}\right)\vee\cdots\vee\left(\bigvee_{i_1,i_2,\cdots,i_k=1}^{t}\frac{\varphi}{i_1,i_2,\cdots,i_k}\right)\vee\cdots\vee\left(\frac{\varphi}{1,2,\cdots,t}\right),$$

其中用记号$\frac{\varphi}{\varnothing}$表示公式

$$\bigwedge_{i=1}^{t}d(z,y_i)>m+n_i\wedge(d(x,z)=m).$$

用记号$\frac{\varphi}{i_1,i_2,\cdots,i_k}$表示公式

$$\bigwedge_{l=1}^{t-k}(d(z,y_{j_l})>m+n_{j_l})\wedge\bigwedge_{l=1}^{k}(d(z,y_{i_l})\leqslant m+n_{i_l})\wedge$$
$$\varphi(x,y_1,\cdots,\hat{y}_{j_1},\cdots,\hat{y}_{j_{t-k}},\cdots,y_t,z)$$

($y_1,y_2,\cdots,y_t$ 中除去 $y_{i_1},y_{i_2},\cdots,y_{i_k}$ 的部分记为 $y_{j_1},y_{j_2},\cdots,y_{j_{t-k}}$).

实际上公式$\frac{\varphi}{i_1,i_2,\cdots,i_k}$中含 x 的项是$\frac{\varphi}{1,2,\cdots,t}$的子公式,不含 x 的项完全可以提到存在量词的前面而不影响量词的消去. 所以只需考虑公式$\frac{\varphi}{1,2,\cdots,t}$,即

$$(d(z,y_1)\leqslant m+n_1)\wedge\cdots\wedge(d(z,y_t)$$
$$\leqslant m+n_t)\wedge\varphi(x,y_1,\cdots,y_t,z).$$

由此,在文章的后半部分,我们将只考虑 $d(z,y_i)\leqslant m+n_i$时的情形.

按照上述分析,当 $d_i,h_0,h_i,L_{i,j}(1\leqslant i,j\leqslant t)$固定后,对应有集合 C 的一组分块 $B(y_i)_{d_i}^{n_i}$,若这些分块能否覆盖集合 C 的性质不依 y_i 的位置不同而变化,我们称公式 $F(d_i,h_0,h_i,L_{i,j}\mid 1\leqslant i,j\leqslant t)$能拟一致覆盖集合 C.

定理 4 每一个和谐的分布公式 $F(d_i,h_0,h_i,L_{i,j}\mid 1\leqslant i,j\leqslant t)$,都能拟一致覆盖集合 C.

证 我们将对 t 采用归纳法完成证明. 首先假设 $d_1\leqslant d_2\leqslant\cdots\leqslant d_t$,否则可以重新排序使之满足此关系. 如图 1 所示,对于给定的 d_i,y_1,顺序称 y_1,z 之间

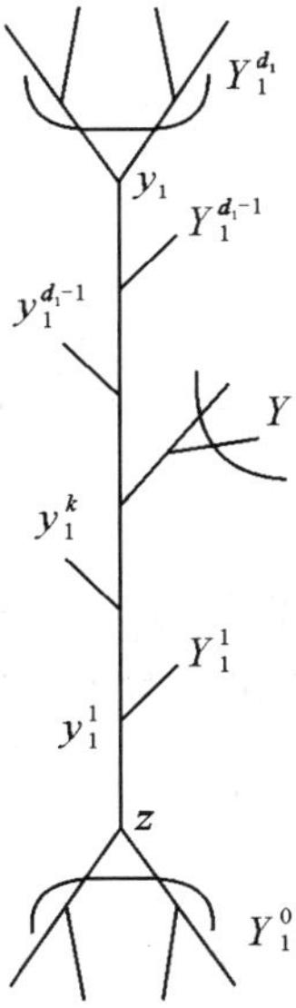

图 1

的 d_1-1 个点为 $y_1^1,\cdots,y_1^{d_1-1}$. 与 y_1^k 有边相连且不在此通路上的节点记为 $Y_1^k(1\leqslant k\leqslant d_1-1)$ 称为 y_1 的标志点. 同时称 z 为标志点 Y_1^0, y_1 为标志点 $Y_1^{d_1}$. 所有与 Y_1^k 相通且与 z 的距离大于 $d(z,Y_1^k)$ 的节点构成一个以 Y_1^k 为根的子树$(1\leqslant k\leqslant d_1)$. 所有与 z 相通且与 y_1 不在同一分支的节点构成关于 Y_1^0 的子树. 若 $y_1\equiv z$, 则整个树构成 Y_1^0, 不妨也分别称这些对应子树为 $Y_1^k(0\leqslant k\leqslant d_1)$. 而且这些子树互不相交, 与 y_1^k $(1\leqslant k<d_1)$ 构成整个树. 子树 $Y_1^k(0\leqslant k\leqslant d_1)$ 将集合 C 分割成 d_1+1 块, 其中有一分块为 $B(y_1)_{d_1}^{n_1}$, 对应子树 Y_1^k.

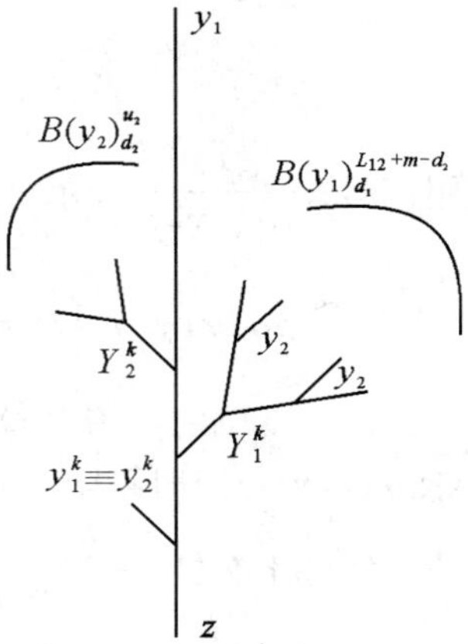

图 2

当 $t=1$ 时, 显然只有一个 $B(y_1)_{d_1}^{n_1}$, 结论成立; 当 $t=2$ 时, 对于 y_2 及给定的 $d_2,h_0,h_1,h_2,L_{12},y_2$ 的位置是否出现在子树 Y_1^k 上是可判定的, 从而有以下 4 种可能(如图 2):

(1) y_2 在子树 Y_1^k 上, 且

$$B(y_2)_{d_2}^{n_2}\subset B(y_1)_{d_1}^{n_1};$$

(2) y_2 在子树 Y_1^k 上, 且

$$B(y_2)_{d_2}^{n_2}\cap B(y_1)_{d_1}^{n_1}=\varnothing;$$

(3) y_2 不在子树 Y_1^k 上, 且 $B(y_1)_{d_1}^{n_1}\subset B(y_2)_{d_2}^{n_2}$;

(4) y_2 不在子树 Y_1^k 上(在另一子树 Y_1^l), 且 $B(y_2)_{d_2}^{n_2}\cap B(y_1)_{d_1}^{n_1}=\varnothing$.

前 3 种情形下, 显然成立; 对于第 4 种情形, 如果 $B(y_2)_{d_2}^{n_2}$ 不在 y_2 所在子树 Y_1^l 中, 则不管 y_2 的取法, 集合 $B(y_2)_{d_2}^{n_2}$ 不变, 从而覆盖的一致性仍然保持; 如果 $B(y_2)_{d_2}^{n_2}$ 在 y_2 所在子树 Y_1^l 中, 则 y_2 的位置影响到集合 $B(y_2)_{d_2}^{n_2}$, 分以下两种情形: (a) y_2 加细了此分块, 即至少分成两个分块, 而 $B(y_2)_{d_2}^{n_2}$ 只占其中一块, 则另一块使得 $C\setminus\bigcup_{i=1}^{2}B(y_i)_{d_i}^{n_i}$ 非空, 从而保证覆盖的一致性; (b) y_2 没有加细此分块, 即此分块就是 $B(y_2)_{d_2}^{n_2}$, 此时 y_2 取法唯一, 不具二歧性.

当 $t=p$ 时, $y_1,y_2,\cdots,y_{p-1}$ 已经取好, 记 $B(y_i)_{d_i}^{n_i}(1\leqslant i<p)$ 所在子树分别为 $Y_{p-1}^{k},\cdots,Y_{p-1}^{k_{p-1}}$, 对于 y_p, 把 $\bigcup_{i=1}^{p-1}B(y_i)_{d_i}^{n_i}$ 作为一个整体考虑, 分析 y_2 时的证明可知, 只有当分块 $B(y_p)_{d_p}^{n_p}$ 在子树 $Y_{p-1}^{k_p}$ 上时, y_p 的位置不同

才影响集合$\bigcup_{i=1}^{p}B(y_i)_{d_i}^{n_i}$，此时依然可以证明覆盖的一致性(同上述(a)(b)的情形).

由上讨论即证得每个和谐的公式$F(d_i,h_0,h_i,L_{i,j}\mid 1\leqslant i,j\leqslant t)$都能拟一致覆盖集合$C$.

由此定理，倘若$C\subseteq\bigcup_{i=1}^{t}B(y_i)_{d_i}^{n_i}$，即集合$C$的$t$个分块$B(y_i)_{d_i}^{n_i}$有限覆盖$C$，则有

$$F(d_i,h_0,h_i,L_{i,j}\mid 1\leqslant i,j\leqslant t)\rightarrow\theta.$$

据此把全部形如$F(d_i,h_0,h_i,L_{i,j}\mid 1\leqslant i,j\leqslant t)$的公式分为两类(以后简记为$F(d_i,h_0,h_i,L_{i,j})$)——能够有限覆盖$C$的和不能有限覆盖$C$的.

定理 5 当$d(z,y_i)\leqslant n_i+m(1\leqslant i\leqslant t)$时

$$\theta\leftrightarrow\bigvee_{\bigcup B(y_i)_{d_i}^{n_i}\supset C}F(d_i,h_0,h_i,L_{i,j}).$$

证 其中一个方向是平凡的，只证另一个方向，即

$$\theta\rightarrow\bigvee_{\bigcup B(y_i)_{d_i}^{n_i}\supset C}F(d_i,h_0,h_i,L_{i,j}).$$

设$\mathfrak{A}$是T的一个模型，如果存在自由变元的解释使模型满足θ，那么这个解释确定一个分布公式$F(d_i,h_0,h_i,L_{i,j})$，而且这个公式所确定的集合$\bigcup_{i=1}^{t}B(y_i)_{d_i}^{n_i}$有限覆盖$C$.

到此为止我们已经完成了量词消去的全过程.

定理 6 完全二叉树的理论T是完备理论.

证 对于一阶语言$\mathcal{L}$中的任何一个句子φ，根据完全二叉树理论的量词消去法，它T-等价于基本公式的布尔组合. 而基本公式集中只有恒真和恒假句子不含有自由变元，所以句子φ或者T-等价于恒真句子或者T-等价于恒假句子，下面两个式子

$$T\models\varphi\quad 和\quad T\models(\neg\varphi)$$

必有一个成立. 即证明完全二叉树的理论T是完备理论.

Abstract The method of quantifier elimination has been one of the powerful tools in the computer science and algebraic model theory. In this article, we deal with the theory of complete binary trees. After giving a set of fomulas as the Basic Formulas, we use the layout formulas and finite covering to prove that the theory of complete binary trees admits quantifier elimination.

Keywords Complete binary trees, Quantifier elimination, Basic formulas, Layout formulas, Finite covering.

参考文献

[1] Rose B I. Rings which admit elimination of quantifiers. J. Symbolic Logic, 1978, 43(1): 92—112.

[2] Berline Ch. Rings admit elimination of quantifiers. J. Symbolic Logic, 1981, 46(1): 56—58.

[3] Berline Ch. Elimination of quantifiers for non semi-simple rings of characteristic *p*. Springer Lecture Notes, 1980, 834: 10—20.

[4] Berline Ch, Cherlin G. QE rings in characteristic p^n. J. Symbolic Logic, 1983, 48(1): 140—162.

[5] Boffa M, Macintyre A, Point F. The quantifier elimination problem for rings without nilpotent elements and for semi-simple rings, Springer Lecture Notes, 1980, 834: 20—30.

[6] Weispfenning V. Quantifier elimination for distributive lattices and measure algebras. Zeitschrift M. L., 1985, 31(1): 249—261.

[7] Weispfenning V. Quantifier elimination for modules. Arch. Math. Logik Grundlag, 1985, 25(1): 1—11.

[8] Chang C C, Keisler H J. Model theory. North-Holland Publ. Co., 1990 (Third Edition).

[9] Wilfrid H. Model theory. Cambridge: Cambridge University Press, 1993.

[10] Wang S Q. Foundation of Model theory. Beijing: Science Press, 1987 (in Chinese).

[11] Shen F X. Introduction of Model Theory. Beijing: Beijing Normal University Press, 1995 (in Chinese).

数学学报中文版,2008,51(2):311～318

完全二叉树理论的计算复杂度[①]

On the Computational Complexity of the Theory of Complete Binary Trees

摘要 完全二叉树的一阶理论已被证明具有量词消去的性质,进而计算了完全二叉树模型中元素的 CB 秩. 本文利用有界 Ehrenfeucht-Fraïssé 博弈研究完全二叉树的一阶理论,证明了此理论的时间计算复杂度上界为 $2^{2^{cn}}$,空间计算复杂度上界为 2^{dn}(其中 n 为输入长度,c,d 为合适的常数).

关键词 完全二叉树的一阶理论,有界 **Ehrenfeucht-Fraïssé** 博弈,计算复杂度.

§0. 引言

树形结构广泛应用于计算机科学的各个领域,例如:逻辑程序设计、数据库和知识表示等. Vorobyov 在文[1]中讨论了由 Herbrand 域导出的有限树的一阶理论的可判定性和复杂性.

近年来,人们对完全二叉树理论的研究不断深入,得到了不少好的成果. 刘吉强等在文[2]中利用分布公式和有限覆盖证明了完全二叉树理论具有量词消去的性质,从而是完全的. 陈磊等在文[3]中计算了完全二叉树模型中元素的 CB 秩,并对其元素进行了划分.

① 收稿日期:2006-07-25;接受日期:2007-07-25.
本文与李志敏和李祥合作.

但是,关于该理论的计算复杂度一直没有相关的结果.无论是从理论还是从实际应用的角度,确定完全二叉树理论的计算复杂度上界是一个重要且有趣的课题.本文运用有界 Ehrenfeucht-Fraïssé 博弈,得到了如下结论:

完全二叉树一阶理论的时间计算复杂度上界为 $2^{2^{cn}}$,空间计算复杂度上界为 2^{dn}(其中 n 为输入长度,c,d 为合适的常数).

§1. 预备知识

为了阅读方便,在本部分不加证明地列举了完全二叉树理论 T 和 Ehrenfeucht-Fraïssé 博弈的有关概念以及相关结果.

§1.1 完全二叉树理论

定义 1 本文所讨论的完全二叉树理论 T,是由一阶语言 $\mathscr{L}=\{E,R\}$ 中以下公理组成的,其中 E 是二元关系符号,R 为常量符号.

(Ⅰ)图的理论

(a)无环:$\forall x\neg(xEx)$;

(b)无向:$\forall xy(xEy\leftrightarrow yEx)$.

(Ⅱ)树的理论

(a)存在唯一的根节点,记为 R.这条公理可表示成如下式子:

$$\exists y_1y_2((y_1\neq y_2)\wedge(REy_1\wedge REy_2)\wedge(\forall_z(REz\to(z\equiv y_1\vee z\equiv y_2))));$$

(b)无圈,也就是:$\neg\sigma_3,\neg\sigma_4,\cdots,\neg\sigma_n,\cdots$,这里 $\sigma_n(n>3)$表示如下公式

$$\exists y_1y_2\cdots y_n\left(\left(\bigwedge_{i\neq j}y_i\neq y_j\right)\wedge(y_1Ey_2)\wedge\cdots(y_nEy_1)\right).$$

(Ⅲ)完全二叉树

$\forall x\neq R$,有且仅有 3 个不同的节点与之相邻

$$\forall x((x\neq R)\to(\exists_{y_1y_2y_3}((xEy_1\wedge xEy_2\wedge xEy_3)\wedge(\forall z(zEx\to (z\equiv y_1\vee z\equiv y_2\vee z\equiv y_3)))))).$$

语言 $\mathscr{L}$ 的满足理论 T 的模型 $\mathfrak{A}=\langle A,E,R\rangle$,被称为完全二叉树模型.其中 A 为树的节点集,xEy 表示节点 x 与 y 有边相连,R 为唯一的一个根节点.

文中常用到如下结论:

(1)任意两个节点之间都有唯一的通路,用 $PW(x,y)$表示节点 x 到

y 的路径.

用 $d(x,y)$记节点 x 到 y 的距离. 事实上

$$d(x,y)=n\leftrightarrow \exists_{z_1z_2\cdots z_{n-1}}\left(\left(\bigwedge_{i\neq j} z_i\neq z_j\right)\wedge\left(\bigwedge_{i=1}^{n-1} x\neq z_i\right)\wedge\left(\bigwedge_{i=1}^{n-1} y\neq z_i\right)\wedge\right.$$

$$\left.(xEz_1)\wedge(z_1Ez_2)\wedge\cdots\wedge(z_{2-1}E_y)\right);$$

用 $\| x \|=h(x)=d(x,R)$表示 x 的范数.

(2)完全二叉树理论的消去集为 $d(x,y)=n,d(R,x)=m,n,m=0,1,\cdots$,其中 $d(x,y)=1$ 和 $d(x,y)=0$ 分别是原子公式 xEy 和 $x\equiv y$(见文[2]).

§1.2 有界 Ehrenfeucht-Fraïssé 博弈

下面的讨论主要围绕着有界 Ehrenfeucht-Fraïssé 博弈展开. 在文[4]和[5]中,Ferrante 与 Rackoff 利用该博弈对适当的模型给出了判定程序. 但本文作者建议读者阅读文献[6],在文[6]中,罗里波的叙述更清晰明确.

设$\mathfrak{A}=\langle A,E,R\rangle$是一个模型. $\forall a\in A$,定义 a 的一个范数 $\| a \|$. 假设有如下句子

$$\phi_0=(Q_1x_1)(Q_2x_2)\cdots(Q_nx_n)\psi(x_1,x_2,\cdots,x_n).$$

对该句子,已经将变量 $x_1,x_2,\cdots,x_k$ 替换成 $a_1,a_2,\cdots,a_k\in A$ 得到如下句子:

$$\phi_k=(Q_{k+1}x_{k+1})\cdots(Q_nx_n)\psi(a_1,a_2,\cdots,a_k,x_{k+1},\cdots,x_n).$$

接下来,将用 a_{k+1} 去替换 x_{k+1}. 希望能使 a_{k+1} 的范数控制在某一特定的范围内. 于是引入控制函数如下

$$H(n-k,k,m)=h\in\mathbf{N},\text{其中}\ \| a_1 \|,\| a_2 \|,\cdots,\| a_k \|\leqslant m.$$

判定程序所需的步数是由 A 中范数不超过某一固定数的元素个数和控制函数 H 决定的,当然期望它们越小越好.

在多数情况下直接讨论 $\bar{a}_k\equiv_n\bar{b}_k$(该式表示 $\bar{a}_k=(a_1,a_2,\cdots,a_k)$与$\bar{b}_k=(b_1,b_2,\cdots,b_k)$满足量词深度不超过 n 的相同的句子集)是相当困难的. 但找一个等价关系 $\bar{a}_kE_{h,k}\bar{b}_k$ 相对要容易一些.

在已知范数,控制函数 H 和等价关系的情况下,可以证两个玩家甲和乙利用模型 M_1 和 M_2 进行如下博弈:

在博弈中，两个玩家交错地从 M_1 或 M_2 中取元素. 规定在每一步中，甲和乙不能在同一模型中取元素，每个玩家都须完成 n 步. 乙已知有 $n+1$ 个等价关系. 博弈开始于 $\varnothing E_{n,0}\varnothing$. 设第 k 步结束时，两玩家选取的元素

$$(a_1,a_2,\cdots,a_k)\in M_1^k \text{ 和 } (b_1,b_2,\cdots,b_k)\in M_2^k$$

满足关系 $\bar{a}_k E_{n-k,k}\bar{b}_k$. 对甲取的任意元素 $a_{k+1}\in M_1$，根据等价关系，乙可以找到一个元素 $b_{k+1}\in M_2$（如果甲选取 b_{k+1}，则乙选取 a_{k+1}）满足

$$\bar{a}_{k+1}E_{n-k-1,k+1}\bar{b}_{k+1}.$$

乙取胜的充要条件是

$$\bar{a}_n E_{0,n}\bar{b}_n \rightarrow \bar{a}_n \equiv_0 \bar{b}_n.$$

乙取胜须满足的条件是在任一步 k，$\bar{a}_k E_{n-k,k}\bar{b}_k$.

因此，有如下定理：

定理 1(Ferrante and Rackoff) 设 A_1，A_2 是两个模型. 假设玩家乙有等价关系 $E_{n,k}$ 与甲进行博弈. 这就是说 A_1，A_2 的元素之间的等价关系 $E_{n,k}$ 满足

(i) $\bar{a}_k E_{0,k}\bar{b}_k \rightarrow \bar{a}_k \equiv_0 \bar{b}_k$，其中 $1\leqslant k\leqslant n$.

(ii) 给定 $\bar{a}_k E_{r+1,k}\bar{b}_k$ 且对任意的 $a_{k+1}\in A_1$，存在 $b_{k+1}\in A_2$，使得 $\bar{a}_{k+1}E_{r,k+1}\bar{b}_{k+1}$，其中 $1\leqslant r\leqslant n$. 交换 a_{k+1} 与 b_{k+1} 在定理中的位置结论不变.

则有：

(i) $\bar{a}_{k+1}E_{r,k+1}\bar{b}_{k+1}\rightarrow \bar{a}_k \equiv_r \bar{b}_k$，对任意的 $\bar{a}_k\in A_1^k$ 且 $\bar{a}_k\in A_2^k$.

(ii) $\mathrm{Th}_n(A_1)=\mathrm{Th}_n(A_2)$（也就是说模型 A_1 与模型 A_2 中量词深度不超过 n 的句子集相等）.

有的情况下 $A_1=A_2=A$ 且对任意选取的元素序列 $a_1,a_2,\cdots,a_k$ 和 $b_1,b_2,\cdots,b_k$，使得 $\bar{a}_k E_{n,k}\bar{b}_k$ 且 $\|b_1\|,\|b_2\|,\cdots,\|b_k\|\leqslant m$. 甲选取任意元素 $a_{k+1}\in A$，乙能找到一个元素 $b_{k+1}\in A$ 满足

$$\bar{a}_{k+1}E_{n-k-1,k+1}\bar{b}_{k+1},\ \|b_{k+1}\|\leqslant H(n-k,k,m).$$

这个条件称为有界条件，此时的博弈叫作有界 Ehrenfeucht-Fraïssé 博弈. 那么下面的定理提供了一个 Th(A) 中量词深度不超过 n 的句子的判定程序.

定理 2(Ferrante and Rackoff) 设 A 是一个模型. 已知有界 Ehrenfeucht-Fraïssé 博弈有等价关系 $E_{n,k}$，元素 a 的范数为 $\|a\|$ 且控制函数为 $H(n,k,m)$，满足如下条件：

(i)如果 $\bar{a}_k E_{n+1,k} \bar{b}_k$ 且 $\| b_1 \|, \| b_2 \|, \cdots, \| b_k \| \leqslant m$,那么对任意的 $a_{k+1} \in A$ 存在 $b_{k+1} \in A$,使得

$$\bar{a}_{k+1} E_{n,k+1} \bar{b}_{k+1} \text{和} \| b_{k+1} \| \leqslant H(n,k,m).$$

(ii)对任意的非负实数 r,范数不超过 r 的元素个数是 $F(r)$.

那么判定量词深度为 n 的一阶句子 ϕ 最多需要 $\prod_{i=1}^{n} F(L_i)$ 步,其中 $L_0=1$ 且

$$L_{i+1}=H(n-i,i,L_i),(i=0,1,\cdots,n-1).$$

这里一步(有时叫大步)的意思是既无变量又无量词的句子 ϕ 的检测程序.

§2. 完全二叉树模型上的有界 Ehrenfeucht-Fraïssé 博弈及其复杂度上界

依上节所叙,首先定义完全二叉树模型上的一个等价关系,使得玩家乙有取胜策略.

§2.1 完全二叉树模型上的等价关系

设 $\mathfrak{A}'=\langle A,E,R\rangle$ 是一个完全二叉树模型,其中 A 为树的节点集,xEy 表示节点 x,y 有边相连,R 为唯一的树根节点.

$T(a,k)$表示以 a 为根且深度为 k 的子树

$$T(a,b,k)=T(a,k)\cup T(b,k),$$

其中 $d(a,b)\leqslant k$.

$T(a,b,k)\clubsuit T(a',b',k)$表示 $T(a,b,k)\cong T(a',b',k)$且 $PW(a,a')\cong PW(b,b')$.

设 $a_1,a_2,b_1,b_2\in A$ 且 $n\in\mathbf{N}$. 用$(a_1,a_2)\approx_n(b_1,b_2)$表示

(1)或者 $d(a_1,a_2)>2^n$ 且 $d(b_1,b_2)>2^n$,

(2)或者 $d(a_1,a_2)=d(b_1,b_2)\leqslant 2^n$,且 $T(a_1,a_2,2^n)\clubsuit T(b_1,b_2,2^n)$.

设 $n\in A$, $k\in\mathbf{N}$, 且 $\bar{a}_k,\bar{b}_k\in\mathbf{N}^k$. 称 $\bar{\mathbf{a}}_k$ 与 $\bar{\mathbf{b}}_k$ 有等价关系 $\mathbf{E}_{n,k}$,记作 $\bar{\mathbf{a}}_k\mathbf{E}_{n,k}\bar{\mathbf{b}}_k$,如果满足条件:

(1)对任意 $i\in[1,k]$,$(R,a_i)\approx_{n+1}(R,b_i)$.

(2)对任意 $i,j\in[1,k]$,$i\neq j$,有$(a_i,a_j)\approx_n(b_i,b_j)$.

§2.2 完全二叉树模型上的有界 Ehrenfeucht-Fraïssé 博弈

根据定理 1 和定理 2,希望下面两个引理在完全二叉树模型上成立.

引理 1 设 $\bar{a}_k,\bar{b}_k\in A^k$ 满足条件 $\bar{a}_k E_{0,k}\bar{b}_k$. 那么 $\forall k\in[1,n]$，有 $\bar{a}_k\equiv_0\bar{b}_k$. 这就是说 $\bar{a}_k$ 和 $\bar{b}_k$ 满足相同的原子公式.

证 根据 $\bar{a}_k E_{0,k}\bar{b}_k$ 的定义，我们有：

(i) $\forall i\in[1,k]$，$(R,a_i)\approx_1(R,b_i)$，也就是：要么 $\|a_i\|>2$，且 $\|b_i\|>2$；

要么 $\|a_i\|=\|b_i\|\leqslant 2$，且 $PW(R,a_i)\cong PW(R,b_i)$.

(ii) $\forall i\neq j\in[1,k]$，有 $(a_i,a_j)\equiv_0(b_i,b_j)$，也就是：要么 $d(a_i,a_j)>1$，且 $d(b_i,b_j)>1$；

要么 $d(a_i,a_j)=d(b_i,b_j)=1$，且 $T(a_i,a_j,1)♣T(b_i,b_j,1)$.

容易证明：

(1)如果 $a_i\equiv a_j$，那么 $b_i\equiv b_j$.

(2)如果 a_iEa_j，那么 b_iEb_j，且 $T(a_i,b_i,1)\cong T(a_j,b_j,1)$.

于是，$\bar{a}_k$ 和 $\bar{b}_k$ 满足相同的原子公式.

引理 2 对任意的 $\bar{a}_k,\bar{b}_k\in A^k$ 满足 $\bar{a}_k E_{n+1,k}\bar{b}_k$，且 $\forall i\in[1,k]$，$\|b_i\|\leqslant m$，任取 $a_{k+1}\in A$，则存在 $b_{k+1}\in A$，使得

$$\bar{a}_{k+1}E_{n,k+1}\bar{b}_{k+1}\text{且}\ \|b_{k+1}\|\leqslant 2^{n+1}+m+1.$$

证 依题设 $\bar{a}_k E_{n+1,k}\bar{b}_k$ 可知

(H_1) $\forall i\in[1,k]$；或者 $\|a_i\|>2^{n+2}$，且 $\|b_i\|>2^{n+2}$；

或者 $\|a_i\|=\|b_i\|\leqslant 2^{n+2}$，且 $PW(R,a_i)\cong PW(R,b_i)$

(H_2) $\forall i;j\in[1,k]$，或者 $d(a_i,a_j)>2^{n+1}$，且 $d(b_i,b_j)>2^{n+1}$；

或者 $d(a_i,a_j)=d(b_i,b_j)\leqslant 2^{n+1}$，且 $T(a_i,a_j,2^{n+1})♣T(b_i,b_j,2^{n+1})$.

对任意给定的 a_{k+1}，我们将找一个 b_{k+1} 满足 $\bar{a}_{k+1}E_{n,k+1}\bar{b}_{k+1}$，也就是，$(H_1)$和$(H_2)$成立，同时下面结论也成立.

(C_1) $\|a_{k+1}\|>2^{n+1}$，且 $\|b_{k+1}\|>2^{n+1}$；

或者 $\|a_{k+1}\|=\|b_{k+1}\|\leqslant 2^{n+1}$，且 $PW(R,a_{k+1})\cong PW(R,b_{k+1})$.

(C_2) $\forall i\in[1,k]$，或者 $d(a_i,a_{k+1})>2^n$，且 $d(b_i,b_{k+1})>2^n$；

或者 $d(a_i,a_{k+1})=d(b_i,b_{k+1})\leqslant 2^n$，且 $T(a_i,a_{k+1},2^n)♣T(b_i,b_{k+1},2^n)$.

分三种情况讨论：

情形 1 $\|a_{k+1}\|\leqslant 2^{n+1}$.

这时，只需取 b_{k+1}，使得 $\|a_{k+1}\|=\|b_{k+1}\|\leqslant 2^{n+1}$，且 R 是 a_{k+1} 与 b_{k+1} 的对称中心. 显然

$$\| a_{k+1} \| = \| b_{k+1} \| \leqslant 2^{n+1} \text{且} PW(R, a_{k+1}) \cong PW(R, b_{k+1}),$$

故条件(C_1)成立.

下面证明条件(C_2)成立,设 $i \in [1, k]$.

子情形 1.1 $\| a_i \| \leqslant 2^{n+2}$.

由条件(H_1)知 $\| a_i \| = \| b_i \|$,且 $PW(R, a_i) \cong PW(R, b_i)$,于是($C_2$)是成立的.

子情形 1.2 $\| a_i \| > 2^{n+2}$.

由条件(H_1)知 $\| b_i \| > 2^{n+2}$.

$$d(a_i, a_{k+1}) \geqslant d(R, a_i) - d(R, a_{k+1}) \geqslant 2^{n+2} - 2^{n+1} > 2^n,$$

并且

$$d(b_i, b_{k+1}) \geqslant d(R, b_i) - d(R, b_{k+1}) \geqslant 2^{n+2} - 2^{n+1} > 2^n$$

于是条件(C_2)也满足.

情形 2 $\| a_{k+1} \| > 2^{n+1}$且存在 $h \in [1, k]$,使得 $a_{k+1} \in T(a_h, 2^n)$,这就是 $d(a_h, a_{k+1}) \leqslant 2^n$. 显然,存在同构映射 ρ,使得 $T(a_h, 2^n) \cong T(b_h, 2^n)$. 这时,可以找到 $b_{k+1} \in T(b_h, 2^n)$满足 $b_{k+1} \equiv \rho(a_{k+1})$.

下面用反证法证明 $\| b_{k+1} \| > 2^{n+1}$.

假设 $\| b_{k+1} \| \leqslant 2^{n+1}$,那么

$$\| b_h \| \leqslant \| b_{k+1} \| + d(b_{k+1}, b_h) \leqslant 2^{n+1} + 2^n \leqslant 2^{n+2}.$$

由条件(H_1)知, $\| a_h \| = \| b_h \|$,又 $T(a_h, 2^n) \cong T(b_h, 2^n)$,从而 $\| a_{k+1} \| = \| b_{k+1} \|$. 这就产生了矛盾. 于是,($C_1$)是成立的.

对于条件(C_2),设 $i \in [1, k]$,若 $i = h$,

$$d(a_h, a_{k+1}) = d(b_h, b_{k+1}) \leqslant 2^n \text{ 且 } T(a_h, a_{k+1}, 2^n) \clubsuit T(b_h, b_{k+1}, 2^n).$$

若 $i \neq h$,考虑下面两种情况:

子情形 2.1 $d(a_i, a_h) > 2^{n+1}$.

由条件(H_2)知 $d(b_i, b_h) > 2^{n+1}$,

$$d(a_i, a_{k+1}) \geqslant d(a_h, a_i) - d(a_h, a_{k+1}) > 2^{n+1} - 2^n > 2^n.$$

同理 $d(b_i, b_{k+1}) > 2^n$.

子情形 2.2 $d(a_i, a_h) \leqslant 2^{n+1}$.

由条件(H_2)知 $d(b_i, b_h) = d(a_i, a_h) \leqslant 2^{n+1}$. 于是有

或者 $d(a_i, a_{k+1}) = d(b_i, b_{k+1}) > 2^n$;

或者 $d(a_i, a_{k+1}) = d(b_i, b_{k+1}) \leqslant 2^n$ 且 $T(a_i, a_{k+1, 2^n}) \clubsuit T(b_i, b_{k+1}, 2^n)$.

条件(C_2)是满足的.

情形 3 $\|a_{k+1}\|>2^{n+1}$,且 $\forall i\in[1,k]$,$d(a_i,a_{k+1})>2^n$.

设 $b_i\in\{b_i\}$满足条件 $\|b_j\|\geqslant\{\|b_i\|\}$.

选择 b_{k+1},使得 $\|b_{k+1}\|=\|b_j\|+2^{n+1}+1\leqslant 2^{n+1}+m+1$,从而得到 $d(b_i,b_{k+1})\geqslant\|b_{k+1}\|-\|b_i\|=2^{n+1}+1+\|b_j\|-\|b_i\|\geqslant 2^n+1>2^n$.

综上所述,可以给出 $\|b_{k+1}\|$ 的上界:

情形 1 $\|b_{k+1}\|\leqslant 2^{n+1}$.

情形 2 $\|b_{k+1}\|\leqslant\|b_h\|+d(b_h,b_{k+1})\leqslant m+2^n$.

情形 3 $\|b_{k+1}\|\leqslant 2^{n+1}+m+1$.

由以上三种情况,可置 $\|b_{k+1}\|\leqslant 2^{n+1}+m+1$.

§2.3 完全二叉树理论的时间复杂度上界

下面运用定理 2 来给出完全二叉树理论的时间复杂度上界:

引理 3 设 $m_0=1$,且 $m_{k+1}=H(n-k,k,m_k)$,$k=0,1,\cdots,n-1$,那么有

$$m_k\leqslant 2^{n+2}+k+2,n\geqslant 2.$$

证 不妨取控制函数 $H(n,k,m)=2^{n+1}+m+1$.

对 k 用数学归纳法,容易证明

$$m_k=2^{n-k+2}(2^k-1)+k+2\leqslant 2^{n+2}+k+2.$$

引理 4 完全二叉树理论中,判定输入长度为 n 的句子 ϕ,不超过 $2^{2^{n+\log(5n)}}$ 步.

证 由引理 1 和引理 2 知,定理 1 与定理 2 的题设条件是满足的,为了判定句子 ϕ,须检测 $\prod_{k=0}^{n-1}F(m_k)$ 无量词公式,这里 $m_0=1$ 且

$$m_{k+1}=H(n-k,k,m_k),k=0,1,\cdots,n-1.$$

由完全二叉树的定义知集合$\{a\in A:\|a\|\leqslant m_k\}$的元素个数是

$$F(m_k)=2^{m_k+1}-1\leqslant 2^{m_k+1}.$$

因此,检测步的总数不超过:

$$\prod_{k=0}^{n-1}(2^{m_k+1})\leqslant\prod_{k=0}^{n-1}2^{(2^{n+2}+k+2)}=2^{\sum_{k=0}^{n-1}(2^{n+2}+k+2)}=2^{(n2^{n+2}+\frac{n(n-1)}{2}+2n)}\leqslant$$

$$2^{(n2^{n+2}+n^2)}\leqslant 2^{5n2^n}=2^{2^{(n+\log(5n))}}.$$

于是有如下定理:

定理 3 完全二叉树理论中,判定输入长度为 n 的句子 ϕ,其时间复杂度为 $2^{2^{cn}}$(c 为合适的常数).

§2.4 完全二叉树理论的空间复杂度上界

空间复杂度主要考虑 Turing 机计算过程中，工作带所消耗的单元数量，它是所需储存的个体数的 log 函数(以 2 为底的对数).

用数学归纳法易证如下引理：

引理 5 在完全二叉树模型中，设句子$(Q_1x_1)(Q_2x_2)\cdots(Q_nx_n)\psi(x_1,x_2,\cdots,x_n)$是前束范式，$k\in[1,n]$，$Q_k\in\{\forall,\exists\}$，$\psi(x_1,x_2,\cdots,x_n)$是无量词句子. 令 $m_k=2^{n+2}+k+2$，那么

$$\mathfrak{U}=\langle A,E,R\rangle\models(Q_1x_1)(Q_2x_2)\cdots(Q_nx_n)\psi(x_1,x_2,\cdots,x_n)$$

当且仅当

$$\mathfrak{U}=\langle A,E,R\rangle\models(Q_1x_1\leq m_1)(Q_2x_2\leq m_2)\cdots(Q_nx_n\leq m_n)\psi(x_1,x_2,\cdots,x_n).$$

这里 $Q_kx_k\leq m_k$，$k\in[1,n]$表示第 k 个量词所辖的变元 x_k 满足

$$\|x_k\|\leqslant m_k.$$

该引理表明，前束范式中量词数量确定了计算复杂度.

由于语句的长度为 n 时，它至多含有 n 个量词，我们不妨设句子 ϕ 的量词深度恰好为 n.

由引理 5 知，在判定一个句子 ϕ 时，只需考虑范数(即树的高度)不超过 $2^{n+2}+n+2$ 的完全二叉树. 用一个邻接矩阵表示它，那么存储这棵树所需的个体不超过 $2^{(2^{n+2}+n+2)}\leqslant 2^{5(2^n)}$.

在判定句子的过程中，其他过程并不改变空间的指数级. 对 n 个量词，其涉及的个体不超过

$$(2^{5(2^n)})^n=2^{5n(2^n)}=2^{2^{n+\log(5n)}},$$

所以所需储存空间不超过 $2^{n+\log(5n)}$.

故有如下定理：

定理 4 对完全二叉树理论，判定量词深度为 n 的句子 ϕ 的空间复杂度为 2^{dn}(d 为合适常数).

§3. 结论与几个问题

本文运用有界 Ehrenfeucht-Fraïssé 博弈给出了判定完全二叉树理论的时间和空间复杂度上界. 这自然会产生如下没有解决的问题：

(1)如何得到复杂度的上确界?

(2)如何得到复杂度的下界?

致谢 第一作者感谢导师李祥教授以及北京师范大学罗里波教授的指导与讨论.

Abstract The first-order theory of a complete binary tree is decidable by the quantifier elimination, we also know the *CB* rank of elements of a complete binary tree. In this paper, by using the bounded Ehrenfeucht-Fraïssé game, we demonstrate that the first order theory of a complete binary tree can be decided within linear double exponential Turing time $2^{2^{cn}}$ and Turing space 2^{dn} (n is the length of input, c and d are suitable constants).

Keywords first-order theory of a complete binary tree, bounded Ehrenfeucht-Fraïssé game; upper bounds for computational complexity.

参考文献

[1] Vorobyov S. On the bounded theories of finite trees. In Asian'96 Computing Science Conference, Lect. Notes Comput. Sci., 1996.

[2] Liu J. Q, Liao D S, Luo L B. Quantifier elimination for complete binary trees. Acta Mathematica Sinica, Chinese Series, 2003, 46(1): 95—102.

[3] Chen L, Shen F X. The *CB* rank of the elements in complete binary trees. Acta Mathematica Sinica, Chinese Series, 2005, 48(2): 245—250.

[4] Ehrenfeucht A. An application of games to the completeness problem for formalized theories. Fund. Math., 1961, 49: 129—141.

[5] Ferrante J, Rackoff C. The computational complexity of Logical theories. Lecture Notes in Mathematics, Vol 718, 1979.

[6] Luo L B. On the computational complexity of the theory of abelian groups. Annals of Pure and Applied Logic, 1988, 37: 205—248.

《上帝创造数》，湖南科技出版社

可计算实数及其在判定问题上的应用①

On Computable Numbers, with an Application to the "Entscheidungs Problem"

"可计算"数可以简单地描述为它的小数表示方式可以用有限的手段来计算的实数. 虽然表面上这篇文章主要是讨论可计算数，几乎是同样容易地可以定义和讨论整数变元或者是实数或者是可计算变元等的可计算谓词和可计算函数. 无论如何在每一种情形所涉及的基本问题是相同的，而我之所以选择可计算数来做详尽的叙述那是因为这样会牵涉最少的烦琐的技巧. 我希望能很快地给出可计算数、函数以及其他的概念之间的关系的说明. 这将包含展开用可计算数来表示的一个变元等的实数函数理论. 根据我的定义，一个数是可计算的，如果它的小数可以用一个机器写出来.

在 § 9, § 10 我给出一些论断是想用来说明可计算数包含了所有通常被认为是可计算的数. 特别地，我指出一些很大的数类是可计算的. 它们包含例如所有的代数数中的实数，贝塞尔函数的零点中的实数，以及数 π, e 等. 但是可计算数并不包含所有的可以定义的数，并且给出了一个可以定义但是不可以计算的数的例子.

虽然可计算数类是如此的庞大并且在很多方面和实数类相似，但是它是可数的. 我在 § 8 考察某些论断，它们好像是在证明相反的结论. 经过正确地运用这些论断中的一个，神奇地得到了和哥德尔** 类似的结论，这些结

---

① 原作者 艾·图灵

[发表在 Proceedings of the London Mathematical Society, 42(2), 1936.]

中译者 罗里波

Gödel. Uber formal unentscheidbare Sätzc der Principia Mathematica und ver-wandter Systeme. *Monashefte Math Phys*, 1931, 38: 173—198.

果有有价值的应用,特别地证明了(§11)希尔伯特的判定问题无解.

在近来的一篇文章中丘奇*引入了“有效可计算”的想法,它是和我的可计算性等价的,但是定义的方式有很大的不同.丘奇也得到了关于判定问题的类似结论.**在这篇文章的一个附录中给出了“可计算性”和“有效可计算性”之间等价的证明框架.

§1. 计算的机器

我们说过可计算数是它们的小数表示可以用有限的手段来计算的数.这一点要更加精确地定义.在我们到达§9之前我们还不对这个定义的合理性作实质上的判断.现在我们只能说这个合理性的判断有赖于人类的记忆库应该是有限制的.

我们可以将一个人在计算一个实数的过程和一个机器相比较,后者只能有有限数个条件 $q_1, q_2, \cdots, q_r$,它们叫作“m-配置”.机器提供一条带子(类似于纸)从机器中运行穿过,并且划分为节(叫作“方格”),每一个方格可以载入一个“记号”.在每一瞬间恰好有一个方格,比如说第 r 个方格,载入有符号 $\mathfrak{S}(r)$是“在机器之内”.我们可以把这个方格叫作是“被扫描的方格”.在被扫描的方格里的记号可以叫作是“被扫描的记号”.可以这样说“被扫描的记号”是唯一的机器“直接认知”的记号.当然利用变换 m-配置可以让机器有效地记住一些机器在这之前曾经“见”过的记号.在每一瞬间机器可能的行为取决于它的 m-配置 q_n 和被扫描的记号 $\mathfrak{S}(r)$.由 $q_n, \mathfrak{S}(r)$所组成的对就叫作配置:因此配置决定机器可能的行为.在某些配置之下,被扫描的记号是空白时(也就是说方格里没有记号时),机器在空白的方格里写下一个新的记号,在另外的配置之下机器擦去被扫描的记号.机器可以变换被扫描的方格,但是只能向左边移动一格或者向右边移动一格.除了这些动作之外可以更换 m-配置.所写下的记号中有一些是来自所计算的实数的小数数字系列.另一些仅仅是“帮助记忆”的草稿.只有这些草稿将要被消掉.

---

* Church A. An unsolvable problem of elementary number theory. *American J. of Math*, 1936, 58: 345—363.

** Church A. A note on the Entscheidungsproblem. *J of Symbolic Logic*, 1936, 1(1): 40—41.

我的论点是这些操作涵盖了用于计算一个数的全部操作. 当读者熟悉了这个机器的理论之后再来说明这个论点会比较容易一些. 所以我在下一节进行理论的展开并且认为大家已经懂得了“机器”“带子”“扫描”等的含义.

§2. 定义

自动机

如果在每一个阶段机器的动作(在§1的意义之下)完全由配置所确定,我们将这种机器叫作“自动机”(或者 a-机器).

为了某些目的我们可能使用机器(选择机或者 c-机器),它的动作只是部分的被配置所决定(所以在§1我们用“可能的行为”的说法). 当这样的一个机器走到一个这种含混的配置时它必须等到一个外在的运算子作出任意的选择才能继续运转下去. 这就是我们使用机器来处理公理化系统的情形. 在这篇文章里我只处理自动机,所以我们常常省略前缀 a-.

计算的机器

如果一个 a-机器打印两类记号,其中第一类(叫作数字)整个地由0和1所组成(其他的符号叫作第二类符号),那么这种机器叫作计算的机器. 如果机器备有一条空白的带子并且从正确的 m-初始配置启动,它所打印出的子序列是第一类的记号叫作是机器所计算的序列. 在这个序列的前面放上一个小数点所得到的二进位序列所代表的实数叫作是机器所计算的数.

在机器运转的任何阶段,扫描方格的数字,带子上所有记号的完全序列,以及 m-配置说是描述了这一阶段的完整的配置. 机器在相邻的两个完整的配置之间的变化叫作是机器的一个动作.

循环的和无循环的机器

如果一个计算的机器写下的第一类记号的数目永不超过一个固定的有限数,它就叫作循环的机器. 不是这样的机器就叫作无循环的机器.

如果一个机器走进了一个不可能继续运行的配置,或者它继续运行但是不再打印更多的第一类记号,而是打印第二类记号,它就是循环的机器. 词语“循环”的重要性将在§8加以解释.

可计算的序列和数

一个序列叫作是可以计算的,如果它可以被一个无循环的机器来计算.

一个数是可计算的,如果它和一个由无循环机器计算出来的数相差一个整数.

为了避免混淆我们使用可计算序列多于可计算数.

§3. 计算的机器的例子

Ⅰ. 可以构造一个机器计算序列 010101…,机器有 4 个 m-配置"b""c""z""e"并且可以打印"0"和"1". 机器的运行状况是用下面的表格来描述,其中"R"表示机器要扫描紧挨着已经扫描过的方格右边的一个方格. "L"的解释类似. "E"是说要擦掉被扫描的记号而"P"表示"打印". 这个表格(以及随之而来的同样类型的表格)理解为一个在表格的第一、第二列所描述的配置机器的运转是依次做表格的第三列上的动作,然后机器进入表格的最后一列所描述的 m-配置. 当在第二列留下空白时,这就理解为第三列和第四列可以是任何记号或者没有记号. 机器从 m-配置 b 开始带子是空白的.

配 置		动 作	
m-配置	记号	运转	最后 m-配置
b	空白	$P0,R$	c
c	空白	R	e
e	空白	$P1,R$	k
k	空白	R	b

如果(和§1 的描述相反)我们允许字母 L 和 R 在运转列出现一次以上,那么我们就可以大大地简化表格.

配 置		动 作	
m-配置	记号	运转	最后 m-配置
b	空白	$P0$	b
b	0	$R,R,P1$	b
b	1	$R,R,P0$	b

Ⅱ. 作为一个稍微困难一点的例子,我们可以构造一个机器用来计算序列 001011011101111011111…机器设有 5 个 m-配置"o""q""p""f""b"和打印"e""x""0""1"的能力. 带子上的前三个记号是"$e\ e$ 0";其他的数字放在跟随着的交替的方格. 紧随的方格我们只打印"x",其他什么也不打

印. 这些字母是用来替我们"保持位置"并且在它们完成任务之后被擦掉. 我们也安排数字的序列在相间的方格中使得没有空格.

配　置		动　作	
m-配置	记号	运转	最后 m-配置
b		$Pe, R, Pe, R, P0, R, R, P0, L, L$	o
o	0	R, Px, L, L, L	o
o	1		q
q	0 或者 1	R, R	q
q	空白	$P1, L$	p
p	x	E, R	q
p	e	R	f
p	空白	L, L	p
f	任意	R, R	f
f	空白	$P0, L, L$	o

为了说明这个机器的工作方式给出下面的表格用以表示开头几个完整的配置. 这些完整的配置是用写下它们在纸带上的记号序列来呈现的,而它们的 m-配置就写在它们所扫描的记号下面. 紧挨着的配置用冒号分隔开来.

```
 :ee0  0:ee0  0:ee0  0:ee0  0  :ee0  01
b   o      q      q      q        p
ee0  0  1:ee0  0  1:ee0  0  1:ee0  0  1:
   p      p        f               f
ee0  0  1:ee0  0  1   :ee0  0  1  0:
      f           f        o
ee0  0  1x0:…
     o
```

这个表格也可以用下面的形式写出来

$$b{:}ee\,o\,0\,0{:}ee\,q\,0\,0\,{:}\cdots \quad ①$$

其中被扫描的符号的左边让出一个空间用以写入 m-配置. 这种形式是不太容易理解,但是后面我们要用它来说明理论上的问题.

规定数字是隔一相间的方式来写是很有用的:我总是用这个办法. 这

个由隔一相间的方格所组成的序列，我把它叫作 F-方格，而其他的序列叫作 E-方格. E-方格上的记号是准备要消去的. 在 F-方格上的记号形成一个连续的序列. 在没有到达终点以前是没有空格的. 每对相邻的 F-方格之间不需要有一个以上的 E-方格. 表面上的需要一个以上的 E-方格可以用足够多品种的记号用来印入 E-方格. 如果一个记号 β 是在一个 F-方格 S 内，而一个记号 α 是在与 S 右侧相邻的方格内，那么 S 和 β 就说是做上了记号 α. 打印这个 α 的过程就叫作将 S(或者 β)做上记号 α.

§4. 简化了的表格

有一些类型的过程几乎在所有的机器中都会用到，并且这些过程在某些机器中会以多种不同的方式来应用. 这些过程包括复制记号的序列，比较序列，消去全部给定的某种形式的记号等. 考虑到这种过程我们可以用“骨架表”的办法来极大地简化表格中的 m-配置. 在骨架表中出现有大写的德文字母和小写的希腊字母. 用一个 m-配置来代替表格中的大写德文字母，再用一个符号来代替表格中的小写希腊字母，就可以得到一个 m-配置的表格.

骨架表仅仅是理解为简记号：它们没有什么本质上的问题. 只要读者知道如何将一个骨架表翻译成原来的表就不需要给出对这种联系的额外定义. 让我们考虑下面的例子：

m-配置	记号	动作	最后 m-配置
$\mathfrak{f}(\mathfrak{E},\mathfrak{B},\alpha)$	ə	L	$\mathfrak{f}_1(\mathfrak{E},\mathfrak{B},\alpha)$
	其他	L	$\mathfrak{f}(\mathfrak{E},\mathfrak{B},\alpha)$
$\mathfrak{f}_1(\mathfrak{E},\mathfrak{B},\alpha)$	α		$\mathfrak{E}$
	非 α	R	$\mathfrak{f}_1(\mathfrak{E},\mathfrak{B},\alpha)$
	空格	R	$\mathfrak{f}_2(\mathfrak{E},\mathfrak{B},\alpha)$.
$\mathfrak{f}_2(\mathfrak{E},\mathfrak{B},\alpha)$	α		$\mathfrak{E}$
	非 α	R	$\mathfrak{f}_1(\mathfrak{E},\mathfrak{B},\alpha)$
	空格	R	$\mathfrak{B}$.

从 m-配置 $\mathfrak{f}(\mathfrak{E},\mathfrak{B},\alpha)$机器寻找最左边的 α 形式的记号，找到后机器进入 m-配置 $\mathfrak{E}$(第一个 α). 如果找不到这样的一个记号，机器进入 m-配置 $\mathfrak{B}$.

如果(譬如说)我们是想将全部的 $\mathfrak{E}$ 替换成 $\mathfrak{q}$，$\mathfrak{B}$ 替换成 $\mathfrak{r}$，α 替换成

x，我们就会有一个关于 m-配置 $\mathfrak{f}(\mathfrak{q},\mathfrak{r},x)$ 的完整的表格. $\mathfrak{f}$ 叫作是"m-配置函数"或者"m-函数".

在一个 m-函数的表达式中，只有那些 m-配置和机器的符号是允许替换的. 那些表达式需要枚举得或多或少清晰一些：它们可能含有像 $\mathfrak{p}(\mathfrak{e},x)$ 这样的表达式；如果有任何的 m-函数可以使用就应该有这种表达式. 如果我们不坚持这样清晰的枚举，而简单地认为机器有某些(枚举出来的)m-配置，并且所有从某些 m-函数中替换 m-配置得来的 m-配置，我们通常会得到无限多个 m-配置；例如我们可能说机器想要得到 m-配置 $\mathfrak{q}$，并且所有 m-配置是由在 $\mathfrak{p}(\mathfrak{C})$ 中用 m-配置替换 $\mathfrak{C}$ 而得到. 那么就有 m-配置 $\mathfrak{q},\mathfrak{p}(\mathfrak{q}),\mathfrak{p}(\mathfrak{p}(\mathfrak{q}))\mathfrak{p}(\mathfrak{p}(\mathfrak{p}(\mathfrak{q})))\cdots$

因此我们对规则做如下的解释. 我们给出机器的 m-配置的名字，大部分使用 m-函数来表示的. 我们也给出一些骨架表. 所有我们需要的就是机器的完整的 m-配置表. 这就可以在骨架表中反复替换而得到.

更多的例子

("→"记号的解释是用来表示"机器进入到…"m-配置.)

$\mathfrak{e}(\mathfrak{C},\mathfrak{B},\alpha)$		$\mathfrak{f}(\mathfrak{e}_1(\mathfrak{C},\mathfrak{B},\alpha),\mathfrak{B},\alpha)$
$\mathfrak{e}_1(\mathfrak{C},\mathfrak{B},\alpha)$	E	$\mathfrak{C}$

从 $\mathfrak{e}(\mathfrak{C},\mathfrak{B},\alpha)$ 消去第一个 α 并且→$\mathfrak{C}$，如果没有这样的 α，→$\mathfrak{B}$.

$\mathfrak{e}(\mathfrak{B},\alpha)$	$\mathfrak{e}(\mathfrak{e}(\mathfrak{B},\alpha),\mathfrak{B},\alpha)$

从 $\mathfrak{e}(\mathfrak{B},\alpha)$ 所有 α 都消去并且→$\mathfrak{B}$.

最后一个例子好像是比起大部分的例子来多少有点难于解释，让我们假设在某个机器的 m-配置清单中出现有 $\mathfrak{e}(\mathfrak{b},x)$(譬如说$=\mathfrak{q}$). 表格是

$\mathfrak{e}(\mathfrak{b},x)$	$\mathfrak{e}(\mathfrak{e}(\mathfrak{b},\mathfrak{x}),\mathfrak{b},x)$

或者

$\mathfrak{q}$	$\mathfrak{e}(\mathfrak{q},\mathfrak{b},x)$

或者，更详细一些：

$\mathfrak{q}$		$\mathfrak{e}(\mathfrak{q},\mathfrak{b},x)$
$\mathfrak{e}(\mathfrak{q},\mathfrak{b},x)$		$\mathfrak{f}(\mathfrak{e}_1(\mathfrak{q},\mathfrak{b},x),\mathfrak{b},x)$
$\mathfrak{e}_1(\mathfrak{q},\mathfrak{b},x)$	E	$\mathfrak{q}$

在这里我们可以替换 $\mathfrak{e}_1(\mathfrak{q},\mathfrak{b},x)$ 以 $\mathfrak{q}'$，然后给出 $\mathfrak{f}$ 的表(使用右替换)，最后得到一个不出现 m-函数的表.

$\mathfrak{pe}(\mathfrak{C},\beta)$		$\mathfrak{f}(\mathfrak{pe}_1(\mathfrak{C},\beta),\mathfrak{C},ə)$
$\mathfrak{pe}_1(\mathfrak{C},\beta)$ { 任意	R,R	$\mathfrak{pe}_1(\mathfrak{C},\beta)$}
空白	$P\beta$	$\mathfrak{C}$
$\mathfrak{l}(\mathfrak{C})$	L	$\mathfrak{C}$
$\mathfrak{r}(\mathfrak{C})$	R	$\mathfrak{C}$

从 $\mathfrak{pe}_1(\mathfrak{C},\beta)$ 在符号序列的末尾机器打印 β 并且→$\mathfrak{C}$.

从 $\mathfrak{f}'(\mathfrak{C},\mathfrak{B},\alpha)$ 它对 $\mathfrak{f}(\mathfrak{C},\mathfrak{B},\alpha)$ 做同样的事情，但是在→$\mathfrak{C}$ 之前向左走.

$\mathfrak{f}'(\mathfrak{C},\mathfrak{B},\alpha)$		$\mathfrak{f}(\mathfrak{l}(\mathfrak{C}),\mathfrak{B},\alpha)$
$\mathfrak{f}''(\mathfrak{C},\mathfrak{B},\alpha)$		$\mathfrak{f}(\mathfrak{r}(\mathfrak{C}),\mathfrak{B},\alpha)$
$\mathfrak{c}(\mathfrak{C},\mathfrak{B},\alpha)$		$\mathfrak{f}'(\mathfrak{c}_1(\mathfrak{C}),\mathfrak{B},\alpha)$
$\mathfrak{c}_1(\mathfrak{C})$	β	$\mathfrak{pe}(\mathfrak{C},\mathfrak{B})$

$\mathfrak{c}(\mathfrak{C},\mathfrak{B},\alpha)$ 机器将标记着 α 的第一个符号写在末尾，然后→$\mathfrak{C}$.

最后一行是代表所有可能的行，它们是由相关的机器的带子上可能出现的符号代入 β 而得到的.

$\mathfrak{ce}(\mathfrak{C},\mathfrak{B},\alpha)$	$\mathfrak{c}(\mathfrak{e}(\mathfrak{C},\mathfrak{B},\alpha),\mathfrak{B},\alpha)$
$\mathfrak{ce}(\mathfrak{B},\alpha)$	$\mathfrak{ce}(\mathfrak{ce}(\mathfrak{B},\alpha),\mathfrak{B},\alpha)$

$\mathfrak{ce}(\mathfrak{B},\alpha)$ 机器按顺序在所有标号为 α 的记号末尾复制，擦掉记号 α，然后→$\mathfrak{B}$.

$\mathfrak{re}(\mathfrak{C},\mathfrak{B},\alpha,\beta)$	$\mathfrak{f}(\mathfrak{re}_1(\mathfrak{C},\mathfrak{B},\alpha,\beta),\mathfrak{B},\alpha)$
$\mathfrak{re}_1(\mathfrak{C},\mathfrak{B},\alpha,\beta)E,P\beta$	$\mathfrak{C}$

$\mathfrak{re}(\mathfrak{C},\mathfrak{B},\alpha,\beta)$ 机器将第一个 α 换成 β 并且→$\mathfrak{C}$，如果没有 α，→$\mathfrak{B}$.

$\mathfrak{re}(\mathfrak{B},\alpha,\beta)$	$\mathfrak{re}(\mathfrak{re}(\mathfrak{B},\alpha,\beta)\mathfrak{B},\alpha,\beta)$

$\mathfrak{re}(\mathfrak{B},\alpha,\beta)$ 机器将所有 α 换成 β：→$\mathfrak{B}$.

$\mathfrak{cr}(\mathfrak{C},\mathfrak{B},\alpha)$	$\mathfrak{c}(\mathfrak{re}(\mathfrak{C},\mathfrak{B},\alpha,a)\mathfrak{B},\alpha)$
$\mathfrak{cr}(\mathfrak{B},\alpha)$	$\mathfrak{cr}(\mathfrak{cr}(\mathfrak{B},\alpha),\mathfrak{re}(\mathfrak{B},a,\alpha),\alpha)$

$\mathfrak{cr}(\mathfrak{B},\alpha)$ 与 $\mathfrak{ce}(\mathfrak{B},\alpha)$ 不同仅在于不消去字母 α. 当字母"a"在带子上不出现时，m-匹配 $\mathfrak{cr}(\mathfrak{B},\alpha)$ 才会用到.

$\mathfrak{cp}(\mathfrak{C},\mathfrak{A},\mathfrak{E},\alpha,\beta)$		$\mathfrak{f}(\mathfrak{cp}_1(\mathfrak{E}_1,\mathfrak{A},\beta),\mathfrak{f}(\mathfrak{A},\mathfrak{E},\beta),\alpha)$
$\mathfrak{cp}_1(\mathfrak{C},\mathfrak{A},\beta)$	γ	$\mathfrak{f}(\mathfrak{cp}_2(\mathfrak{C},\mathfrak{A},\gamma),\mathfrak{A},\beta)$
$\mathfrak{cp}_2(\mathfrak{C},\mathfrak{A},\gamma)$	{ γ	$\mathfrak{C}$
	非 γ	$\mathfrak{A}$

第一个标记为 α 的记号和第一个标记为 β 的记号进行比较. 如果 α

和 β 都不存在，→$\mathfrak{E}$ 如果两个符号相同，→$\mathfrak{C}$，余下的情况，→$\mathfrak{A}$.

$\mathfrak{cpe}(\mathfrak{C},\mathfrak{A},\mathfrak{E},\alpha,\beta)$	$\mathfrak{cp}(\mathfrak{e}(\mathfrak{e}(\mathfrak{C},\mathfrak{C},\beta),\mathfrak{C},\alpha),\mathfrak{A},\mathfrak{E},\alpha,\beta)$

$\mathfrak{cpe}(\mathfrak{C},\mathfrak{A},\mathfrak{E},\alpha,\beta)$ 与 $\mathfrak{cp}(\mathfrak{C},\mathfrak{A},\mathfrak{E},\alpha,\beta)$ 不同，是在如果查出是相同时，第一个标记为 α 的记号和第一个标记为 β 的记号都要消掉.

$\mathfrak{cpe}(\mathfrak{A},\mathfrak{E},\alpha,\beta)$	$\mathfrak{cpe}(\mathfrak{cpe}(\mathfrak{A},\mathfrak{E},\alpha,\beta),\mathfrak{A},\mathfrak{E},\alpha,\beta)$

$\mathfrak{cpe}(\mathfrak{A},\mathfrak{E},\alpha,\beta)$. 标记为 α 的记号序列和标记为 β 的记号序列进行比较，如果它们相似，→$\mathfrak{E}$. 否则→$\mathfrak{A}$，有些标记为 α,β 的记号要消去.

$\mathfrak{q}(\mathfrak{C})$	任意	R	$\mathfrak{q}(\mathfrak{C})$
	空白	R	$\mathfrak{q}_1(\mathfrak{C})$
$\mathfrak{q}_1(\mathfrak{C})$	任意	R	$\mathfrak{q}(\mathfrak{C})$
	空白		$\mathfrak{C}$
$\mathfrak{q}(\mathfrak{C},\alpha)$			$\mathfrak{q}(\mathfrak{q}_1(\mathfrak{C},\alpha))$
$\mathfrak{q}_1(\mathfrak{C},\alpha)$	α	L	$\mathfrak{C}$
	非 α		$\mathfrak{q}_1(\mathfrak{C},\alpha)$

机器找到最后的形为 α 的记号，然后→$\mathfrak{C}$

$\mathfrak{pe}_2(\mathfrak{C},\alpha,\beta)$	$\mathfrak{pe}(\mathfrak{pe}(\mathfrak{C},\beta),\alpha)$

$\mathfrak{pe}_2(\mathfrak{C},\alpha,\beta)$. 机器在末尾打印 $\alpha\beta$.

$\mathfrak{ce}_2(\mathfrak{B},\alpha,\beta)$	$\mathfrak{ce}(\mathfrak{ce}(\mathfrak{B},\beta),\alpha)$
$\mathfrak{ce}_3(\mathfrak{B},\alpha,\beta,\gamma)$	$\mathfrak{ce}(\mathfrak{ce}_2(\mathfrak{B},\beta,\gamma),\alpha)$

$\mathfrak{ce}_3(\mathfrak{B},\alpha,\beta,\gamma)$ 机器首先复制下标号为 α 的记号于末尾，然后复制标号为 β 的记号，最后再复制标号为 γ 的记号，复制完成之后消去 α,β,γ.

$\mathfrak{e}(\mathfrak{C})$	ə	R	$\mathfrak{e}_1(\mathfrak{C})$
	非 ə	L	$\mathfrak{e}(\mathfrak{C})$
$\mathfrak{e}_1(\mathfrak{C})$	任意	R,E,R	$\mathfrak{e}_1(\mathfrak{C})$
	空白		$\mathfrak{C}$

从 $\mathfrak{e}(\mathfrak{C})$ 所有作为标记的记号都予以消去. →$\mathfrak{C}$.

§5. 可计算序列的枚举

一个可计算序列 γ 是被一个计算 γ 的机器的一个描述所确定. 所以§3 中的序列 001011011101111011111…就由计算它的表格所确定. 事实上任何可计算序列都可以被一个这样的表格所描述.

将这种表格设定一种标准的形式会是很有用的. 首先我们假设这种表格是用同第一个表格相同的一种形式给出的,如前面§3的例子Ⅰ. 这就是说在动作列上的数据总是下面的形式之一 $E{:}E,R{:}E,L{:}Pa{:}\ Pa,R{:}$ $Pa,L{:}R{:}L{:}$或者什么数据也没有. 表格总是可以用引入更多的 m-匹配来作成这种形式. 现在让我们给出数字于 m-匹配,把它们像在§1那样叫作 $q_1,q_2,\cdots,q_R$. 初始的 m-匹配总是叫作 q_1. 我们也给出数字于记号 $S_1,S_2,\cdots,S_m$,并且特别地,空格 $=S_0$, $0=S_1$, $1=S_2$. 表格的各行现在具有下面的形式.

m-匹配	记号	动作	最后 m-匹配	
q_i	S_j	PS_k,L	q_m	(N_1)
q_i	S_j	PS_k,R	q_m	(N_2)
q_i	S_j	PS_k	q_m	(N_3)

像下面的行

q_i	S_j	E,R	q_m

可以写成

q_i	S_j	PS_0,R	q_m

像下面的行

q_i	S_j	R	q_m

可以写成

q_i	S_j	PS_j,R	q_m

这样一来所有原来表中的行都可以写成$(N_1)(N_2)(N_3)$的形式之一.

从每一个(N_1)形式的行,我们形成表达式 $q_iS_jS_kLq_m$,从每一个(N_2)形式的行,我们形成表达式 $q_iS_jS_kRq_m$,从每一个(N_3)形式的行,我们形成表达式 $q_iS_jS_kNq_m$.

让我们写下所有的如上形成的表达式并且用分号';'将它们分隔开来. 用这个办法我们得到关于机器的一个完整的描述. 在这个描述中,我们将 q_i 换成字母 D 跟着 i 个重复的字母 A,将 S_j 换成字母 D 跟着 j 个重复的字母 C. 这个对机器的新的描述可以叫作是标准的描述(S. D). 它完全是由字母"A""C""D""L""R""N"以及符号";"来组成.

如果最后我们将"A"换成"1","C"换成"2","D"换成"3","L"换成"4","R"换成"5","N"换成"6",";"换成"7"我们得到一个完全用阿拉伯数字来组成的关于机器的描述. 这些数字所组成的整数叫作是机器的一个描述数(D. N). D. N 唯一地决定 S. D 以及机器的结构. D. N 是 n 的机器可以描述为 $\mathscr{M}(n)$.

对每一个可计算序列至少对应着一个描述数,然而没有描述数会对应到一个以上的可计算序列. 可计算序列和数就这样是可枚举的了.

让我们找到 §3 机器I的描述数. 当我们重新命名 m-匹配它的表格成了:

q_1	S_0	PS_1,R	q_2
q_2	S_0	PS_0,R	q_3
q_3	S_0	PS_2,R	q_4
q_4	S_0	PS_0,R	q_1

同样的机器也可以加上无关紧要的行而得到另外的表格,譬如说:

q_1	S_1	PS_1,R	q_2

我们的第一个标准形式就是

$$q_1S_0S_1Rq_2;q_2S_0S_0Rq_3;q_3S_0S_2Rq_4;q_4S_0S_0Rq_1;.$$

标准描述是

DADDCRDAA;*DAADDRDAAA*;*DAAADDCCRDAAAA*;*DAAAADDRDA*;

描述数是

31332531173113353111731113322531111731111335317

下面的数也是一个描述数

3133253117311335311173111332253111173111133531731323253117

一个数是一个非循环机器的描述数叫作是符合要求的数. 在 §8 证明了没有一个一般的过程来确定一个给定的数是否符合要求.

§6. 万能计算机

发明一个单个的机器,它可以用来计算任意可计算序列是可能的. 如果为这个机器 $\mathfrak{U}$ 提供一条带子,开始时在它上面写上某一个计算的机器 $\mathscr{M}$ 的 S. D,那么机器 $\mathfrak{U}$ 就会计算和机器 $\mathscr{M}$ 相同的序列. 在这一节里我简略地解释这个机器的动作. 下一节专门来给出机器 $\mathfrak{U}$ 的完整的表格.

让我们首先假设我们有一个机器 $\mathscr{M}'$,它将 $\mathscr{M}$ 的完整的匹配写在相继的 F-方格上. 这些应该是表示成像在与 §3 机器 II 的第二种描述表达(C)相同的形式,把所有符号写成一行. 或者,更好一些,(像在 §5)将这

个描述用更换 m-匹配成“D”接上适当数目的重复的“A”，更换数字成“D”接上适当数目的重复的“C”. 字母“A”和“C”的数目符合 § 5 的选择，特别地使得，“0”替换成“DC”，“1”替换成“DCC”，而空格替换成“D”. 这些替换是在完整的匹配，像在(C)那样放在一起之后. 如果我们先做替换就会产生困难. 在每一个完整的匹配中空格就会被替换成“D”，使得这个完整的匹配就不可能表示成一个有限的符号序列.

如果在 § 3 机器 Ⅱ 的描述中我们把“o”替换成“DAA”，“e”替换成“$DCCC$”，“q”替换成“$DAAA$”，那么序列(C)变成：

$$DA:DCCCDCCCDAADCDDC:DCCCDCCCDAAADCDDC:\cdots \quad ②$$

(这些是在 F-方格上的记号.)

不难看出，如果 $\mathscr{M}$ 能被构造出来，那么 $\mathscr{M}'$ 也能. $\mathscr{M}'$ 的运转方式可以依靠把它的动作(也就是说 $S.D$)写在它自己的某个地方(也就是说在 $\mathscr{M}'$ 内)；每一步可以按照这些规则来执行. 我们只看这些规则是能够被取出或者是交换成别的，并且我们得到一些东西很类似于万能机器.

有一样东西是缺少的：在现在机器 $\mathscr{M}'$ 不打印数字. 我们可以改进这一点，在完整的匹配的每一对相邻的方格中打印出现在新的而不是老的完整匹配中的数字. 那么②变成

$$DDA:0:0:DCCCDCCCDAADCDDC:DCCC\cdots \quad ③$$

不是全部很明显的 E-方格会留下足够的空间用以做“粗糙的工作”，但是事实上是对的.

像在②那样在冒号之间的字母序列可以用来当作完整的匹配的标准描述. 当这些字母像在 § 5 里那样被替换成数字时，我们将会有数字的完整匹配的描述，它可以叫作描述数.

§ 7. 万能机器的详细描述

下面给出关于这个万能机器运转的表格. 机器功能的 m-匹配出现在表格的第一列和最后一列，再加上那些当我们写出 m-函数的非简化表格时所出现的所有 m-匹配. 例如 $\mathfrak{e}(\mathfrak{anf})$ 出现在表格，是一个 m-函数. 它的非简化表格(见 § 5)是

$\mathfrak{e}(\mathfrak{anf})$	e	R	$\mathfrak{e}_1(\mathfrak{anf})$
	非 e	L	$\mathfrak{e}(\mathfrak{anf})$

$\mathfrak{e}_1(\mathfrak{anf})$	任意	R,E,R	$\mathfrak{e}_1(\mathfrak{anf})$
	空白		$\mathfrak{anf}$

由此得出 $\mathfrak{e}_1(\mathfrak{anf})$是 $\mathfrak{A}$ 的一个 m-匹配.

当 $\mathfrak{A}$ 准备好开始工作的时候,运行穿过的纸带在一个 F-方格上带着符号 e 并且也在相邻的 E-方格;在这以后,只在 F-方格,来了机器的 $S.D$ 跟着成对的冒号"∷"(一个单个的记号,在 F-方格). 这 $S.D$ 含有一些用分号分隔开的指令.

每一条指令由五个部分组成

(i)"D"跟着一序列的字母"A". 这描述了有关的 m-匹配.

(ii)"D"跟着一序列的字母"C". 这描述了被扫描的记号.

(iii)"D"跟着另外一个序列的字母"C",这描述了记号里面被扫描的记号是要改变的.

(iv)"L""R""N",描述机器是向左,向右,或者原地不动.

(v)"D"跟着一序列的字母"A". 这描述了最后的 m-匹配.

机器 $\mathfrak{A}$ 是要能够打印"A""C""D""0""1""u""v""w""x""y""z". 它的 $S.D$ 由";""A""C""D""L""R""N"形成.

附属的骨架表

$\mathfrak{con}(\mathfrak{C},\alpha)$	非 A	R,R	$\mathfrak{con}(\mathfrak{C},\alpha)$
	A	$L,P\alpha,R$	$\mathfrak{con}_1(\mathfrak{C},\alpha)$
$\mathfrak{con}_1(\mathfrak{C},\alpha)$	A	$R,P\alpha,R$	$\mathfrak{con}_1(\mathfrak{C},\alpha)$
	D	$R,P\alpha,R$	$\mathfrak{con}_1(\mathfrak{C}_2,\alpha)$

$\mathfrak{con}(\mathfrak{C},\alpha)$从一个 F-方格譬如说 S,开始,把紧靠着 S 右边描述一个匹配的记号序列 C 做上标记 α. $\rightarrow\mathfrak{C}$.

$\mathfrak{con}_2(\mathfrak{C},\alpha)$	C	$R,P\alpha,R$	$\mathfrak{con}_2(\mathfrak{C},\alpha)$
	非 C	R,R	$\mathfrak{C}$

$\mathfrak{con}(\mathfrak{C},\alpha)$. 在最后的匹配中机器所扫描的方格是在 C 的最后方格的右边的第四个方格. C 不做标记.

$\mathfrak{A}$ 的表格

$\mathfrak{b}$		$\mathfrak{f}(\mathfrak{b}_1,\mathfrak{b}_2,∷)$
$\mathfrak{b}_1$	$R,R,P:,R,R,PD;R,R,PA$	$\mathfrak{anf}$

$\mathfrak{b}$. 机器在"∷"之后的方格打印":DA". $\rightarrow\mathfrak{anf}$.

anf	q(anf$_1$, :)
anf$_1$	con(tom, y)

anf. 机器对最后一个完整的匹配中的匹配做上标号 y. →tom

tom	;	R, Pz, L	con(tmp, x)
	z	L, L	tom
	非 z 非;	L	tom

tom 机器找到最后一个没有标记 z 的分号，对这个分号标上 z 并且以匹配 x 跟着它. →tmp.

tmp	cpe(e(tom, x, y), sim, x, y)

tmp 机器比较标号为 x 的序列和标号为 y 的序列. 如果它们相同，将所有字母 x 和 y 消去. →sim. 否则→tom.

anf. 从长远来看，与最后一个匹配相关联的最后一个指令已经找到. 在这以后它可以被认为是跟在最后一个分号后面做有标记 z. →sim.

sim		f′(sim$_1$, sim$_1$, z)
sim$_1$		con(sim$_2$,)
sim$_2$	A	sim$_3$,
	非 A R, Pu, R, R, R	sim$_2$
sim$_3$	非 A L, Py	c(mt, z)
	A L, Py, R, R, R	sim$_3$

sim 机器给指令做标记. 将有关动作的指令中准备执行的部分做上标号 u，以及最后与 y 构成 m-匹配. 字母 z 消去了.

mt			q(mt$_1$, :)
mt$_1$	非 A	R, R	mt$_1$,
	A	L, L, L, L	mt$_2$
mt$_2$	非 C	R, Px, L, L, L	mt$_2$,
	:		mt$_4$,
	D	L, L, L, L	mt$_3$
mt$_3$	非:	R, Pv, L, L, L	mt$_3$,
	:		mt$_4$
mt$_4$			con(l(l(mt$_5$)))

mf₅	{任意 空格	R,Pw,R P:	mf₅, sh

mt. 最后一个完整的匹配分成四段来标记. 匹配不加以标记. 在它前面的记号以 x 标记. 完整的匹配中剩下的部分分为两段. 其中第一段以 v 为标记,第二段以 w 为标记,整个完成之后打印一个冒号. →sh.

sh			f(sh₁,inst,u)
sh₁		L,L,L	sh₂
sh₂	{D 非 D	R,R,R,R	sh₂ sh₃ inst
sh₃	{C 非 C	R,R	sh₄, inst
sh₄	{C 非 C	R,R	sh₅, pe₂(inst,0,:)
sh₅	{C 非 C		inst pe₂(inst,1,:)

sh. 检查指令(标记了 u). 如果找到了这些指令,它们指示打印 0 或者打印 1,那么在末尾打印 0:或者 1:.

inst			q(l(inst₁),u)
inst₁	α	R,E	insh₁(α)
inst₁(L)			ce₅(ov,v,y,x,u,w)
inst₁(R)			ce₅(ov,v,x,u,y,w)
inst₁(N)			ce₅(ov,v,x,y,u,w)
ov			e(anf)

inst. 写下下一个完整的匹配,执行有标号的指令. 字母 v,y,x,u,w 被擦掉. →anf.

§ 8.　对角线过程的应用

有人认为证明实数是不可数的论断* 会同样证明可计算数和序列也是不可数的. 也可能认为可计算数的极限也是可计算的,很明显的这是成

* Cf Hobson E W. Theory of functions of a real variable. 2nd ed. ,1921.

立的只在可计算数的序列是由某些规则来定义的.

或者我们可以应用对角线过程."如果可计算序列是可数的,设 α_n 是第 n 个可计算序列,并且 $\phi_n(m)$ 是 α_n 的第 m 个数字. 设 β 是一个数字序列,它的第 n 个数字是 $1-\phi_n(n)$. 因为 β 是可计算的,存在一个数 K 使得 $1-\phi_n(n)=\phi_K(n)$ 对所有 n 都成立. 令 $n=K$,我们得到 $1=2(\phi_K(K))$. 这样一来 1 是一个偶数. 这是不可能的. 因此可计算序列是不可数的."

这个推断的错误在于假设 β 是可计算的. 它或许会是对的,如果能够用有限的方法来枚举可计算序列,但是问题是枚举可计算序列是等价于发现一个给定的数是否是一个无循环机器的 D. N,而我们并没有一般的过程来在有限多步之内做这件事. 实际上经过正确地运用对角线过程论断,我们可以证明这样的过程是不存在的.

最简单和最直接的证明就是指出,如果这样的过程存在,那么就有一个计算 β 的机器. 这个证明虽然完全没有问题,但是它会使读者感觉到"一定是弄错了". 我将要给出的证明就没有这方面的缺陷,并且还对"无循环"设想的重要性做一些深入的观察. 它与构造 β 没有关系,但是与构造 β' 有关. 它的第 n 个数字是 $\phi_n(n)$.

让我们假设存在一个这样的过程;那就是说,我们可以发明一个机器 $\mathfrak{D}$ 使得,当提供任意计算机器 $\mathscr{M}$ 的 S. D,它就试验这个 S. D,如果 $\mathscr{M}$ 是循环的机器,那么就给这个 S. D 标上记号 u,如果 $\mathscr{M}$ 是无循环的机器,那么就给这个 S. D 标上记号 s. 将机器 $\mathfrak{D}$ 和 $\mathfrak{U}$ 组合起来我们可以构造一个机器 $\mathfrak{H}$ 用以计算序列 β'. 机器 $\mathfrak{H}$ 需要一条带子. 我们可以假设它的 E-方格所使用的记号与 F-方格所使用的记号完全不同,并且当它得到了它的结论时所有机器 $\mathfrak{D}$ 的工作痕迹都要消去.

机器 $\mathfrak{H}$ 的动作分为若干个段落. 在开头的 $N-1$ 个段落,除了其他事情之外,机器写下 $1,2,\cdots,N-1$ 并且利用机器 $\mathfrak{D}$ 试验,其中有譬如说 $R(N-1)$ 个数被发现,无循环机器的 D. N'. 在第 N 个段落机器 $\mathfrak{D}$ 试验数 N. 如果 N 满足要求,也就是说如果它是无循环机器的 D. N,那么令 $R(N)=1+R(N-1)$ 前 $R(N)$ 个数字都是某个机器的 D. H,也就计算出来了. 第 $R(N)$ 个数字(就是 N)可以写下来作为 $\mathfrak{H}$ 所计算的序列 β' 的一个数字. 如果 N 不满足要求,那么 $R(N)=R(N-1)$ 并且机器前进到第 $(N+1)$ 个段落的运作.

从 $\mathfrak{H}$ 的构造过程我们可以看出 $\mathfrak{H}$ 是无循环的. 每一个 $\mathfrak{H}$ 的运作段落都会在有限多步结束. 这是因为由我们关于 $\mathfrak{D}$ 的假设,确定 N 是否满足要求会在有限多步之内完成. 如果 N 不满足要求,那么第 N 个段落已经结束. 如果 N 满足要求,那么机器 $\mathscr{M}(N)$的 $D.N$ 就是 N. $\mathscr{M}(N)$就是无循环的机器,因此它的第 $R(N)$个数字可以在有限多步之内计算完成. 当这个数字计算完成并且写了下来作为序列 β' 的第 $R(N)$个数字,第 N 个段落结束. 所以 $\mathfrak{H}$ 是无循环的.

现在设 K 是 $\mathfrak{H}$ 的 $D.N$. 在第 K 个计算段落机器 $\mathfrak{H}$ 作什么运作呢?他要试验 K 是不是符合要求,给出一个是"s"或者"u"的判决. 由于 K 是 $\mathfrak{H}$ 的 $D.N$ 并且 $\mathfrak{H}$ 是无循环的,判决不可能是"u". 另一方面判决也不可能是"s". 因为如果这样,那么在 $\mathfrak{H}$ 的第 K 个段落的动作中,$\mathfrak{H}$ 被要求计算 $D.N$ 是 K 的机器的开头 $R(K-1)+1=R(K)$个数字,然后写下第 $R(K)$个数字作为 $\mathfrak{H}$ 的计算序列中的一个数字. 计算前 $R(K)-1$ 个数字可以没有什么问题来执行. 但是计算第 $R(K)$个数字的指令中却有"计算 H 所计算的前 $R(K)$个数字并且写下第 $R(K)$个数字"的条文. 所以第$R(K)$个数字将永远找不出来. 也就是说 $\mathfrak{H}$ 是循环的机器,矛盾于我们在上一节的发现,也矛盾于"s"的判决. 所以两种判决都不可能并且不可能有机器 $\mathfrak{D}$.

我们可以更进一步指出不存在机器 $\mathscr{E}$,对它当给出一个任意的机器 $\mathscr{M}$ 的 S,D 时,机器是否能打印一个给定的符号(譬如说 0).

我们将首先证明,如果存在这样一个机器 $\mathscr{E}$,那么就存在一个一般的过程来决定是否存在一个机器,$\mathscr{M}$ 打印 0 无限多次,设 $\mathscr{M}_1$ 是一个机器它和 $\mathscr{M}$ 打印同样的序列. 所不同的是当机器 $\mathscr{M}$ 在打印第一个 0 的位置停留时,$\mathscr{M}_1$ 确打印 $\bar{0}$. $\mathscr{M}_2$ 是要将开头的两个 0 替换成 $\bar{0}$ 等这样下去. 所以如果 $\mathscr{M}$ 是打印

$$ABA01AAB0010AB\cdots$$

那么 $\mathscr{M}_1$ 打印

$$ABA\bar{0}1AAB0010AB\cdots$$

并且 $\mathscr{M}_2$ 打印

$$ABA\bar{0}1AAB\bar{0}010AB\cdots$$

现在设 $\mathscr{F}$ 是一个机器,当它被提供机器 $\mathscr{M}$ 的 $S.D$ 时,将连续地写下机器 $\mathscr{M}$,$\mathscr{M}_1$ 和 $\mathscr{M}_2$ 的 $S.D$,…(有这样的一个机器). 我们将 $\mathscr{F}$ 和 $\mathscr{E}$ 组合起来得到一个新的机器,$\mathscr{G}$. 在 $\mathscr{G}$ 的动作中首先用 $\mathscr{F}$ 来写下 $\mathscr{M}$ 的 $S.D$,然后

用 $\mathscr{E}$ 来试验它，如果它发现 $\mathscr{M}$ 永远不打印 0，那么写下：0：；然后 $\mathscr{F}$ 写下 $\mathscr{M}_1$ 的 $S.D$，并且再对这个 $S.D$ 进行试验，如果它发现 $\mathscr{M}_1$ 永远不打印 0，那么写下：0：；…如此进行下去. 现在让我们用 $\mathscr{E}$ 来试验 $\mathscr{G}$. 如果发现 $\mathscr{G}$ 永远不打印 0，那么 $\mathscr{M}$ 打印 0 无限多次；如果发现 $\mathscr{G}$ 有时打印 0，那么 $\mathscr{M}$ 不会打印 0 无限多次.

类似地存在一个一般的过程用以确定是否 $\mathscr{M}$ 打印 1 无限多次. 将这些过程组合起来我们可以得到一个过程用以试验是否机器 $\mathscr{M}$ 打印无限多个数字，也就是说我们有一个确定机器 $\mathscr{M}$ 是否无循环的过程. 所以机器 $\mathscr{E}$ 不存在.

贯穿于整个小节中使用了表述“存在一个一般的过程用以确定…”它等价于“存在一个机器用以确定…”. 它们的用法是合理的当且仅当我们关于“可计算”的定义是合理的. 因为每一个“一般过程”的问题都可以表示成关于一个一般过程用来确定一个整数 n 是否具有性质 $G(n)$ 的问题. [例如 $G(n)$ 的含义可能是“n 是可满足的”或者“n 是一个可证明公式的哥德尔表示式”]，并且这是等价于计算一个数，它的第 n 位数字是 1，如果 $G(n)$ 成立，或者 0，如果它不成立.

§9. 可计算数的扩充

还没有尝试证明“可计算数”包括所有自然被认为是可计算的数. 所有可以给出的论断基本上仅限于直观的愿望，并且由于这个原因在数学上是不太满意的. 论断的真实问题是“什么过程在计算一个数中可以被执行?”

我将要用到的论断有三种类型.

(a)直接来源于直观.

(b)两个定义等价的证明(在新定义有更大的直观来源的情形).

(c)给出一大类的数，它们是可计算的.

一旦所有可计算数都是“可计算的”得到认可，同样特征的其他几个命题也就得出. 特别地，它推断出，如果有一个一般的过程用以决定一个希尔伯特函量演算是可以证明的，那么这个决定可以由一个机器来执行.

Ⅰ.[(a)型]这个论断仅仅是 §1 的思想的精确化

计算通常是在纸上写某些记号来做. 我们可以假设这张纸是像儿童的算术书那样被划分为方格. 在初等算术里纸的二维特征是有时被使用.

但是这种用法通常是可以避免的，并且我想大家会同意纸的二维特征对计算不是非要不可的. 那么我假设计算是在一条划分为方格的一维纸带上执行. 我还将假设可以打印的记号的数目是有限的. 如果我们允许无限多个记号，那么就有差别极其微小的记号. * 这个对符号数目的限制的效果并不很严重. 总是可能在单个符号的地方用符号序列. 所以一个阿拉伯数字例如 17 或者 999999999999999 通常地是当作一个单个符号来处理. 类似地在任何欧洲语言，字是当作一个单个符号来处理(无论如何中文尝试拥有可数无限多个记号). 从我们的观点看来单个和复合的符号的区别在于复合的符号如果太过于长，不可能一眼就看清. 这是根据经验得来的. 我们不能一瞥就说出 9999999999999999 和 999999999999999 是否相同.

在任何时刻计算机的行为是由它所观察的记号，以及它在该时刻的"思想状态"来确定的. 我们可以假设在一个瞬间计算机可以观察的记号或者方格数目有一个界限 B. 如果它想观察更多一些，它必须用继续的观察. 我们也假设被考虑的思想状态的数目也是有限的. 这一点的理由和限制记号的数目具有相同的特征. 如果我们允许无限多个思想状态，它们之中有些就会"非常接近"以至于混淆. 再次，这个限制不至于严重地影响计算，因为使用更多复杂的思想状态可以用在带子上多写些记号来避免.

让我们想象将计算机所做的运转分解成"简单的动作"，它们是如此的初等使得不容易再想把它们进一步分开. 每一个这样的运作是由计算机的物理系统和它的带子一些变化来组成的. 如果我们知道在纸带上的记号序列，它们正在被计算机(可能是按一定的顺序)所观察，以及计算机的思想状态，我们就知道系统的状态. 我们可以假设在一个简单的动作中不改变超过一个以上的记号. 任何其他变化可以设定为几个这一类简单变化. 关于它的记号可能会以这种方式更换的方格的情况与关于被观察的方格相同. 因此，不失一般性，记号被更换的方格总是被观察的方格.

除了改变这些记号之外，简单的运作必须包括改变被观察方格的分

---

* 如果我们理解一个符号是照字面在方格上打印，我们可以假设方格是 $0<x<1, 0<y<1$. 符号是定义成为方格内的点集，也就是说打印机油墨所占领的集合. 如果这个集合限制是可测的我们可以定义两个符号之间的"距离"是变换一个符号到另一个符号的费用. 如果移动一个单位面积的打印机油墨一个单位距离是一个单位，那么在 $x=2$ $y=0$ 有无限多的油墨供应. 考虑到这个拓扑，符号形成一个条件紧致空间.

布. 新的被观察方格必须被计算机很快地认出. 我想假设它们只能是离最接近的直接前面的被观察方格的距离不超过一个固定总数的方格. 让我们说每一个新的被观察方格是在直接前面的被观察方格 L 个方格之内.

关于"直接可认出", 可能会想有其他类型的方格是直接可认出的. 特别地, 标记有特殊记号的方格可以认为是直接可认出的. 现在如果这些方格仅仅被单个符号所标记, 这种符号只有有限多个, 并且我们不想添加这些有标记的方格到被观察的方格来搞乱我们的理论. 如果, 从另一方面讲, 它们用一序列记号来标记, 我们不能将认知过程看成是一个简单的过程. 这是一个需要加以说明的基本点. 在大多数的数学文章中方程式和定理是编号的. 通常地这种号码不会超过(譬如说)1000. 因此可能从它的号码一眼就认出这个定理, 但是如果文章很长, 我们可能用到定理157767733443477; 而在文章的后面我们可能发现"…所以(应用定理157767733443477)我们得到…", 为了弄清哪一个是有关的定理我们必须逐个符号进行比较, 可能还要用铅笔来逐个剔除符号以免某一个符号被计算两次. 如果不管还有人认为存在其他的"直接可认出"的方格, 只要不搞乱我的论点这些方格可以由某些过程来找到, 而我这种类型的机器有这个能力, 就可以了. 这个想法在后面Ⅲ展开.

因此简单的运作必须包含:

(a)在被观察的方格中的一个改变记号.

(b)将一个被观察方格改变成为在前面被观察的某一个方格距离 L 个方格之内的另一个方格.

有可能这些变化必须连带地改变思想状态. 所以最一般的单个运作必须看成是下面的一个:

(A)可能的记号变化(a)连同一个可能的思想状态的变化.

(B)可能的被观察的方格变化(b)连同一个可能的思想状态的变化.

实际执行的运作是由计算机的思想状态和被观察的方格来确定的. 特别地它们确定运作被执行之后的思想状态.

我们现在可以构造一个机器用以做这个计算机的工作. 每一个计算机的思想状态对应于机器的一个 m-匹配. 机器扫描方格 B 对应于 B 方格被计算机所观察. 再任意动作机器能改变一个在被扫描的方格上的记号, 或者能改变任意一个被扫描的方格到另外一个方格距离其他的一个被扫描的方格不超过 L 个方格. 在这个动作做完,并且跟着的匹配是由被扫描的记号和 m-匹配所确定. 刚才描述的机器和 § 2 所定义的机器没有很大的区别,并且对应于任意一个这种类型的机器可以构造一个计算同样序列的计算的机器,那就是说用计算机来计算序列.

Ⅱ.[类型(b)]

如果希尔伯特函量演算的概念* 是修正了的,也是系统化了的,并且也是只用到有限数个记号,它使构造一个找到所有演算中可以证明的公式的自动机** $\mathscr{K}$ 成为可能. ***

现在假设 α 是一个序列,并且让我们以 $G_\alpha(x)$ 来表示命题"α 的第 x 个数字是 1",所以 $-G_\alpha(x)$ 的含义是"α 的第 x 个数字是 0". 进一步假设我们能够找到一个性质的集合,它定义序列 α 并且它也能用 $G_\alpha(x)$,命题函数 $N(x)$ 和 $F(x,y)$ 来表示,其中 $N(x)$ 的含义是"x 是一个非负整数",$F(x,y)$ 的含义是"$y=x+1$". 当我们将所有这些命题用合取连接起来,我们将得到一个公式,譬如说 $\mathfrak{A}$. 它定义了 α. $\mathfrak{A}$ 的各项中必须包含皮亚诺公理的各个必要部分,也就是说

$$\exists u N(u) \& (x)(N(x) \to (\exists y) F(x,y)) \& (F(x,y) \to N(y)).$$

我们把它简单记作 P.

当我们说"$\mathfrak{A}$ 定义 α",我们的意思是 $-\mathfrak{A}$ 不是一个可以证明的公式,并且也有,对每一个 n 下面的两个公式 A_n 或者 B_n 之一是可以证明的. ****

$$\mathfrak{A} \& F^{(n)} \to G^{\alpha} u^{(n)} \qquad ④$$

$$\mathfrak{A} \& F^{(n)} \to -G^{\alpha} u^{(n)} \qquad ⑤$$

其中 $F^{(n)}$ 的含义是 $F(u,u') \& F(u',u'') \& \cdots F(u^{(n-1)},u^{(n)})$.

这样我就说 α 是一个可计算序列:计算 α 的机器 $\mathscr{K}_\alpha$ 可以从 $\mathscr{K}$ 经过一

---

* 通篇中概念"函量演算"的含义是有限制的希尔伯特函量演算.

** 最自然的办法是首先构造一个选择机器(§ 2)来做这个工作. 然后容易构造所要求的自动机器. 我们总是可以假设选择就是在 0 与 1 之间选择. 每一个证明就将会被一个选择序列 $i_1, i_2, \cdots, i_n$ ($i_1=0$ 或者 1, $i_2=0$ 或者 1, $\cdots$, $i_n=0$ 或者 1), 所确定, 并且所以数 $2^n+i_1 2^{n-1}+i_2 2^{n-2}+\cdots+i_n$ 完全确定了这个证明. 这个自动的机器成功地实行了证明 1, 证明 2, 证明 3, …

*** 作者发现了这个机器的描述.

**** 公式的否定是写在一个式子的前面,而不是在上面加横线.

个相当简单的修正而得到.

我们将 $\mathscr{K}_\alpha$ 的动作划分为段落. 第 n 个段落要找到 α 的第 n 个数字. 在第 $n-1$ 个段落结束之后两个冒号被打印在所有记号之后,并且所有以后的工作完全是做在这两个冒号的右边. 第一步是写下字母 A,接着写公式④,然后 B 接着写⑤. 然后机器 $\mathscr{K}_\alpha$ 开始做机器 $\mathscr{K}$ 的工作,每当找到一个可以证明的公式,就拿它来与④⑤进行比较. 如果它和④一样,那么就打印数字"1",并且第 n 段落的工作结束. 如果它和⑤一样,那么就打印数字"0",第 n 段落的工作也结束. 如果它和两者都不同,那么 $\mathscr{K}$ 的工作从它中断的地方重新开始,迟些或者早些就会到达一个公式④或者⑤;这可以从我们关于 α,$\mathfrak{A}$ 的假设,以及 $\mathscr{K}$ 的特性推断出来. 所以第 n 个段落最终一定会结束,$\mathscr{K}_\alpha$ 是无循环的机器,α 是可计算的.

也可以证明用这种方法、用公理来定义的数包含了所有的可计算数. 这个可以用函量演算的条款描述计算的机器来完成.

需要记住,我们将比较特别的含义附着于词语"$\mathfrak{A}$ 定义 α". 可计算数不包括所有(在普通意义之下)的可定义数. 设 δ 是一个序列,它的第 n 个数字是 1 或者 0,代表 n 是不是可满足的. 由 §8 一个定理的直接推论,δ 是不可计算的. 它是(到目前为止我们所知道的)每一个 δ 所指定的数字都是可计算的,但不是由统一的过程来计算的. 当足够多的 δ 的数字计算完成之后,为了得到更多的数字需要一个从本质上是新的方法.

Ⅲ. 这可以被看成是一个Ⅰ的化简或者是Ⅱ的推论

像在Ⅰ那样,我们假设计算是在一条纸带上执行;但是我们避免引进"思想状态",而是考虑它的更加物化和确定的副本. 总是可以让计算机中断它的工作走开,并且整个忘掉它,然后在以后的某一个时候回来继续工作. 如果它想这样做,它必须留下一条注释指令(用一种标准的方式写下来)解释怎样继续进行工作. 这条注释就是"思想状态"的副本. 我们将假设计算机就是这样断断续续地工作,它永不在一个座位上做多于一步. 注释指令必须能够使它执行一步并且写下下一步的注释. 所以计算进行的任何阶段完全由注释指令和带上的记号来确定. 那时,系统的状态可以用一条单个表示式(记号的序列)和注释指令来描述,这个表示式是由一些带上的记号跟着一个 Δ(假设在其他地方没有用到过)再跟着一个注释指令. 这个表示式可以叫作"状态公式". 我们知道这个状态公式在任意给定的阶段是被做上一步的状态公式所确定，我们假设这两个公式之间的关系是可以在函量演算中表示的. 换一句话说我们假设有一个公理 $\mathfrak{A}$，它用任何阶段的状态公式对于前一阶段的状态公式的关系来表示管理计算机动作的规则，如果是这样，我们可以构造一个机器来写下相继的状态公式，并且以此来计算所要求的数.

§ 10. 可计算数的大类的例子

从定义一个整数变元的可计算函数和一个可计算变元的可计算函数开始会是有用的. 定义一个整数变元的可计算函数有很多等价的方法. 可能最简单的办法如下. 如果 γ 是一个可计算序列，其中 0 出现无限多次 *，并且 n 是一个整数，那么让我们定义 $\xi(\gamma,n)$ 作为 γ 的在第 n 个 0 和第 $n+1$ 个 0 之间的 1 的数目. 那么 $\phi(n)$ 是可以计算的，如果 $\phi(n)=\xi(\gamma,n)$，对所有 n 和某些 γ. 一个等价的定义是这样的. 设 $H(x,y)$ 的含义为 $\phi(x)=y$. 那么，如果我们能够发现一个无矛盾的公理 $\mathfrak{A}_\phi$，使得 $\mathfrak{A}_\phi \rightarrow P$，并且对每一个整数 n，存在一个整数 N，使得

$$\mathfrak{A}_\phi \& F^{(N)} \rightarrow H(u^{(n)}, u^{(\phi(n))}).$$

并且使得，如果 $m \neq \phi(n)$，那么对某些 N'，

$$\mathfrak{A}_\phi \& F^{(N')} \rightarrow -H(u^{(n)}, u^{(m)}).$$

那么 ϕ 可以说是一个可计算函数.

我们不能一般地定义一个可计算实数函数，因为没有一个一般的方法来描述一个实数，但是我们可以定义一个可计算变元的可计算函数. 如果 n 是可以满足的，设 γ_n 为 $\mathscr{M}(n)$所计算的数，并且设

$$\alpha_n = \tan\left(\pi\left(\gamma_n - \frac{1}{2}\right)\right).$$

γ_n 只有两种可能 $\gamma_n=1$ 或者 $\gamma_n=0$，两种都能使 $\alpha_n=0$. 因此，只要 n 跑过符合要求的数，α_n 就跑过可计算数. ** 现在设 $\phi(n)$ 是一个可计算函数，它可以被证明是对任意符合要求的变元，它的函数值是符合要求的. *** 那么

* 如果 $\mathscr{M}$ 计算 γ，那么 $\mathscr{M}$ 是否打印 0 无限多次的问题和 $\mathscr{M}$ 是否无循环的问题具有同样的特征.

** 一个函数 α_n 可以有很多其他的不同方法定义来跑过可计算数.

*** 虽然不可能找到一个一般的过程来确定一个给定数是符合要求的，总是可以证明某些数的类是符合要求的.

由 $f(\alpha_n)=\alpha_{\phi(n)}$. 定义的函数 f 是可计算函数，并且所有一个可计算变元的可计算函数都可以表示成这种形式.

类似地多个变元的可计算函数，一个整数变元的可计算值函数等都可以给出定义.

我将宣布一些关于可计算性的定理，但是只给出(ii)和与(iii)类似的一个定理的证明.

(i)一个整数变元或者可计算变元的可计算函数也是可计算的.

(ii)用可计算函数来递归地定义的任意一个整数变元函数是可计算的. 也就是说，如果 $\phi(m,n)$是可计算的，并且 r 是一个整数，那么 $\eta(n)$是可计算的，其中

$$\eta(0)=r,$$
$$\eta(n)=\phi(n,\eta(n-1)).$$

(iii)如果 $\phi(m,n)$是一个含有两个整数变元的可计算函数，那么 $\phi(n,n)$是一个 n 的可计算函数.

(iv)如果 $\phi(n)$是一个可计算函数，它的值总是 0 或者 1，那么序列，它的第 n 个数字是 $\phi(n)$是可计算的.

戴得金定理的一般形式，如果我们将其中"实数"通篇改为"可计算数"，是不成立的. 但是它的下面形式是成立的：

(v)如果 $G(\alpha)$是一个可计算数的命题函数并且

$$(a)\quad \exists\alpha\exists\beta\{G(\alpha)\&(-G(\beta))\}.$$
$$(b)\quad \{G(\alpha)\&(-G(\beta))\}\rightarrow\alpha<\beta.$$

并且存在一个确定 $G(\alpha)$的值一般的过程，那么存在一个可计算数 ξ 使得

$$G(\alpha)\rightarrow\alpha\leqslant\xi,\qquad -G(\alpha)\rightarrow\alpha\geqslant\xi.$$

换句话说，这个定理成立对任意段落的可计算数，它们使得存在一个确定一个数属于哪一类的一般过程.

由于这个对戴得金定理的限制，我们不能说一个有界上升序列的可计算数的极限仍然是可计算的. 这一点可以考虑下面的序列来理解

$$-1,-\frac{1}{2},-\frac{1}{4},-\frac{1}{8},-\frac{1}{16},\frac{1}{2},\cdots$$

另一方面，(v)使我们能够证明

(vi)如果 α 和 β 是可计算的，并且 $\alpha<\beta$，并且 $\phi(\alpha)<0<\phi(\beta)$，其中 $\phi(\alpha)$是一个可计算上升连续函数，那么存在一个唯一的可计算数 γ，满足

$\alpha<\gamma<\beta$ 并且 $\phi(\gamma)=0$.

可计算收敛

我们说一个可计算数的序列 β_n 叫作是可计算地收敛的，如果有一个可计算变元 ε 的可计算整值函数 $N(\varepsilon)$ 使得我们可证明，如果 $\varepsilon>0$ 并且 $n>N(\varepsilon)$，并且 $m>N(\varepsilon)$，那么 $|\beta_n-\beta_m|<\varepsilon$.

这样我们就可以证明：

(vii)一个幂级数的系数形成一个可计算数的可计算序列是在所有它的可计算区间内部的可计算点收敛的.

(xiii)可计算收敛序列的极限是可计算的. 并且带有这明显的"一致收敛"的定义：

(ix)一个可计算函数的一致可计算收敛的可计算序列的极限是一个可计算函数. 所以

(x)一个幂级数它的系数形成一个可计算函数的可计算序列的和是在所有它的收敛区间内部的可计算函数.

由(viii)以及 $\pi=4\left(1-\frac{1}{3}+\frac{1}{5}-\cdots\right)$ 我们可以推演出 π 是可计算的.

由 $e=1+1+\frac{1}{2!}+\frac{1}{3!}+\cdots$ 我们可以推演出 e 是可计算的.

由(vi)我们可以推演出所有实代数数是可计算的.

由(vi)以及(x)我们可以推演出所有贝塞尔函数的实的零点是可计算的.

(ii)的证明.

设 $H(x,y)$ 的含义是"$\eta(x)=y$"，并且设 $K(x,y,z)$ 的含义是"$\phi(x,y)=z$"，$\mathfrak{A}_\phi$ 是 $\phi(x,y)$ 的公理. 我们以 $\mathfrak{A}_\eta$ 表示下面的式子：

$$\begin{aligned}&\mathfrak{A}_\phi \& P \& (F(x,y)\to G(x,y)) \& (G(x,y) \& G(y,z)\to G(x,z))\\&\quad \& (F^{(r)}\to H(u,u^{(r)})) \& (F(v,w) \& H(v,x)\\&\quad \& K(w,x,z)\to H(w,z))\\&\quad \& [H(w,z) \& G(z,t)\vee G(t,z)\to(-H(w,t))].\end{aligned}$$

我将不给出 $\mathfrak{A}_\eta$ 的和谐性证明. 这样的一个证明可以利用希尔伯特和博尼斯所著"Crundlagen der Mathematik(Berlin，1934)，p. 209 et seq"中的办法构造出来. 从含义上看和谐性也是清楚的.

假设对某些 n. N 我们已经证明了

$$\mathfrak{A}_{\eta}\&F^{(N)}\rightarrow H(u^{(n-1)},u^{(\eta(n-1))}).$$

那么对于某一个 M,

$$\mathfrak{A}_{\phi}\&F^{(M)}\rightarrow K(u^{(n)},u^{(\eta(n-1))},u^{(\eta(n))})$$

$$\mathfrak{A}_{\eta}\&F^{(M)}\rightarrow F(u^{(n-1)},u^{(n)})\&H(u^{(n-1)},u^{(\eta(n-1))})\\ \&K(u^{(n)},u^{(\eta(n-1))},u^{(\eta(n))}),$$

并且

$$\mathfrak{A}_{\eta}\&F^{(M)}\rightarrow[F(u^{(n-1)},u^{(n)})\&H(u^{(n-1)},u^{(\eta(n-1))})\\ \&K(u^{(n)},u^{(\eta(n-1))},u^{(\eta(n))})\rightarrow H(u^{(n)},u^{(\eta(n))})].$$

所以

$$\mathfrak{A}_{\eta}\&F^{(M)}\rightarrow H(u^{(n)},u^{(\eta(n))}).$$

也有

$$\mathfrak{A}_{\eta}\&F^{(r)}\rightarrow H(u,u^{(\eta(0))}).$$

所以对每一个 n 这种形式的某些公式

$$\mathfrak{A}_{\eta}\&F^{(M)}\rightarrow H(u^{(n)},u^{(\eta(n))}).$$

是可以证明的. 也有,如果 $M'\geqslant M$ 并且 $M'\geqslant m$ 并且 $m\neq\eta(u)$,那么

$$\mathfrak{A}_{\eta}\&F^{(M')}\rightarrow G(u^{(\eta(n))},u^{(m)})\vee G(u^{(m)},u^{(\eta(n))}).$$

并且

$$\mathfrak{A}_{\eta}\&F^{(M')}\rightarrow[\{G(u^{(\eta(n))},u^{(m)})\vee G(u^{(m)},u^{(\eta(n))})\\ \&H(u^{(n)},u^{(\eta(n))})\}\rightarrow(-H(u^{(n)},u^{(m)}))].$$

所以

$$\mathfrak{A}_{\eta}\&F^{(M')}\rightarrow(-H(u^{(n)},u^{(m)})).$$

因此我们关于可计算函数的第二个定义的条件得到满足. 结论是 η 是一个可计算函数.

(iii)的一个修正形式的证明.

假设给出一个机器 $\mathfrak{N}$,它启动时带子的 F 方格子上开头的两个记号是 əə,跟着是一个任意多个"F"字母所组成在 F-方格上的序列,并且 m-匹配是 b,将计算序列 γ_n,依赖于数字 n 和"F"字母. 如果 $\phi_n(m)$是 γ_n 的第 m 个数字,那么序列 β 的第 n 个数字是 $\phi_n(n)$是可以计算的.

我们假设 $\mathfrak{N}$ 的表已经用这种方式写出,每一行只有一个动作出现在动作列. 我们也假设 Ξ,Θ,$\overline{0}$ 和 $\overline{1}$ 不在表中出现,并且我们将通篇的 ə 替换成 Θ,0 替换成 $\overline{0}$,并且 1 替换成 $\overline{1}$. 更多的替换在如下的表格列出

符号行 $\mathfrak{A}$	α	$P\bar{0}$	$\mathfrak{B}$
替换成 $\mathfrak{A}$	α	$P\bar{0}$	$\mathfrak{re}(\mathfrak{B},\mathfrak{u},h,k)$

以及

符号行	$\mathfrak{A}$	α	$P\bar{1}$	$\mathfrak{B}$
替换成	$\mathfrak{A}$	α	$P\bar{1}$	$\mathfrak{re}(\mathfrak{B},\mathfrak{v},h,k)$

表上再加上其他的行

$\mathfrak{u}$		$\mathfrak{pe}(\mathfrak{u}_1,\mathfrak{0})$
$\mathfrak{u}_1$	$R,Pk,R,P\Theta,R,P\Theta$	$\mathfrak{u}_2$
$\mathfrak{u}_2$		$\mathfrak{re}(\mathfrak{u}_3,\mathfrak{u}_3,k,h)$
$\mathfrak{u}_3$		$\mathfrak{pe}(\mathfrak{u}_2,F)$

还有类似的行,以 v 替换 u,以 1 替换 0,再加上下面的行

$\mathfrak{c}$	$R,P\Xi,R,Ph$	$\mathfrak{b}$.

我们这样就得到机器 $\mathfrak{N}'$ 的表,它计算 β,初始的 m-匹配是 $\mathfrak{c}$,并且初始的扫描符号是第二个 ə.

§11.　在判定问题上的应用

§8 的结果有一些非常重要的应用. 特别的,它们可以用来证明希尔伯特判定问题是无解的. 目前我们把自己限定于这个特殊问题的证明上. 对这个问题的形成请读者参看希尔伯特和阿克曼所著 Grundzüge der Theoretischen Logik(柏林,1931),第三章.

因此我提议来证明不可能有确定一个给定的函量演算 K 的公式 $\mathfrak{A}$ 是否可以证明的一般过程,也就是说没有这样的机器,如果向它提供任意的一个公式 $\mathfrak{A}$,机器会最终判定这个公式是否是可以证明的.

我也许应该提醒我要证明的内容与众所周知的歌德尔* 的结果有很大的不同. 歌德尔(在数学原理的形式主义之内)证明了存在命题 $\mathfrak{A}$ 使得 $\mathfrak{A}$ 或者 $-\mathfrak{A}$ 都是不可以证明的. 作为这个的一个结论,数学原理的(或者 K 的)和谐性也不可能在形式主义之内加以证明. 另一方面,我要证明没有能告诉我们一个给定的公式 $\mathfrak{A}$ 是在 K 之内可以证明的,或者也可以同样地说,由 K 添加一条额外的公理 $-\mathfrak{A}$ 所组成的系统是否和谐也是不可以证明的.

* Loc. cit.

如果歌德尔所证明的反面能够加以证明，也就是说，如果对每一个$\mathfrak{A}$，$\mathfrak{A}$或者$-\mathfrak{A}$是可以证明，那么我们就可以得到判定问题的直接解. 因为我们可以发明一个机器$\mathscr{K}$，它将一个接着一个地证明所有可以证明的公式. 迟早$\mathscr{K}$就会达到$\mathfrak{A}$或者$-\mathfrak{A}$，如果它达到$\mathfrak{A}$，我们就知道$\mathfrak{A}$是可以证明的. 如果它达到$-\mathfrak{A}$，那么，由于K是和谐的（希尔伯特和阿克曼，p. 65）我们就知道$\mathfrak{A}$是不可以证明的.

由于K中没有整数证明看起来是有些冗长. 基本的设想是很直接的.

对应于每一个计算的机器$\mathscr{M}$我们构造一个公式$Un(\mathscr{M})$并且我们证明，如果我们有一个一般的方法来确定$Un(\mathscr{M})$是可以证明的，那么就有一个一般的方法来确定$\mathscr{M}$在某一个时候会打印0.

证明中所用的命题的函数的解释如下：

$R_{S_i}(x,y)$解释为"在（$\mathscr{M}$的）完整匹配x里，在方格y的记号是S".

$I(x,y)$解释为"在（$\mathscr{M}$的）完整匹配x里，在方格y被扫描".

$K_{q_m}(x,y)$解释为"在（$\mathscr{M}$的）完整匹配x里，m-匹配是q_m".

$F(x,y)$解释为"y是x的直接后继元素".

$Inst\{q_iS_jS_kLq_l\}$是下面式子的简记号

$$(x,y,x',y')\{(R_{S_j}(x,y)\&I(x,y)\&K_{q_i}(x)\&F(x,x')\&F(y',y))$$
$$\rightarrow(I(x',y')\&R_{S_k}(x',y)\&K_{q_l}(x')$$
$$\&(z)[F(y',z)\vee(R_{S_j}(x',z)\rightarrow R_{S_k}(x',z))])\}.$$

$Inst\{q_i,S_j,S_k,Rq_l\}$和$Inst\{q_i,S_j,S_k,Nq_l\}$是其他的一些类似地构造的表示式.

让我们把$\mathscr{M}$的描述按§6的第一种标准型表示出来. 这个描述由一些诸如"$q_iS_jS_kLq_l$"（或者与R或者N代替L）的表示来组成. 让我们构成所有相应的表示式类似于$Inst\{q_iS_jS_kLq_l\}$并做它们的逻辑和式. 这个式子我们把它记作$Des(\mathscr{M})$.

公式$Un(\mathscr{M})$成了

$$(\exists u)[N(u)\&(x)(N(x)\rightarrow(\exists x')F(x,x'))$$
$$\&(y,z)(F(y,z)\rightarrow N(y)\&N(z))\&(y)R_{S_0}(u,y)$$
$$\&I(u,u)\&K_{q_1}(u)\&Des(\mathscr{M})]\rightarrow(\exists s)(\exists t)[N(s)\&N(t)\&R_{S_1}(s,t)].$$

$[N(u)\&\cdots\&Des(\mathscr{M})]$可以简记为$A(\mathscr{M})$.

当我们按上面所提议的含义来替换时，我们发现$Un(\mathscr{M})$可以做这样

的解释"在 $\mathscr{M}$ 的某一个完整匹配中 S_1（就是 0）在带子上出现"和这一点相对应我证明

(a)如果 S_1 在 $\mathscr{M}$ 的某一个完整匹配的带子上出现，那么 $Un(\mathscr{M})$ 是可以证明的.

(b)如果 $Un(\mathscr{M})$ 是可以证明的，那么 S_1 在 $\mathscr{M}$ 的某一个完整匹配的带子上出现.

当这些做完了以后定理的剩余部分就是明显的了.

引理 1 如果 S_1 在 $\mathscr{M}$ 的某一个完整匹配的带子上出现，那么 $Un(\mathscr{M})$ 是可以证明的.

我们要指出怎样证明 $Un(\mathscr{M})$. 假设在第 n 个完整匹配的带子上出现的记号序列是 $S_{r(n,0)}, S_{r(n,1)}, \cdots, S_{r(n,n)}$ 后面的记号全部都是空格，被扫描的记号是第 $i(n)$ 个，并且 m-匹配是 $q_{k(n)}$. 然后我们就可以形成这个命题.

$$R_{S_{r(n,0)}}(u^{(n)}, u) \& R_{S_{r(n,1)}}(u^{(n)}, u') \& \cdots \& R_{S_{r(n,n)}}(u^{(n)}, u^{(n)})$$
$$\& I(u^{(n)}, u^{(i(n))}) \& K_{q_{k(n)}}, (u^{(n)})$$
$$\& (y) F((y, u') \vee F(u, y) \vee F(u', y) \vee \cdots \vee F(u^{(n-1)}, y) \vee R_{S_0}(u^{(n)}, y)).$$

对它我们可以简记为 CC_n.

和以前一样 $F(u,u')\&F(u',u'')\&\cdots\&F(u^{(r-1)},u^{(r)})$ 简记为 $F^{(r)}$.

我们将证明所有形如 $A(\mathscr{M})\&F^{(n)}\to CCn$（简记为 CF_n）是可以证明的. CF_n 的含义是"$\mathscr{M}$ 的第 n 个完整匹配是如此如此"，这里"如此如此"就是实际上的 $\mathscr{M}$ 的第 n 个完整匹配. 所以 CF_n 应该是可以证明的.

CF_0 肯定是可以证明的，因为在完整的匹配中所有的记号都是空格，m-匹配是 q_1，并且扫描的方格是 u，也就是说 CC_0 是

$$(y) R_{S_0}(u, y) \& I(u, u) \& K_{q_1}(u).$$

由此 $A(\mathscr{M}\to CC_0)$ 是明显的.

下一步我们要说 $CF_n\to CF_{n+1}$ 对每一个 n 都是可以证明的. 根据由第 n 个动作过渡到第 $n+1$ 个动作是机器是向左、向右移动或者是在原地停留要考虑三种情形. 我们假设应用第一种情形，也就是说机器向左移动. 其他应用的情形类似地可以得出. 如果 $r(n,i(n))=a$, $r(n+1,i(n+1))=c$, $k(i(n))=b$，并且 $k(i(n+1))=d$. 那么 $Des(\mathscr{M})$ 必定含有 $Inst\{q_aS_bS_dLq_c\}$ 作为它的一个项，也就是说

$$Des(M) \to Inst\{q_aS_bS_dLq_c\}.$$

所以

$$A(\mathscr{M})\&F^{(n+1)}\rightarrow Inst\{q_aS_bS_dLq_c\}\&F^{(n+1)}.$$

但是

$$Inst\{q_aS_bS_dLq_c\}\&F^{(n+1)}\rightarrow(CC_n\rightarrow CC_{n+1}).$$

是可以证明的，所以

$$A(\mathscr{M})\&F^{(n+1)}\rightarrow(CC_n\rightarrow CC_{n+1})$$

以及

$$(A(\mathscr{M})\&F^{(n)}\rightarrow CC_n)\rightarrow(A(\mathscr{M})\&F^{(n+1)}CC_{n+1})$$

也就是说

$$CF_n\rightarrow CF_{n+1}.$$

都是可以证明的，

CF_n 对每一个 n 是可以证明的. 现在这个引理假设 S_1，在某一个完整的匹配的某个地方，在 $\mathscr{M}$ 所打印的记号序列中出现；那就是，对某些整数 N,K,CC_N 有 $R_{S_1}(u^{(N)},u^{(K)})$ 作为它的一个项，因此 $CC_N\rightarrow R_{S_1}(u^{(N)},u^{(K)})$ 是可以证明的. 这样我们就得到

$$CC_N\rightarrow R_{S_1}(u^{(N)},u^{(K)})$$

以及

$$A(\mathscr{M})\&F^{(N)}\rightarrow CC^N.$$

我们还得到

$$(\exists u)A(\mathscr{M})(\exists u)(\exists u')\cdots(\exists u^{(N')})(A(\mathscr{M})\&F^{(N)}),$$

其中 $N'=\max(N,K)$. 并且因此得到

$$(\exists u)A(\mathscr{M})\rightarrow(\exists u)(\exists u')A!\ -(\exists u^{(N')})R_{S_1}(u^{(N)},u^{(K)}),$$

$$(\exists u)A(\mathscr{M})\rightarrow(\exists u^{(N)})(\exists u^{(K)})R_{S_1}(u^{(N)},u^{(K)}),$$

$$(\exists u)A(\mathscr{M})\rightarrow(\exists s)(\exists t)R_{S_1}(s,t),$$

也就是说 $Un(\mathscr{M})$ 得到了证明.

这就证明了引理 1.

引理 2 如果 $Un(\mathscr{M})$ 是可以证明的，那么 S_1 在 $Un(\mathscr{M})$ 的某个完整匹配的带子上出现.

如果我们在一个可以证明的公式中以命题函数替换函数变元，我们得到一个真命题. 特别地，如果我们替换在本节开头 $Un(\mathscr{M})$ 含义的表格，我们得到一个真命题，它的含义是“S_1 在 $Un(\mathscr{M})$ 的某个完整匹配的带子上出现.”

现在我们来证明判定问题是不可以解决的. 假设不是这样的. 那么就存在一个一般的(机械)过程用以确定 $Un(\mathscr{M})$ 是否可以证明的. 由引理 1 和 2,这蕴涵存在一个确定 $\mathscr{M}$ 是否会打印 0 的过程,并且由 § 8,这是不可能的. 所以判定问题是不可以解决的.

考虑到对带有限制量词系统的公式的大量判定问题是有解的特殊情形,将公式 $Un(\mathscr{M})$ 表示成所有量词放在前面是很有用的. $Un(\mathscr{M})$ 实际上是可以表示成如下形式.

$$(u)(\exists x)(w)(\exists u_1)\cdots(\exists u_n)\mathfrak{B}, \qquad ⑥$$

其中 $\mathfrak{B}$ 不含有量词,并且 $n=6$. 利用一个不重要的改进我们可以得到一个具有形式⑥的公式保有所有 $Un(\mathscr{M})$ 的性质而 $n=5$.

附录

可计算性和有效可演算性

1936 年 8 月 28 日添加

关于所有有效可演算(λ-可定义)的序列是可以计算的定理以及它的逆定理在下面给出一个轮廓性的证明. 对合式公式(W. F. F.)以及“逆定理”理解为像丘奇和克林尼所用过的那样的含义. 在下面的第二个证明中承认有几个公式存在,但是没有给出证明;这些公式可以直接地构造出来. 想了解详情请参看克林尼的文章“A theory of positive integers ill formal logic”,American Journal of Math. ,57 (1935),153—173,219—244.

代表一个整数 n 的 W. F. F. 记作 N_n. 我们说一个序列 γ 它的第 n 个数字 $\phi_\gamma(n)$ 是 λ-可定义的或者是有效地可演算的,如果 $1+\phi_\gamma(u)$ 是一个 n 的 λ-可定义函数,也就是说,如果有一个 W. F. F. ,M_γ 使得对所有整数 n,

$$\{M_\gamma\}(N_n)\mathrm{conv}N_{\phi_\gamma(n)+1},$$

也就是说$\{M_\gamma\}(N_n)$可以转化成 $\lambda xy.x(x(y))$ 或者转化成 $\lambda xy.(x(y))$ 根据 λ 的第 n 个数字是 1 或者 0.

为要证明每一个 λ-可计算序列 γ 是可计算的,我们需要指出怎样构造一个机器来计算 γ:为了使用机器在转换的演算中做一个简单的修改是方便的. 这个改变包括用 $x,x',x'',\cdots$ 来代替 $a,b,c,\cdots$ 我们现在构造一个机器 $\mathscr{L}$,当它提供公式 M_γ 时,能写下序列 γ. $\mathscr{L}$ 的构造是和泛函数演算中证明所有可以证明的公式的机器 $\mathscr{K}$ 有些类似. 我们首先构造一个选择

型机器 $\mathscr{L}_1$，如果提供一个 W. F. F.，譬如说它是 M，经过适当的操作，它能得到任意一个 M 可以转换成的公式. 然后 $\mathscr{L}_1$ 又可以改进成为一个自动机器 $\mathscr{L}_2$. 它能够将所有 M 能够转换成的公式逐个地产生出来(请参看 p. 154 脚注[*]). 机器 $\mathscr{L}$ 把机器 $\mathscr{L}_2$ 包含在内作为一个部件. 当提供公式 M_γ 时 $\mathscr{L}$ 的动作分成若干段落，其中第 n 个段落是专门用来找到 γ 的第 n 个数字. 这个第 n 个段落的第一步是形成公式 $\{M_\gamma\}(N_n)$. 然后这个公式就提供给 $\mathscr{L}_2$，它将公式逐个地转换成为各种其他的公式. 每一个可以转换成的公式最终总会出现，并且每一个公式只要找到就拿来和 N_2 也就是 $\lambda x[\lambda x'[\{x\}(\{x\}(x'))]]$，以及 N_1 也就是 $\lambda x[\lambda x'[\{x\}(x')]]$，i. e. 进行比较. 如果它和这两个式子中的第一个式子相同，那么机器打印数字 1，并且第 n 个段落结束. 如果它和这两个式子中的第二个式子相同，那么机器打印数字 0，并且第 n 个段落结束，如果这两种情形都不是，那么 $\mathscr{L}_2$ 恢复工作. 由假设 $\{M_\gamma\}(N_n)$ 可以转换成为 N_2 或者 N_1；因此第 n 段落最终一定会结束，也就是说 γ 的第 n 个数字最终一定会写下来.

想要证明每一个可计算序列 γ 是 λ-可定义的，我们必须说明如何找到一个公式 M_γ，使得对所有整数 n，

$$\{M_\gamma\}(N_n)\operatorname{conv}N_{1+\phi_\gamma(n)}.$$

设 $\mathscr{M}$ 为一个机器，它能计算 γ 并且让我们用数作为工具取出 $\mathscr{M}$ 的一些完整匹配的描述，例如我们可以使用 § 6 中的完整匹配的 D. N 作为描述. 设 $\xi(n)$ 为这个 $\mathscr{M}$ 的第 n 个完整匹配的 D. N. 机器 $\mathscr{M}$ 的表给我们一个如下形式的 $\xi(n+1)$ 和 $\xi(n)$ 之间的关系

$$\xi(n+1)=\rho_\gamma(\xi(n)),$$

其中 ρ_γ 是一个非常受限制的函数形式，虽然通常地不是很简单：它是由 $\mathscr{M}$ 的 $\mathscr{M}$ 表所确定的. ρ_γ 是 λ 可定义的(我省略了这一点的证明)，也就是说有一个 W. F. F. A_γ 使得，对所有整数 n，

$$\{A_\gamma\}(N_{\xi(n)})\operatorname{conv}N_{\xi(n+1)}.$$

设 $U=\lambda u[\{\{u\}(A_\gamma)\}(N_r)]$. 其中 $r=\xi(0)$；那么对所有整数 n，

$$\{U_\gamma\}(N_n)\operatorname{conv}N_{\xi(n)}.$$

可以证明有一个公式 V 使得

$$\{\{V\}(N_{\xi(n+1)})\}(N_{\xi(n)})\begin{cases}\operatorname{conv}N_1, & \text{如果从第 } n \text{ 个到第 } n+1 \text{ 个完整匹配，打印 } 0,\\ \operatorname{conv}N_2, & \text{打印 } 1,\\ \operatorname{conv}N_3, & \text{其他.}\end{cases}$$

设 W_γ 表示

$$\lambda u[\{\{V\}(\{A_\gamma\}(\{U_\gamma\}(u)))\}(\{U_\gamma\}(u))]$$

使得对每一个整数 n，

$$\{\{V\}(N_{\xi(n+1)})\}(N_{\xi(n)})\mathrm{conv}\{W_\gamma\}(N_n),$$

并且设 Q 是一个公式使得

$$\{\{Q\}(W_\gamma)\}(N_s)\mathrm{conv}N_{r(z)},$$

其中 $r(s)$是第 s 个整数 q 对它$\{W_\gamma\}(N_q)$可以转换成为 N_1 或者 N_2. 所以如果 $M_\gamma=\lambda w[\{W_\gamma\}(\{\{Q\}(W_\gamma)\}(w))]$它就得到所要求的性质 *.

美国新泽西州普林西顿大学研究生学院

更正

可计算数及其在判定问题上的应用

艾·图灵

在一篇题为“可计算数及其在判定问题上的应用”的文章中作者给出了一个“演算功能的判定问题是不可解”** 的证明. 这个证明中包含有一个形式的错误***，它将在这里得到改正：在同一文章的一些语句也需要得到修改，虽然它们从含义上看其实不是错误的.

$Inst\{q_iS_jS_kLq_l\}$的式子应该改为

$$(x,y,x',y')\{(R_{S_j}(x,y)\&I(x,y)\&K_{q_i}(x)\&F(x,x')\&F(y',y))$$

$$\rightarrow(I(x',y')\&R_{S_k}(x',y)\&K_{q_l}(x')\&F(y',z)\vee[(R_{S_0}(x,z)\rightarrow R_{S_0}(x',z))$$

$$(R_{S_1}(x,z)\rightarrow R_{S_1}(x',z))\&\cdots\&(R_{S_u}(x,z)\rightarrow R_{S_u}(x',z))])\}.$$

---

* 在一个可计算序列的 λ-可定义性的完整证明中最好是用一个对我们的装置更加容易地掌握的描述来代替完整匹配的数字化的描述来修正这个办法. 让我们选择某些整数来代表机器的记号和 m-匹配. 假设在一个完整匹配中代表带子上连续记号的数字是 $s_1s_2\cdots s_n$，并且第 n 个记号被扫描，并且 m-匹配有数字 t；这样我们就可以用这个公式来代表完整匹配$[[N_{s_1},N_{s_2},\cdots,N_{s_{m-1}}],[N_t,N_{s_m}],[N_{s_{m+1}},\cdots,N_{s_n}]]$，其中$[a,b]$的含义是 $\lambda u[\{\{u\}(a)\}(b)]$，$[a,b,c]$的含义是 $\lambda u[\{\{\{u\}(a)\}(b)\}(c)]$，等.

** 作者感谢伯尼斯指出这个错误.

*** Proc. London Math. Soc, Ser. 2. Vol. 43,. No. 2 198.

The Journal of Symbolic Logic, 1983, 48(3): 539～542

可数齐次模型的模型数[①]

On the Number of Countable Homogeneous Models

The number of homogeneous models has been studied in [1] and other papers. But the number of countable homogeneous models of a countable theory T is not determined when dropping the GCH. Morley in [2] proves that if a countable theory T has more than $\aleph_1$ nonisomorphic countable models, then it has $2^{\aleph_0}$ such models. He conjectures that if a countable theory T has more than $\aleph_0$ nonisomorphic countable models, then it has $2^{\aleph_0}$ such models. In this paper we show that if a countable theory T has more than $\aleph_0$ nonisomorphic countable *homogeneous* models, then it has $2^{\aleph_0}$ such models.

We adopt the conventions in [1]～[3]. Throughout the paper T is a theory and the language of T is denoted by L which is countable.

Lemma 1 *If a theory T has more than $\aleph_0$ types, then T has $2^{\aleph_0}$ nonisomorphic countable homogeneous models.*

Proof Suppose that T has more than $\aleph_0$ types. From [2, Corollary 2.4] T has $2^{\aleph_0}$ types. Let σ be a T type with n variables, and $T' = T \cup \{\sigma(c_1, c_2, \cdots, c_n)\}$, where $c_1, c_2, \cdots, c_n$ are new constants. T' is consistent and has a countable model $(\mathfrak{A}, a_1, a_2, \cdots, a_n)$. From [3, Theorem 3.2.8] the reduced model $\mathfrak{A}$ has a countable homogeneous elementary extension $\mathfrak{B}$.

① Received January 18, 1980; revised November 24, 1980.

σ is realized in $\mathfrak{B}$. This shows that every type σ is realized in at least one countable homogeneous model of T. But each countable model can realize at most $\aleph_0$ types. Hence T has at least $2^{\aleph_0}$ countable homogeneous models. On the other hand, a countable theory can have at most $2^{\aleph_0}$ nonisomorphic countable models. Hence the number of nonisomorphic countable homogeneous models of T is $2^{\aleph_0}$.

In the following, we shall use the languages L_α ($\alpha=0,1,2$) defined in [2]. We give a brief description of them. For a countable theory T, let K be the class of all models of T. $L=L_0$ is countable. If σ is a type of L, then $\wedge\sigma$ is a formula of $L_{\omega_1\omega}$. L_1 is the smallest regular language generated by the set L_0 with all $\wedge\sigma$ added to it. If L_1 is countable, let τ be a type of L_1. $\wedge\tau$ is a formula of $L_{\omega_1\omega}$, L_2 is the smallest regular language generated by the set L_1 with all $\wedge\tau$ added to it.

Lemma 2 *Suppose that* $|L_1|=\aleph_0$. *Then we have*:

(1) *Corresponding to every countable homogeneous model* $\mathfrak{A}$ *of* T, *there is an* L_1 *type* $\tau_{\mathfrak{A}}$ *such that if* $\mathfrak{A}\not\cong\mathfrak{B}$, *then* $\tau_{\mathfrak{A}}\neq\tau_{\mathfrak{B}}$.

(2) *If* $|L_2|<2^{\aleph_0}$, *then T has at most* $\aleph_0$ *nonisomorphic countable homogeneous models*.

Proof Let $\mathfrak{A}\not\cong\mathfrak{B}$ be two countable homogencous models of T. From [3, Proposition 3. 2. 9 (iii)], they realize different sets of L_0 types. We may suppose without loss of generality that $\sigma(v_1, v_2, \cdots, v_n)$ is an L_0 type realized by $\mathfrak{A}$ but not by $\mathfrak{B}$.

Consider the sentence φ of L:

$$\varphi=(\exists v_1, v_2, \cdots, v_n)(\wedge\sigma(v_1, v_2, \cdots, v_n)).$$

Obviously, $\mathfrak{A}\models\varphi$, $\mathfrak{B}\models\neg\varphi$. Let $\tau_{\mathfrak{A}}, \tau_{\mathfrak{B}}$ be the L_1 types with no variables determined by $\mathfrak{A}, \mathfrak{B}$, respectively. Then $\varphi\in\tau_{\mathfrak{A}}$, $\neg\varphi\in\tau_{\mathfrak{B}}$. Hence $\tau_{\mathfrak{A}}\neq\tau_{\mathfrak{B}}$. Therefore, the number of nonisomorphic countable homogeneous models is at most the number of the L_1 types with no variables. This proves (1).

If $|L_2|<2^{\aleph_0}$. From [2, Corollary 2. 4] $|L_2|=\aleph_0$. Then the number of types of L_1 is also $\aleph_0$. And by (1) the number of nonisomorphic countable homogeneous models is at most $\aleph_0$. (2) is proved.

Lemma 3 *Suppose that* $|L_1|=\aleph_0$, $|L_2|=2^{\aleph_0}$. *If T has more than* $\aleph_0$ *nonisomorphic countable homogeneous models, then T has* $2^{\aleph_0}$ *such models.*

Proof Let $\sigma_i(v_1, v_2, \cdots, v_{n_i})(i<\omega)$ be the L_0 types and $S(L_1, K)$ the set of all L_1 types. From [2, Theorem 2.3], $S(L_1, K)$ is an analytic subset of 2^{L_1}. Consider the L_1 sentences

$$\varphi_{ij}=(\forall u_1, \cdots, u_{n_i+1}, v_i, \cdots, v_{n_i})$$
$$[(\bigwedge\sigma_i(u_1, \cdots, u_{n_i}))\wedge(\bigwedge\sigma_i(v_1, \cdots, v_{n_i}))\wedge(\bigwedge\sigma_j(u_1, \cdots, u_{n_i+1}))\rightarrow$$
$$(\exists v_{n_i+1})(\bigwedge\sigma_j(v_1, \cdots, v_{n_i+1}))]\ (\text{for } n_j=n_i+1),$$
$$\varphi_{ij}=(\forall v)(v\equiv v)(\text{for the other cases}).$$

Let X be the set of all sentences of L_1.

$$U_{ij}=\{s: s\subseteq X, \varphi_{ij}\in s\}\quad (i, j<\omega)$$

are basic subsets of 2^X. Hence $U=\bigcap_{i,j}U_{ij}$ is a Borel subset of 2^X. And $V=U\cap S(L_1, K)$ is an analytic subset of 2^X.

From the proof of Lemma 2 there is an L_1 type τ corresponding to each countable homogeneous model $\mathfrak{A}$ of T such that $\mathfrak{A}\not\cong\mathfrak{B}$ implies $\tau_{\mathfrak{A}}\neq\tau_{\mathfrak{B}}$. From the homogeneity of $\mathfrak{A}$ we have $\varphi_{ij}\in\tau\ (i, j<\omega)$, $\tau\in U_{ij}$, $\tau\in U$, $\tau\in V$, and $|V|>\aleph_0$. V is an Uncountable analytic subset. The power of V is $2^{\aleph_0}$.

Corresponding to each $\tau\in V$, there is a model $\mathfrak{B}\in K$, $\mathfrak{B}\models\tau$. From [4, p. 22], the downward Löwenheim-Skolem theorem of a countable set of sentences in $L_{\omega_1\omega}$, τ has a countable model $\mathfrak{C}$. The homogeneity of $\mathfrak{C}$ follows from $\varphi_{ij}\in\tau$. $\mathfrak{C}$ is a T model because $T\subset\tau$. Also $\tau_{\mathfrak{A}}\neq\tau_{\mathfrak{B}}$ implies $\mathfrak{A}\not\cong\mathfrak{B}$, hence T has $2^{\aleph_0}$ nonisomorphic countable homogeneous models.

Theorem *For any countable theory T, the number of nonisomorphic countable homogeneous models of T is either* $\leqslant\aleph_0$ *or* $=2^{\aleph_0}$.

Proof Immediate from the above lemmas.

Corollary *If all countable models of a complete theory T are homogeneous, then the number of countable models of T is either* $2^{\aleph_0}$, *or* $\aleph_0$ *or* 1.

Proof From [5] (indicated by the referee) we know that the number of countable models of T cannot be a finite number $n\neq 1$. The corollary is implied from the theorem.

We can easily find complete theories to have the number of nonisomorphic countable homogeneous models $=1$, $\aleph_0$, or $2^{\aleph_0}$. For finite n, in [3, Exercise 5. 1. 16], there are complete theories which have exactly 2 or 3 nonisomorphic countable homogeneous models. We will generalize these results to $n=2^r3^s$ in another paper.

References

[1] Keisler H J and Morley M. On the number of homogeneous models of a given power. Israel Journal of Mathematics, 1967, 5: 73—78.

[2] Morley M. The number of countable models. The Journal of Symbolic Logic, 1970, 35(1): 14—18.

[3] Chang C C and Keisler H J. Model theory. North-Holland, Amsterdam, 1971.

[4] Keisler H J. Model theory for infinitary logic. North-Holland, Amsterdam, 1971.

[5] Benda M. Remarks on countable models. Fundamenta Mathematicae. 1974, 81: 107—119.

The Journal of Symbolic Logic 1983,48(3):700～704

自由群的 τ-理论是不可判定的①

The τ-Theory for Free Groups is Undecidable

Abstract In this paper we give short proofs to the undecidability of the τ-theory for free groups and other relevant theories.

§ 1. The τ-theory for free groups

An inducitive τ-theory for free groups is given in [1]. We quote it briefly here. The language L consists of three symbols, the constant symbol"1", a unary operation symbol "τ"and a binary operation symbol "·". The intended interpretation of $\tau(x)$ will be the inverse of the terminal syllable of the reduced word presenting the element x. The axioms of $\mathscr{F}$ are as follows.

(F1) $\tau(1)=1$,

(F2) $\tau(x)\cdot\tau\tau(x)=1$,

(F3) $\tau(x\cdot\tau\tau(x))=\tau(x)$,

(F4) $y=1\vee\tau\tau(y)=\tau(x)\vee\tau(x\cdot\tau\tau(y))=\tau(y)$,

(F5) $x\cdot 1=x$,

(F6) $x\cdot(y\cdot\tau\tau(z))=(x\cdot y)\cdot\tau\tau(z)$,

(F7) $A(1)\ \&\ (\forall xy)(A(x)\ \&\ \tau(x\cdot\tau\tau(y))=\tau(y)\Rightarrow A(x\cdot\tau\tau(y)))$
$\Rightarrow(\forall x)A(x)$, for every well formed formula of L.

① Received June 6, 1981; revised October 13, 1981.

We also need the following abbreviations for positive integers $n \geqslant 2$.

$$X_n \equiv_{df} (\tau\tau(x_1)=x_1 \& \cdots \& \tau\tau(x_n)=x_n \\ \& (\forall z)(\tau(z)=x_1 \vee \tau(z)=\tau(x_1) \vee \cdots \vee \\ \tau(z)=x_n \vee \tau(z)=\tau(x_n) \vee z=1)),$$

$$D_n \equiv_{df} (\exists x_1, x_2, \cdots, x_n)\ X_n,$$

$$E_n \equiv_{df} D_n \& \sim D_{n-1},$$

$$A_n \equiv_{df} X_n(a_1, a_2, \cdots, a_n) \mathop{\&}_{1 \leqslant i < j \leqslant n} (a_i \neq a_j \& a_i \neq \tau(a_j)).$$

Dyson in [1] suggested many problems, among which we have:

Problem PA. Are the theories $\mathscr{F}(A_n)$ model complete?

Problem PE. Are the theories $\mathscr{F}(E_n)$ complete?

Conjecture $\mathscr{F}$. Every sentence or its negation is either an $\mathscr{F}$-theorem or else $\mathscr{F}$-equivalent to a disjunction of sentences of the form E_n.

We will prove that they all have negative answers.

At first we write down a list of sentences which are not very diff-cult to prove within theory $\mathscr{F}$.

Lemma 1 *The following sentences are valid in theory* $\mathscr{F}$.

(F8) $(x \cdot y) \cdot z = x \cdot (y \cdot z)$,

(F9) $(\forall x)(\exists!\ y)(x \cdot y = y \cdot x = 1)$,

(F10) $\tau y \neq \tau(z)\ \&\ y \neq 1 \Rightarrow \tau(z \cdot \tau(y)) = \tau\tau(y)$.

Define cycle multiplication in $\mathscr{F}$:

$$z = x \circ y \equiv_{df} z = x \cdot y \ \&\ (\forall u)(y \cdot u = 1 \Rightarrow \tau(x) \neq \tau(u)).$$

Lemma 2 *Concerning the cycle multiplication the following sentences are valid in* $\mathscr{F}$.

(F11) $x = 1 \vee \tau(z \circ x) = \tau(x)$,

(F12) $(x \circ y) \circ z = x \circ (y \circ z)$, *when at least one side exists*,

(F13) $x \circ y = x \circ z \Rightarrow y = z$,

(F14) $x \circ z = y \circ z \Rightarrow x = y$,

(F15) $x \circ y = u \circ v \ \&\ \tau\tau(y) = y \ \&\ \tau\tau(v) = v \ \&\ y \neq 1 \ \&\ v \neq 1 \Rightarrow x = u \ \&\ y = v$,

(F16) $\tau\tau(x) = x \vee (\exists yz)(z \neq 1 \ \&\ x = y \circ z \ \&\ \tau\tau(z) = z)$,

(F17) $x \circ y \circ z = u \circ v \ \&\ \tau\tau(y) = y \Rightarrow (\exists r)(x \circ y \circ r = u) \vee (\exists s)(s \circ y \circ z = v)$.

With the help of the above lemmas we can prove our main theorem.

Theorem 1 *$\mathscr{F}(A_n)(n\geqslant 2)$ is essentially undecidable.*

Proof From [4, p. 86], we know that the elementary theory $\mathscr{S}_2$ for the free semigroup with two generators is essentially undecidable. We now turn to interpret the theory $\mathscr{S}_2$ into the theory $\mathscr{F}(A_n)$ relatively. Let us begin with the definitions of $\mathscr{S}_2$ (which assigns the elements of a free semigroup with two generators in a free group $\mathscr{G}$) as a new predicate symbol.

$S_2(x)\equiv_{\mathrm{df}} x\neq 1\ \&\ (\forall uvw)(x=u\circ v\circ w\ \&\ \tau\tau(v)=v\ \&\ v\neq 1\Rightarrow v=a_1\vee v=a_2)$.

$S_2(x)$ is written as $x\in S_2$.

We have to verify relatively the axioms of the free semigroup with two generators. These axioms are chosen from [4]. They are

$\Gamma_1\cdot(\forall xyz)((x\circ y)\circ z)=x\circ(y\circ z))$,

$\Gamma_6\cdot(\forall xyz)(x\circ y=x\circ z\Rightarrow y=z)$,

$\Gamma_7\cdot(\forall xyz)(x\circ z=y\circ z\Rightarrow x=y)$,

$\Gamma_9\cdot(\exists ab)((\forall xy)(x\circ a\neq y\circ b)\ \&\ (\forall z)(z=a\vee z=b\vee(\exists x)(z=x\circ a)\vee(\exists y)(z=y\circ b)))$.

The cycle multiplication is closed in S_2, for if $x, y\in S_2$, then from the definition of S_2 we know that every single syllable (the element satisfying $\tau\tau(z)=z\ \&\ z\neq 1$) of the reduced words x and y is a_1 or a_2. Hence the same is true for $x\circ y$, and $x\circ y\in S_2$. The associative law Γ_1 and the cancellation laws Γ_6, Γ_7 are valid in S_2 relatively because they are valid in $\mathscr{F}(A_n)$. Let $z\in S_2$, $z\neq a_1$ and $z\neq a_2$. z can be decomposed as $z=(z\cdot\tau(z))\circ\tau\tau(z)\circ 1$. Here $1\neq z\cdot\tau(z)\in S_2$. From the definition of S_2 we know that $\tau\tau(z)=a_1$ or $\tau\tau(z)=a_2$. Hence

$$z=(z\cdot\tau(z))\circ a_1\quad\text{or}\quad z=(z\cdot\tau(z))\circ a_2.$$

This proves that Γ_9 is valid in S_2 relatively. From [4, p. 16, Theorem 3] we conclude that the theory $\mathscr{F}(A_n)(n\geqslant 2)$ is essentially undecidable.

Theorem 2 *For $n\geqslant 2$ we have:*

(i) *$\mathscr{F}(E_n)$ is essentially undecidable.*

(ii) $\mathscr{F}$ *is undecidable.*

(iii) $\mathscr{F}(E_n)$ *is incomplete.*

(iv) $\mathscr{F}(A_n)$ *is not model complete.*

Proof (i) $\mathscr{F}(A_n)$ *is an inessential extension of* $\mathscr{F}(E_n)$. From [4, p. 16, Theorem 4], $\mathscr{F}(E_n)$ is essentially undecidable.

(ii) $\mathscr{F}(E_n)$ is a finite extension of $\mathscr{F}$. From [4, p. 17, Theorem 5], $\mathscr{F}(\mathscr{E}_n)$ is undecidable.

(iii) $\mathscr{F}(E_n)$ is undecidable and axiomatizable. From [4, Theorem 1] $\mathscr{F}(E_n)$ is incomplete.

(iv) Let Δ_n be the diagram of the free group with n generators. $\mathscr{F}(A_n)\cup\Delta_n$ is axiomatizable and undecidable (the proof is similar to above). Hence by [4, p. 14, Theorem 1], $\mathscr{F}(A_n)\cup\Delta_n$ is not complete and by [5, p. 111, Proposition 3. 1. 7], $\mathscr{F}(A_n)$ is not model complete.

Theorem 3 *The complete τ-theory T for a free group with at least two generators is undecidable.*

§ 2. The length function and the predicate for equality of length

In [2][3] Dyson suggested two theories for free groups. One of them is the theory of free groups with a length function. The other is the theory of free groups with a binary predicate for equality of syllable length. She has proved that the two theories are undecidable. but her proofs are rather long. We will give short proofs for them here. We only discuss the theory of free groups with a binary predicate "$\sim$" as the symbol for equality of length. The undecidability of the theory of free groups with a length function is an immediate consequence of this discussion. Let T ($=T_G$) be the set of all sentences in the language $L=\langle 1, \cdot, \sim\rangle$ that are true in G. Our proof is divided into three lemmas.

Lemma 3 *The letters of the free group are definable in T.*

Proof Let

$\lambda(x,y)\equiv_{\mathrm{df}} x\cdot y\neq y\cdot x \;\&\; x\sim y \;\&\; (\forall z)((\exists uvw)(u\cdot x=x\cdot u$

$\& \ u \sim z \ \& \ v \cdot y = y \cdot v \ \& \ v \sim z \ \& \ xwyx^{-1}w^{-1}y^{-1} = 1 \ \& \ w \sim z) \vee (\exists t)(t \cdot t \sim z))$.

In any free group G, two elements x, y satisfy $\lambda(x, y)$ if and only if x, y are letters. For the "if" part, suppose that x, y are letters and z is any element of G. The length of x (also y) is 1. If the length of z is odd, then we can find a w such that $xwyx^{-1}w^{-1}y^{-1} = 1$ and $w \sim z$. If the length of z is even, then we can find a t, such that $t \cdot t \sim z$. For the "only if" part, suppose that x, y satisfy $\lambda(x, y)$ but x is not a letter. Let z be a letter. From the definition of $\lambda(x, y)$ there exists a u such that $u \sim z, u \cdot x = x \cdot u$ From [6, p. 10, Proposition 2. 17], we know that $x = u^m$ for some m. Similarly $y = v^n$ for some n. Assume that $m, n > 0$, otherwise we change u or v or both into their inverses. From $x \cdot y \neq y \cdot x$ we have $u \cdot v \neq v \cdot u$. Hence u, v are distinct letters and $m = n \geqslant 2$ (because $x \sim y$ and x is not a letter, neither is y). From

$$u^m w v^n u^{-m} w^{-1} v^{-n} = 1$$

we have $w = u^{-m}(v^{-n}u^{-m})^k$ (cf. [7]), but for any k, $u^{-m}(v^{-n}u^{-m})^k$ cannot be a letter, contradicts $w \sim z$. We also know that t cannot exist such that $t \cdot t \sim z$. Hence x is a letter and so is y.

Lemma 4 *The cycle multiplication is definable in T.*

Proof Let

$$u = v \circ w \equiv_{\mathrm{df}} u = v \cdot w \ \& \ (\exists xyrs)(\lambda(x, y) \ \& \ r \cdot x = x \cdot r \ \& \ s \cdot y = y \cdot s \ \& \ v \sim r \ \& \ w \sim s \ \& \ u \sim r \cdot s).$$

From the definition $u = v \cdot w$ cannot change the length by cancellation. Hence it is cycle multiplication.

Lemma 5 *The "τ" operator is definable in T.*

Proof Let

$$y = \tau(x) \equiv_{\mathrm{df}} (x = 1 \ \& \ y = 1) \vee (\exists uvw)(x = u \circ v \ \& \ v \cdot y = 1 \ \& \ \lambda(v, w)).$$

$\tau(x) = y$ is the inverse letter of the terminal syllable of x.

Theorem 4 (Dyson) *The theory T for a free group G with a predicate for equality of syllable length is undecidable.*

Proof Let T be the set of all L sentences which hold in G. Define the τ operator in T. Let T' be the set of all $L'(=\langle 1, \cdot, \tau\rangle)$ sentences. From Theorem 3, T' is undecidable, hence T is undecidable.

References

[1] Huber-Dyson V. An inductive theory for free products of groups. Algebra Universalis, 1979, 9: 35—44.

[2] Huber-Dyson V. The undecidability of the theory of free groups with a length function. University of Calgary, Mathematics Research Paper, 1974, 221: 1—26.

[3] Huber-Dyson V. Talking about free groups in naturally enriched languages. Communications in Algebra, 1977, 11(5): 1 163—1 191.

[4] Tarski A, Mostowski A and Robinson R M. Undecidable theories. Studies in Logic, North-Holland, Amsterdam, 1953.

[5] Chang C C and Keisler H J. Model theory. North-Holland, Amsterdam, 1973.

[6] Lyndon R C and Schupp P E. Combinatorial group theory. Springer-Verlag, Berlin and New York, 1977.

[7] Lo L. A discussion on the equations in free groups. Scientia Sinica, 1981, 25 (2): 161—170.

Annals of Pure and Applied Logic, 1988, 37(2): 205～248

可换群理论的计算复杂性①

On the Computational Complexity of the Theory of Abelian Groups

This paper is dedicated to my parents.

We develop a series of Ehrenfeucht games and prove the following results:

(i) The first order theory of the divisible and indecomposable p-group, the first order theory of the group of rational numbers with denominators prime to p and the first order theory of a cyclic group of prime power order can be decided in $2^{2^{cn \log n}}$ Turing time.

(ii) The first order theory of the direct sum of countably many infinite cyclic groups, the first order theory of finite Abelian groups and the first order theory of all Abelian groups can be decided in $2^{2^{dn}}$ Turing space.

§ 0. Introduction

Every mathematician encounters Abelian groups. We are interested in their elementary theory. The first person to give a decision procedure for the theory of Abelian groups was W. Szmielew in 1950. She introduced four classes of basic sentences and proved that every first-order sentence is equivalent to a Boolean combination of the basic sentences.

① Received 26 April 1984; revised July 1986.
Communicated by A. Prestel.

The basic sentences are of simple structure and easy to check. In her paper [8] she also gave a complete treatment of the elementary subclasses of Abelian groups. In a paper of P. Eklof and E. Fischer another proof of the existence of a decision procedure was given using a model-theoretic method. Neither of these papers discussed the computational complexity of their procedures. After examining Szmielew's procedure carefully we find that her algorithm consumes Turing time

$$2^{2^{\cdot^{\cdot^{\cdot^{2^{cn}}}}}}$$

with height $n-1$ because in some step of quantifier elimination there may happen to be a sentence of the form

$$(\forall x)(\exists y)\phi(x,y)$$

which is equivalent to

$$(\forall x)\bigvee_i \bigwedge_j \psi_{ij}(x),\ \neg(\exists x)\bigwedge_i \bigvee_j \neg\psi_{ij}(x) \quad \text{and} \quad \neg(\exists x)\bigvee_k \bigwedge_l \neg\psi_{kl}(x).$$

The size of the last sentence may be exponential in that of the first. Therefore the procedure of Szmielew is not elementary. In [3] Ferrante and Rackoff determined the computational complexity of many theories. They stated results about the upper bounds of the computational complexities of the theory of a direct sum of countably many infinite cyclic groups and the theory of finite Abelian groups. The complete proofs of them were given in [7]. Their results are as follows:

The first-order theory of a direct sum of countably many infinite cyclic groups and the theory of finite Abelian groups can be decided in $2^{2^{2^{cn}}}$ Turing steps and $2^{2^{2^{dn}}}$ Turing space units for some constants c and d.

Yuri Gurevich conjectured that the first-order theory of all Abelian groups is elementary and suggested that the author work on this topic. This conjecture is confirmed in this paper.

We work on the theory of Abelian groups and its complete extensions and improve the above results. Our main tools in this paper are Ehrenfeucht games. Two kinds of Ehrenfeucht games are used. One of them consists of the Ehrenfeucht games with an equivalence relation producing a winning strategy for player Ⅱ. This kind of games are mainly used in proving elementary equivalences between two groups. The other kind of games are Ehrenfeucht games with norms and used in giving decision procedures for some groups. In Section 1 we describe the features of these games. In Sections 2～4 we prove that there is an Ehrenfeucht game with norm in each of the following groups: The divisible and indecomposable p-group, the group of rationals with denominators prime to p and the cyclic group of order p^w. These games give the decision procedures and allow us to establish the upper bounds for the theories of the corresponding groups. The games for the above three groups are different but their upper bounds are more or less the same. In Sections 4～7 we prove several reduction theorems. According to these theorems we can reduce the base p of a group, the size of a group and the number of summands of a direct sum of groups without altering their satisfiable sentences with quantifier-depth not more than n. Finally in Sections 8～10 we establish the upper bounds for the theories of Abelian groups, finite Abelian groups and a direct sum of infinitely many infinite cyclic groups. Here are our main results:

(i) The first-order theory of the divisible and indecomposable p-group, the first-order theory of the group of rational numbers with denominators prime to p and the first-order theory of a cyclic group of prime power order can be decided in $2^{2^{cn \log n}}$ Turing time.

(ii) The first-order theory of the direct sum of countably many infinite cyclic groups, the first-order theory of finite Abelian groups and the first-order theory of all Abelian groups can be decided in $2^{2^{dn}}$ Turing space.

§ 1. Preparations

In this section we introduce some notation, give a brief look at Ehrenfeucht games and prove a reduction theorem for the theory of Abelian groups. For the convenience of the reader we sketch proofs of some theorems which are available in other places.

§ 1.1 Notations

The following notations are used in this paper. Most of them are commonly used by other authors.

AG: The set of all Abelian groups.

FG: The set of all finite Abelian groups.

Z: The additive group of integers.

Q: The additive group of rational numbers.

$Q_p=\{r/s: r,s\in Z,(s,p)=1\}$.

We use the same symbols for rings and groups if there is no danger of confusion.

$C_{p,n}$: The cyclic group of order p^n.

D_p: The divisible and indecomposable p-group.

$\mathbf{N}$: Natural numbers.

$\bar{a}_k=(a_1,a_2,\cdots,a_k)$; $\bar{x}_k=(x_1,x_2,\cdots,x_k)$.

$\bar{a}_k\equiv_n\bar{b}_k$: $\bar{a}_k$ and $\bar{b}_k$ satisfy the same set of sentences with quantifier-depth less than or equal to n.

$[r]$=The greatest integer not exceeding r.

$\log x=\log_2 x$.

$A+B$: The direct sum of two groups A and B.

Th(A) or Th(C): The-first order theory of a model A or a class of models C.

$\mathrm{Th}_n(A)$ or $\mathrm{Th}_n(C)$: The sentences in Th(A) or Th(C) with quantifier-depth less than or equal to n.

A^*: The direct product of countably many groups isomorphic to A.

We consider a language with equality and a single predicate expressing that $x+y=z$.

§ 1.2 The Ehrenfeucht games

Two versions of Ehrenfeucht games are used in our paper. They are the Ehrenfeucht game G_n and the Ehrenfeucht game with norm.

We give a brief look at the Ehrenfeucht game G_n, which was discussed in detail in [1]. The procedure of the game is as follows. Let A

and B be two Abelian groups and $n\in \mathbf{N}$. We define the Ehrenfeucht game $G_n(A,B)$. There are two players Ⅰ and Ⅱ, and n turns. At turn i, player Ⅰ chooses any element of one of A and B, and player Ⅱ chooses any element of the other. Thus pairs $(a_1,b_1),(a_2,b_2),\cdots,(a_n,b_n)\in A\times B$ are chosen. We say that player Ⅱ wins if, for all $i,j,k\in\{1,2,\cdots,n\}$,

$$a_i+a_j=a_k\leftrightarrow b_i+b_j=b_k.$$

Theorem 1.1 (**Ehrenfeucht**) *If player Ⅱ has a winning strategy in the game $G_n(M_1,M_2)$, then $M_1\equiv_n M_2$. This means that for any first-order sentence*

$$\phi=(Q_1x_1)(Q_2x_2)\cdots(Q_nx_n)\psi(x_1,x_2,\cdots,x_n)$$

where ψ is a quantifier free formula with at most n variables we have

$$\phi \text{ is satisfied in } M_1\leftrightarrow\phi \text{ is satisfied in } M_2.$$

§ 1.3 The game $\mathbf{FR}_n(A,B)$

Ferrante and Rackoff gave a version of Ehrenfeucht game in [3] which led to a decision procedure for a suitable model. This procedure involves a series of testings of a sentence which may be true or false when we substitute a sequence of elements of the model for the bound variables. To constrain such a sequence of elements in a certain range we need to associate with each element a norm.

Let m be a model. For every element $a\in M$ the norm of a is denoted by $\|a\|$ which is usually a natural number. The definition of a norm need not satisfy the properties of a Banach space. For example in the model Z, we could define

(1) $\|a\|=|a|$, the absolute value of a, or

(2) $\|0\|=0$, $\|a\|=[\log a]$ (for $a\neq 0$), or

(3) $\|0\|=0$, $\|a\|=1$ (for $a\neq 0$).

The first two definitions are better for our purpose but the third definition is not so good because the set $\{x\in Z: \|x\|\leqslant r\}$ is too big. In some cases we can even use the third definition.

Suppose that in the sentence

$$\phi_0=(Q_1x_1)(Q_2x_2)\cdots(Q_nx_n)\psi(x_1,x_2,\cdots,x_n)$$

we have already substituted $a_1, a_2, \cdots, a_k \in M$ for variables $x_1, x_2, \cdots, x_k$ containing a sentence

$$\phi_k = (Q_{k+1}x_{k+1})\cdots(Q_n x_n)\psi(a_1, a_2, \cdots, a_k, x_{k+1}, \cdots, x_n).$$

The next step is to substitute a_{k+1} for x_{k+1}. We want to control the norm of a_{k+1} in a certain range. So we introduce a bounding function

$$H(n-k, k, m) = h \in \mathbf{N}$$

where $\| a_1 \|, \| a_2 \|, \cdots, \| a_k \| \leqslant m$.

The number of elements of M with norms less than a fixed number and the bounding function will determine the number of steps needed for the decision procedure, so we always want to keep them as low as possible.

In a lot of models the discussion of the relation $\bar{a}_k \equiv_n \bar{b}_k$ is rather difficult. It happens that a stronger (finer) relation $\bar{a}_n E_{h,k} \bar{b}_k$ is sometimes simpler to use because the definition of the latter may be simpler.

With the help of a norm, a bounding function and an equivalence relation we can now play our game $\mathrm{FR}_n(M_1, M_2)$.

We are given two models M_1 and M_2 and two players Ⅰ and Ⅱ. In the game $\mathrm{FR}_n(M_1, M_2)$ each player makes n moves. This time player Ⅱ has $n+1$ equivalence relations in mind. The game begins at $\varnothing E_{n,0} \varnothing$. Suppose that at the end of the kth move the players have chosen $(a_1, a_2, \cdots, a_k) \in M_1^k$ and $(b_1, b_2, \cdots, b_k) \in M_2^k$ satisfying $\bar{a}_k E_{n-k,k} \bar{b}_k$. For any $a_{k+1} \in M_1$ chosen by player Ⅰ, according to the equivalence relation player Ⅱ can find an element $b_{k+1} \in M_2$ (similarly if player Ⅰ chooses b_{k+1}, then player Ⅱ chooses a_{k+1}) such that $\bar{a}_{k+1} E_{n-k-1,k+1} \bar{b}_{k+1}$. Player Ⅱ wins the game if and only if

$$\bar{a}_n E_{0,n} \bar{b}_n \rightarrow \bar{a}_n \equiv_0 \bar{b}_n.$$

Hence the winning strategy for player Ⅱ is to satisfy the equivalence relation $E_{h,k}$ at each step k.

One sees easily that the game $\mathrm{FR}_n(M_1, M_2)$ is a version of the game $G_n(M_1, M_2)$. Hence we have the following theorem.

Theorem 1.2 (Ferrante and Rackoff) *Let* A, B *be two models. Suppose that there is an equivalence relation* $E_{n,k}$ *for player* Ⅱ *to play the game* $\mathrm{FR}_n(A, B)$. *This means that an equivalence relation* $E_{n,k}$ *is de-*

fined between the elements of A, B *such that*

(i) $\bar{a}_k E_{0,k} \bar{b}_k \rightarrow \bar{a}_k \equiv_0 \bar{b}_k$ *for* $1 \leqslant k \leqslant n$.

(ii) *Given* $\bar{a}_k E_{r+1,k} \bar{b}_k$ *and for every* $a_{k+1} \in A$ *there exists a* $b_{k+1} \in B$ *such that* $\bar{a}_{k+1} E_{r,k+1} \bar{b}_{k+1}$ *for* $1 \leqslant r \leqslant n$. *And vice versa.*

Then

(i) $\bar{a}_{k+1} E_{r,k+1} \bar{b}_{k+1} \rightarrow \bar{a}_k \equiv_r \bar{b}_k$ *for any* $\bar{a}_k \in A^k$ *and* $\bar{b}_k \in B^k$.

(ii) $\mathrm{Th}_n(A) = \mathrm{Th}_n(B)$.

In some cases $M_1 = M_2 = M$ and for any chosen sequences $a_1, a_2, \cdots, a_k$ and $b_1, b_2, \cdots, b_k$ where $\bar{a}_k E_{n,k} \bar{b}_k$ and $\| b_1 \|, \| b_2 \|, \cdots, \| b_k \| \leqslant m$ and for any a_{k+1} chosen by player Ⅰ in M player Ⅱ can find $b_{k+1} \in M$ such that

$$\bar{a}_{k+1} E_{n-k-1,k+1} \bar{b}_{k+1} \quad \text{and} \quad \| b_{k+1} \| \leqslant H(n-k, k, m).$$

Then we have a decision procedure for the sentences of $\mathrm{Th}(M)$ with quantifier-depth n as in the following theorem.

Theorem 1.3 (Ferrante and Rackoff) *Let* A *be a model. Let there be an Ehrenfeucht game* $\mathrm{FR}_n(A, A)$ *with equivalence relation* $E_{n,k}$, *norm* $\| a \|$ *and bounding function* $H(n,k,m)$ *satisfying the following conditions:*

(i) *If* $\bar{a}_k E_{n+1,k} \bar{b}_k$ *and* $\| b_1 \|, \| b_2 \|, \cdots, \| b_k \| \leqslant m$, *then for every* $a_{k+1} \in A$ *there exists a* $b_{k+1} \in A$ *such that* $\bar{a}_{k+1} E_{n,k+1} \bar{b}_{k+1}$ *and* $\| b_{k+1} \| \leqslant H(n,k,m)$.

(ii) *For every non-negative real number* r *the number of elements* $a \in A$ *with norm* $\| a \| \leqslant r$ *is* $F(r)$.

Then the number of steps needed for deciding a first-order sentence ϕ *with quantifier-depth* n *is at most* $\prod_{i=1}^{n} F(L_i)$, *where* $L_0 = 1$ *and* $L_{i+1} = H(n-i, i, L_i)$ *for* $i = 0, 1, \cdots, n-1$. *Here a step (sometimes we call it a big step) means a testing procedure of an instance of* ϕ *without variables and quantifiers.*

§1.4 General reductions for AG

Sometimes we need to reduce a complicated elementary n-equivalence between two Abelian groups to some simpler ones. The following reduction theorem is useful when the equivalence is proved by an Ehrenfeucht game.

Theorem 1.4 *Let A, B and C be Abelian groups. If player* Ⅱ *has a winning strategy in the Ehrenfeucht game* $G_n(A,B)$, *then player* Ⅱ *has a winning strategy in the game* $G_n(A+C,B+C)$.

Proof Suppose that at the end of the kth move players have chosen

$$a_1+c_1\leftrightarrow b_1+c_1, a_2+c_2\leftrightarrow b_2+c_2, \cdots, a_k+c_k\leftrightarrow b_k+c_k.$$

The winning strategy for player Ⅱ in $G_n(A+C,B+C)$ is to choose the element $b_{k+1}+d_{k+1}$ for $a_{k+1}+c_{k+1}$ (or conversely) such that b_{k+1} is chosen for a_{k+1} according to the winning strategy of player Ⅱ in game $G_n(A,B)$ and $d_{k+1}=c_{k+1}$. □

Now we discuss Abelian groups. To check the validity of a sentence ϕ in all Abelian groups is rather difficult because there are Abelian groups of arbitrary cardinality. By the Löwenheim-Skolem Theorem we can reduce this problem to considering only countable models. A better reduction theorem was given by Eklof and Fischer in [2].

Theorem 1.5 (Eklof and Fischer) *Let SG be the class of all finite direct sums of groups of the following types*

$$D_p, Q_p, C_{p,w} \text{ and } Q$$

where $p=2,3,5,\cdots$, $w=0,1,2,\cdots$ *and the number of summands of type Q is* 0 *or* 1. *Then* Th(*SG*)=Th(*AG*).

To simplify our discussion we give a theorem that eliminates the summand *Q*.

Theorem 1.6 *Let* SG_1 *be the class of all finite direct sums of groups of the following types:*

$$D_p, Q_p \quad \text{and} \quad C_{p,w}$$

where $p=2,3,5,\cdots$ *and* $w=0,1,2,\cdots$ *Then* $\mathrm{Th}(SG_1)=\mathrm{Th}(AG)$.

Proof We prove that if ϕ is true in a group $G\in AG$, then ϕ is true in a group $H\in SG_1$.

From [8, p. 261, Theorem 4.21] we know that ϕ is equivalent to a Boolean combination ψ of basic sentences

$$R^{(i)}[p,k,m] \quad \text{and} \quad K[n]$$

where $i=1,2,3$, p is a prime, and $k,m,n\geqslant 1$. The meanings of these sentences are as follows.

(i)$R^{(1)}[p,k,m]$ means that there are m independent elements with respect to p^k and of order p^k.

(ii)$R^{(2)}[p,k,m]$ means that there are m strongly independent elements with respect to p^k.

(iii)$R^{(3)}[p,k,m]$ means that there are m strongly independent elements with respect to p^k and of order p^k.

(iv)$K[n]$ means that for all x in the group $nx=0$.

Now come back to the Boolean combination ψ. Let

$$R^{(i,j)}[p_j,k_i,m_j],K[n_l] \quad ①$$

where $1\leqslant j\leqslant r, 1\leqslant l\leqslant s$, be a complete list of basic sentences which occur in ψ. Expand ψ in conjunctive normal form

$$\psi\leftrightarrow\bigvee_u\bigwedge_v\psi_{uv}$$

where ψ_{uv} is one of the sentences in ① or its negation.

Now suppose that ψ is true in $G\in AG$. Then $\bigwedge_v\psi_{uv}$ is true in G for some u. From [2, Theorem 2. 10] we know that there is a direct product $G_1\in SG$ such that $\bigwedge_v\psi_{uv}$ is true in G_1.

If G_1 doesn't have a summand Q, then let $H=G_1$ and ϕ is true in H.

If $G_1=G_2-Q$ where G_2 is a direct sum without summand Q, then we know that none of the $K[n_l]$ is a factor ψ_{uv} because the elements of Q do not satisfy $K[n_l]$. However $\neg K[n_l]$ might be a ψ_{uv}. Choose a prime number q greater than all p_j and all n_l. Consider the group

$$H=G_2+C_{q,1}.$$

H is a member of SG_1. H and G_1 have the same invariants for $p_1,p_2,\cdots,p_r$ and $K[n_l]$ is not true for $1\leqslant l\leqslant s$ in both G_1 and H. Hence ψ is true in H. Therefore ϕ is true in H. □

§ 2. The game $\mathrm{FR}_n(D_p, D_p)$ and the upper bound for $\mathrm{Th}(D_p)$

§ 2. 1 The main definitions

Since D_p is an Abelian p-group, every element of D_p is uniquely divisible by every integer s not divisible by p. Thus D_p is a Q_p-module.

Lemma 2.1 *D_p is a Q_p-module.*

Division by multiples of p is a difficult point. We can first reduce this to discussing $(1/p^n)a$. For an equation $p^n b = a$ there are exactly p^n solutions for any $a \in D_p$. We are not satisfied with knowing what elements are in the solution set of the equation; sometimes we need to specify each solution. This is given in the next two definitions.

Definition 2.2 θ denotes the identity of D_p.

Definition 2.3 Choose a sequence of elements of D_p

$$a_0 = \theta, a_1, a_2, \cdots, a_n, \cdots$$

such that $a_n = pa_{n+1}$ for $n = 0, 1, 2, \cdots$ Let this sequence be fixed throughout this section and let

$$\frac{\theta}{p^n} = \frac{1}{p^n}\theta$$

denote a_n.

Group D_n is the union of an ascending chain of cyclic groups C_0, C_1, $\cdots$ where C_k has order p^k. We may choose generators a_k for the C_k such that each $a_k = Pa_{k+1}$.

The main reason of using 'θ' instead of '0' to denote the zero element of D_p is to leave '0' as the zero element of Q_p. And it is inconvenient to think about '$0/p^n \neq 0$'.

Lemma 2.4 *Each C_k is a Q_p-submodule of D_p:*

$$\frac{x}{y} \cdot \frac{r}{s} \cdot \frac{\theta}{p^t} = L\frac{\theta}{p^t}.$$

Lemma 2.5 *For each $a \in D_p$ and $n \in \mathbf{N}$ the solutions of the equation*

$$p^n x = a$$

form a coset $y + C_n$.

Proof Let b be a generator of C_n. It is easy to check that $y, y + b, \cdots, y + (p^n - 1)b$ are distinct solutions. Suppose that y is a fixed solution of $p^n x = a$ and x is another solution, then

$$p^n(x - y) = p^n x - p^n y = a - a = \theta.$$

Hence

$$x-y=lb, \quad l=0,1,\cdots,p^n-1.$$

$$x=y+lb. \qquad \square$$

The definition of a norm is flexible but the following two requirements seem to be reasonable.

(a) Only a finite number of elements have the same norm and this number does not depend on p.

(b) $\|\theta/p^n\| < \|\theta/p^{n+1}\|$.

Now we give our definition of norm for the elements of D_p and prove some useful properties.

Definition 2.6 Let $x\in D_p$ have order p^n. Then x can be written in the form

$$x=\frac{q}{r}\cdot\frac{\theta}{p^n} \quad \text{where } q,r\in\mathbf{Z},(p,q)=(p,r)=1.$$

Define

$$\|x\|=\text{Min}\{\text{Max}\{q,p^{n-1}r\}\}$$

where the minimum is taken over pairs (q,r) such that $x=(q/r)\cdot(\theta/p^n)$ and $(p,q)=(p,r)=1$.

By this definition we can list the elements of D_p in terms of their norms. We give the following examples:

Example 1 If p is very large, then

$$\|\theta\|=0,$$

$$\left\|\frac{\pm 1}{p}\theta\right\|=1,$$

$$\left\|\frac{\pm 2}{p}\theta\right\|=\left\|\frac{\pm 1}{2p}\theta\right\|=2,$$

$$\left\|\frac{\pm 3}{p}\theta\right\|=\left\|\frac{\pm 2}{3p}\theta\right\|=\left\|\frac{\pm 1}{3p}\theta\right\|=3.$$

Example 2 Let $p^n=17$, then

$$\left\|\pm x\frac{\theta}{p}\right\|=1 \quad \text{for } x=1,$$

$$\left\|\pm x\frac{\theta}{p^n}\right\|=2 \quad \text{for } x=2 \text{ or } x=8=\frac{1}{2},$$

$$\left\|\pm x\frac{\theta}{p^n}\right\|=3 \quad \text{for } x=3, x=6=\frac{1}{3}, x=5=-\frac{2}{3} \text{ or } x=7=-\frac{3}{2},$$

$$\left\| \pm x \frac{\theta}{p^n} \right\| = 4 \quad \text{for } x=4.$$

Example 3 Let $p^n=25$, then

$$\left\| \pm x \frac{\theta}{p^n} \right\| = 5 \quad \text{for } x=1,2,3 \text{ or } 4,$$

$$\left\| \pm x \frac{\theta}{p^n} \right\| = |x| \quad \text{for } x=6,7,8 \text{ or } 9,$$

$$\left\| \pm x \frac{\theta}{p^n} \right\| = 10 \quad \text{for } x=12=-\frac{1}{2} \text{ or } x=11=-\frac{3}{2}.$$

Example 4 Let $p^n=27$, then

$$\left\| \pm x \frac{\theta}{p^n} \right\| = 9 \quad \text{for } x=1,2,3,4,5,7 \text{ or } 8,$$

$$\left\| \pm x \frac{\theta}{p^n} \right\| = |x| \quad \text{for } x=10,11 \text{ or } 13.$$

Lemma 2.7 (1) *If* $x=(y/z)\cdot(\theta/p^n)$, *where* $(y,p)=(z,p)=1$, *then* $p^{n-1}\leqslant \|x\| < p^n$.

(2) *If* $x=(y_1/z_1)\cdot(\theta/p^n)$, $x_2=(y_2/z_2)\cdot(\theta/p^m)$ *and* $m<n$, *then* $\|x_2\| < \|x_1\|$.

(3) *If* $\|x\|\leqslant m$, *then there are integers* y, z *such that*

$$x=\frac{y}{z}\cdot\frac{\theta}{p^u}, \quad |y|, |zp^{u-1}|\leqslant m \quad \text{and} \quad u-1\leqslant[\log_p m].$$

(4) *If* $x=(y/z)\cdot(\theta/p^n)$ *and* $\|x\|=\mathrm{Max}\{|y|,|zp^{n-1}|\}$, *then* $|z|<p$.

(5) *For a fixed* $m>0$, *there are at most* $2m^2-1$ *elements with norms less than or equal to* m.

(6) *For a fixed* $m>0$ *there exists an element* x *such that* $m<\|x\|\leqslant 2m^2$.

Proof (1) If $x=(q_i/r_i)\cdot(\theta/p^n)$, where $(p,q_i)=(p,r_i)=1$, then any other forms of x can not have smaller n. Hence $\|x\|\geqslant p^{n-1}$.

(2) From (1) we have $\|x_2\|\leqslant p^m\leqslant\|x_1\|$.

(3) Because $p^{u-1}\leqslant m$, we have $u-1\leqslant[\log_p m]$.

(4) From (1) and $\|x\|=\mathrm{Max}\{|y|,|zp^{n-1}|\}$ we have $|zp^{n-1}|\leqslant\|x\|<p^n$. Hence $|z|<p$.

(5) If $\|x\|\leqslant m$, then there are q, r and n such that

$$x=\frac{q}{r}\cdot\frac{\theta}{p^n} \quad \text{and} \quad \mathrm{Max}\{|q|,|rp^{n-1}|\}\leqslant m.$$

If there is a p^u such that $m<p^u\leqslant m^2$, then from (1) there are at most p^u elements with norms smaller than p^u. All the other elements have norms at least p^u. Hence (5) is true.

If there is no p^u such that $m<p^u\leqslant m^2$, then both m and m^2 are smaller than p. Consider $x\neq\theta$. There are at most $2(m-1)$ possibilities for the numerator $q\neq 0$ and m possibilities for the denominator rp^n. Including θ we have at most $m(2(m-1))+1$ elements with norms smaller than or equal to m. Hence the total number of these elements is less than $2m^2$.

(6) Let $t=[\log_p(2m^2-1)]+1$. Consider the following elements

$$\theta,\frac{1}{p^t}\theta,\cdots,\frac{2m^2-1}{p^t}\theta. \qquad ②$$

They are distinct elements (because $2m^2-1<p^t$) and not all having norms less than or equal to m (because of (5) at most $2m^2-1$ elements can have norms less than or equal to m while list ② has $2m^2$ elements). Hence some of them have norms greater than m.

For $1\leqslant i\leqslant 2m^2-1$ in list ② we have

$$\left\|\frac{i}{p^t}\theta\right\|\leqslant\mathrm{Max}\{\|i\|,p^{t-1}\}$$

where $i\leqslant 2m^2-1$ and

$$p^{t-1}=p^{[\log_p(m^2-1)]}\leqslant p^{[\log_p(2m^2-1)]}\leqslant 2m^2-1.$$

Hence $\|(i/p^t)\theta\|\leqslant 2m^2-1$ for $i=0,1,\cdots,2m^2-1$.

Therefore we can find an x such that $m<\|x\|<2m^2$. □

$E_{n,k}$ is going to be defined without involving any quantifiers.

Definition 2.8 The coefficient set of $E_{n,k}$ is a subset of Q_p:

$$V_n=\left\{\frac{r}{s}\in Q_p: r,s\in Z,(p,s)=1 \text{ and } |r|,|s|\leqslant 2^{2^{4n}}\right\}$$

Definition 2.9 Let $n,k\in N$. We say that $\bar{a}_k$ is equivalent to $\bar{b}_k$ and denoted by $\bar{a}_k E_{n,k}\bar{b}_k$ if and only if

$$\sum_{i=0}^{k} v_i a_i=\theta\leftrightarrow\sum_{i=0}^{k} v_i b_i=\theta$$

is true for every sequence $(v_0,v_1,\cdots,v_k)\in V_n^{k+1}$ where

$$a_0=b_0=\frac{1}{p^t}\theta \quad \text{and} \quad t=[\log_p 2^{2^{4n}}].$$

Definition 2.10 The bounding function is defined as

$$H(n,k,m)=2[(k+1)(m2^{2^{4n}})^{k+2}]^2 \text{ for } m>0, \qquad H(n,k,0)=1.$$

§ 2.2 The game $\mathbf{FR}_n(D_p,D_p)$

The quantifier elimination process starts from the innermost part of a formula. It finds an equivalent quantifier free formula $\psi(x_1,x_2,\cdots,x_n)$ for every formula $(\exists x)\phi(x,x_1,\cdots,x_n)$ with ϕ quantifier-free. This allows us to eliminate all of the quantifiers one by one. The main part of Szmielew's [8] was to give such a process. The Ehrenfeucht game starts from the outermost part of the formula. It changes a formula with quantifiers to an equivalent set of formulas with fewer quantifiers. After taking away the quantifiers we obtain a set of formulas. The validity of the original formula depends on the truth value of each of quantifier free formulas.

Lemma 2.11 *For any $\bar{a}_k,\bar{b}_k\in D_p^k$ satisfying $\bar{a}_k E_{n+1,k}\bar{b}_k$ with norms $\|b_1\|,\|b_2\|,\cdots,\|b_k\|\leqslant m$ and any $a_{k+1}\in D_p$ there exists a $b_{k+1}\in D_p$ with norm $\|b_{k+1}\|\leqslant H(n,k,m)$ such that $\bar{a}_{k+1}E_{n,k+1}\bar{b}_{k+1}$.*

Proof Define a set

$$A=\left\{\sum_{i=0}^{k} v_i a_i : v_i\in V_n, i=0,1,\cdots,k, t=[\log_p 2^{2^{4n}}], a_0=\frac{\theta}{p^t}\right\}.$$

Our proof is divided into two cases.

Case 1 There is a $v_{k+1}\in V_n-\{0\}$ such that

$$v_{k+1}a_{k+1}=-\sum_{i=0}^{k} v_i a_i\in A.$$

Let $v_i=r_i/s_i, i=0,1,\cdots,k+1, (s_i,p)=1$. From the definition of V_n we know that $|r_i|,|s_i|<2^{2^{4n}}$. Suppose that $r_{k+1}=p^u r,(r,p)=1$. We choose a suitable $v_{k+1}a_{k+1}\in A$ so that u reaches the minimum. Denote $s_{k+1}=s$. From $|p^u r|=|r_{k+1}|\leqslant 2^{2^{4n}}$ we know that $u\leqslant t$.

$$v_{k+1}a_{k+1}=-\sum_{i=0}^{k} v_i a_i, \qquad \frac{rp^u}{s}a_{k+1}=-\sum_{i=0}^{k}\frac{r_i}{s_i}a_i,$$

$$p^u a_{k+1}=-\sum_{i=0}^{k}\frac{sr_i}{rs_i}a_i, \qquad (rs_i,p)=1. \qquad ③$$

Let b_{k+1} be one of the elements satisfying

$$p^u b_{k+1} = -\sum_{i=0}^{k} \frac{sr_i}{rs_i} b_i. \qquad ④$$

We claim that for any $u_i \in V_n, i=0,1,\cdots,k+1$

$$\sum_{i=0}^{k+1} u_i a_i = \theta \to \sum_{i=0}^{k+1} u_i b_i = \theta$$

Let $u_i = x_i/y_i$ where $x_i, y_i \in \mathbf{Z}, (y_i, p)=1$ and $|x_i|, |y_i| \leqslant 2^{2^{4n}}$ for $i=0, 1, \cdots, k+1$.

Factor x_{k+1} into two parts

$$x_{k+1} = p^w x, \quad y_{k+1} = y \quad \text{and} \quad (x,p)=1.$$

From the minimality of u we know that $w \geqslant u$. Hence we derive the following identities:

$$-u_{k+1} a_{k+1} = \sum_{i=0}^{k} u_i a_i,$$

$$\frac{p^{w-u}x}{y}(-p^u a_{k+1}) = \sum_{i=0}^{k} \frac{x_i}{y_i} a_i,$$

$$\frac{p^{w-u}x}{y}\left(\sum_{i=0}^{k} \frac{sr_i}{rs_i} a_i\right) = \sum_{i=0}^{k} \frac{x_i}{y_i} a_i,$$

$$\sum_{i=0}^{k}\left(\frac{p^{w-u}xsr_i}{yrs_i} - \frac{x_i}{y_i}\right) a_i = \theta,$$

$$\sum_{i=0}^{k}\left(\frac{p^{w-u}xsr_i y_i - yrs_i x_i}{yrs_i y_i}\right) a_i = \theta. \qquad ⑤$$

For $i=1,2,\cdots,k$ we estimate the coefficients of ⑤:

$$|yrs_i y_i| \leqslant (2^{2^{4n}})^4 < 2^{2^{4(n+1)}},$$

$$|p^{w-u}xsr_i y_i - x_i yrs_i| \leqslant 2 \cdot 2(2^{2^{4n}})^5 \leqslant 2^{2^{4n}5+1} \leqslant 2^{2^{4n}6} < 2^{2^{4(n+1)}}.$$

We also know that $(yrs_i y_i, p)=1$. Hence

$$\frac{p^{w-u}xsr_i y_i - yrs_i x_i}{yrs_i y_i} \in V_{n+1}$$

for $i=1,2,\cdots,k$. According to the definition, a_0 in $E_{n,k+1}$ is not the same a_0 as in $E_{n+1,k}$. The latter one we now write as a_0'. Hence we have

$$a_0 = \frac{1}{p^t}\theta, \qquad a_0' = \frac{1}{p^{t'}}\theta,$$

where

$$t=[\log_p 2^{2^{4n}}] \quad \text{and} \quad t'=[\log_p 2^{2^{(4n+1)}}].$$

From Lemmas 2.4 and 2.5 we can find an integer l such that

$$\frac{p^{w-u}xsr_0y_0-yrs_0x_0}{yrs_0y_0}a_0=p^{t-t'}la_0,$$

where

$$|p^{t'-t}l|<p^{t'}\leqslant 2^{2^{4n+4}}.$$

Hence we also have

$$p^{t'-t}l\in V_{n+1}.$$

Now ⑤ can be written as

$$p^{t'-t}la_0+\sum_{i=1}^{k}\frac{p^{w-u}xsr_iy_i-yrs_ix_i}{yrs_iy_i}a_i=\theta. \qquad ⑥$$

All $k+1$ coefficients of ⑥ are in V_{n+1}. Therefore from $\bar{a}_k E_{n+1,k}\bar{b}_k$ we have

$$p^{t'-t}lb'_0+\sum_{i=1}^{k}\frac{p^{w-u}xsr_iy_i-yrs_ix_i}{yrs_iy_i}b_i=\theta. \qquad ⑦$$

Actually $a_0=b_0, a'_0=b'_0$. Hence ⑦ can be transformed back to ⑧.

$$\sum_{i=0}^{k}\frac{p^{w-u}xsr_iy_i-yrs_ix_i}{yrs_iy_i}b_i=\theta. \qquad ⑧$$

Deriving upward from ⑤ to ④ and changing b_i into a_i we obtain

$$\sum_{i=0}^{k+1}u_ib_i=\theta. \qquad ⑨$$

We have finished half of the proof of equivalence. Conversely, suppose we have ⑨ where $u_i=x_i/y_i\in V_n$ for $i=0,1,\cdots,k+1$. Again let

$$u_{k+1}=\frac{x_{k+1}}{y_{k+1}}=\frac{p^w x}{y} \quad \text{and} \quad (x,p)=1.$$

We need to prove that $\sum_{i=0}^{k+1}u_ia_i=\theta$. We claim first that $w\geqslant u$ because $w<u$ leads to a contradiction.

From ⑨ we have

$$-\frac{p^w x}{y}b_{k+1}=\sum_{i=0}^{k}\frac{x_i}{y_i}b_i, \quad p^w b_{k+1}=-\sum_{i=0}^{k}\frac{yx_i}{xy_i}b_i.$$

From the definition of b_{k+1} we have

$$p^u b_{k+1}=-\sum_{i=0}^{k}\frac{sr_i}{rs_i}b_i.$$

Hence

$$p^{u-w}\left(-\sum_{i=0}^{k}\frac{yx_i}{xy_i}b_i\right)=-\sum_{i=0}^{k}\frac{sr_i}{rs_i}b_i,$$

$$\sum_{i=0}^{k}\left(\frac{sr_i}{rs_i}-\frac{p^{u-w}yx_i}{xy_i}\right)b_i=\theta. \tag{⑩}$$

The coefficients of ⑩ can be estimated as we did for ⑤. Hence from $\bar{b}_k E_{n+1,k}\bar{a}_k$ we can transform ⑩ back to its corresponding formula

$$p^{u-w}\left(-\sum_{i=0}^{k}\frac{yx_i}{xy_i}a_i\right)=-\sum_{i=0}^{k}\frac{sr_i}{rs_i}a_i.$$

From ③ we have

$$p^u a_{k+1}=-\sum_{i=0}^{k}\frac{sr_i}{rs_i}a_i.$$

The right-hand side $-\sum_{i=0}^{k}(sr_i/rs_i)a_i$ is p^{u-w} times both $-\sum_{i=0}^{k}(yx_i/xy_i)a_i$ and $p^w a_{k+1}$. We can use Lemma 2. 5 to find an $L(0\leqslant L<p^{u-w})$ such that

$$p^w a_{k+1}=Lb-\sum_{i=0}^{k}\frac{yx_i}{xy_i}a_i. \tag{⑪}$$

where $b=(1/p^{u-w})\theta$. Because $(x,p)=(y,p)=1$ we can write

$$b=\frac{\theta}{p^{u-w}}=\frac{y}{x}\left(\frac{x}{y}\cdot\frac{1}{p^{u-w}}\theta\right).$$

Hence formula ⑪ becomes

$$\begin{aligned}p^w a_{k+1}&=-\sum_{i=1}^{k}\frac{yx_i}{xy_i}a_i-\left(\frac{yx_0}{xy_0}a_0-\frac{y}{x}\cdot\frac{Lx}{y}\cdot\frac{1}{p^{u-w}}\theta\right)\\&=-\sum_{i=1}^{k}\frac{yx_i}{xy_i}a_i-\frac{y}{x}\left(\frac{x_0}{y_0}\cdot\frac{1}{p^t}\theta-\frac{Lx}{y}\cdot\frac{1}{p^{u-w}}\theta\right)\\&=-\sum_{i=1}^{k}\frac{yx_i}{xy_i}a_i-\frac{y}{x}L'\frac{1}{p^t}\theta\end{aligned} \tag{⑫}$$

where $t=[\log_p 2^{2^{4n}}]$ and $u-w\leqslant t$. The subset generated by $(1/p^t)\theta$ is a Q_p-submodule with exactly p^t elements. Hence we can find an integer L' with $0\leqslant L'<p^t$ for formula ⑫. Therefore we have

$$p^w a_{k+1}=-\frac{y}{x}\left(\sum_{i=1}^{k}\frac{x_i}{y_i}a_i+L'a_0\right)\quad\text{and}$$

$$\frac{p^w x}{y}a_{k+1}=\sum_{i=1}^{k}\frac{x_i}{y_i}a_i+L'a_0. \tag{⑬}$$

In formula ⑬, x_i/y_i is from ⑨, and

$$|p^w x| < |p^u x| \leqslant 2^{2^{4n}} \quad \text{and} \quad |y|, |L'| < 2^{2^{4n}}.$$

All coefficients of ⑬ are in V_n, but $w<u$. This contradicts the minimality of u. Therefore we have $w \geqslant u$.

By a method similar to the first half we can prove that

$$\sum_{i=0}^{k+1} u_i b_i = \theta \quad \text{implies} \quad \sum_{i=0}^{k+1} u_i a_i = \theta.$$

Hence we have $\bar{a}_{k+1} E_{n,k+1} \bar{b}_{k+1}$.

Now we need to check that the norm of b_{k+1} satisfies the requirements. To estimate the norm we change the b_i back to their forms related to θ. From Lemma 2. 7 we can write

$$b_i = \frac{e_i}{f_i} \cdot \frac{\theta}{p^{l(i)}}$$

where

$$l(i) \geqslant 0, \quad (e_i, p) = (f_i, p) = 1 \quad \text{and} \quad |e_i|, |f_i p^{l(i)-1}| \leqslant m.$$

Putting these back in a formula similar to ③ gives

$$p^u b_{k+1} = \sum_{i=0}^{k} \frac{sr_i}{rs_i} b_i = \sum_{i=0}^{k} \frac{sr_i e_i}{rs_i f_i} \cdot \frac{\theta}{p^{l(i)}}.$$

We can choose b_{k+1} so that

$$\begin{aligned} b_{k+1} &= \sum_{i=0}^{k} \frac{sr_i e_i}{rs_i f_i} \cdot \frac{\theta}{p^{l(i)+u}} \\ &= \frac{\sum_{i=0}^{k} s_0 f_0 \cdots s_{i-1} f_{i-1} r_i e_i p^{l-l(i)} s_{i+1} f_{i+1} \cdots s_k f_k}{rp^{l+u} s_0 f_0 \cdots s_k f_k} \theta = \frac{\sum}{\prod} \theta. \end{aligned}$$

where $l = \mathrm{Max}\{l(i)\}$. Hence

$$\| b_{k+1} \| \leqslant \mathrm{Max}\{|\sum|, |\prod|\}.$$

Recall the upper bounds of the entries:

$$p^{l-l(i)} \leqslant p^l \leqslant 2^{2^{4n}}, \quad |r_i|, |s_i| \leqslant 2^{2^{4n}} \quad \text{and} \quad |e_i|, |f_i| \leqslant m.$$

Hence we have

$$|\sum| \leqslant (k+1)(2^{2^{4n}})^{k+2} m^{k+2} = (k+1)(m 2^{2^{4n}})^{k+2},$$

$$|\prod| \leqslant (m 2^{2^{4n}})^{k+2}.$$

Therefore

$$\| b_{k+1} \| \leqslant (k+1)(m 2^{2^{4n}})^{k+2} \leqslant H(n,k,m).$$

Case 2 There is no $v_{k+1} \in V_n - \{0\}$ such that $v_{k+1} a_{k+1} \in A$. As in the calculation in Case 1 we know that

$$v_{k+1} b_{k+1} \in B = \left\{ \sum_{i=0}^{k} v_i b_i : v_i \in V_n \right\}$$

only if $\| b_{k+1} \| \leqslant (k+1)(m2^{2^{4n}})^{k+2}$.

Hence if b_{k+1} is chosen to satisfy

$$\| b_{k+1} \| > (k+1)(m2^{2^{4n}})^{k+2},$$

then $v_{k+1} b_{k+1} \notin B$ for all $v_{k+1} \in V_n - \{0\}$. By Lemma 2.7 we can find a b_{k+1} which satisfies

$$\| b_{k+1} \| \leqslant 2[(k+1)(m2^{2^{4n}})^{k+2}]^2 \leqslant H(n,k,m).$$

For this b_{k+1} we can easily derive the following equivalence

$$\sum_{i=0}^{k+1} v_i a_i = \theta \leftrightarrow v_{k+1} = 0 \quad \text{and} \quad \sum_{i=0}^{k} v_i a_i = \theta$$

$$\leftrightarrow v_{k+1} = 0 \quad \text{and} \quad \sum_{i=0}^{k} v_i b_i = \theta \leftrightarrow \sum_{i=0}^{k+1} v_i b_i = \theta.$$

Therefore we obtain finally $\bar{a}_{k+1} E_{n,k+1} \bar{b}_{k+1}$. Our lemma is proved. □

§2.3 The upper bound for D_p

Lemma 2.12 *Let* $m_0 = 1$ *and* $m_{k+2} = H(n-k, k, m_k)$ *for* $k = 0, 1, \cdots, n-1$. *Then we have*

$$m_k \leqslant 2^{2^{4(n+k)\log(n+1)}} \quad \text{for } n \geqslant 3.$$

Proof We proceed by induction. Suppose that

$$m_k \leqslant 2^{2^{4(n+k)\log(n+1)}}.$$

We prove the m_{k+1}. By Definition 2.10,

$$m_{k+1} \leqslant 2(k+1)^2 (2^{2^{4(n+k)\log(n+1)}} 2^{2^{4n}})^{2(k+2)}$$

$$= 2(k+1)^2 \cdot 2^{(2^{4(n+k)\log(n+1)} + 2^{4n})2(k+2)}.$$

It is easy to see that

$$2(k+2) = 2^{\log(2(k+2))} = 2^{1+\log(k+2)}.$$

Hence

$$m_{k+1} \leqslant 2(k+1)^2 2^{(2^{4(n+k)\log(n+1)} + 2^{4n})2^{1+\log(k+2)}}$$

$$\leqslant 2(k+1)^2 2^{(2^{4(n+k)\log(n+1)+1} 2^{1+\log(k+2)})}$$

$$\leqslant 2(k+1)^2 2^{2^{4(n+k)\log(n+1)+2+\log(k+2)}}.$$

Since $n \geqslant 3, k \leqslant n-1$ we have

$$2(k+2)^2 \leqslant 2(2n)^2 = 8n^2 \leqslant 8 \cdot 2^n$$
$$\leqslant 2^{n+3} \leqslant 2^{2^{4(k+n)\log(n+1)+2+\log(k+2)}}.$$

Hence

$$m_{k+1} \leqslant 2^{2^{4(n+k)\log(n+1)+3+\log(n+1)}}.$$

From $\log(n+2) \geqslant 2$ we have

$$3+\log(n+1) \leqslant 4\log(n+1).$$

Therefore we obtain

$$m_{k+1} \leqslant 2^{2^{4(n+k+1)\log(n+1)}}. \qquad \square$$

The following theorem gives an upper bound for the complexity of the theory of D_p.

Theorem 2.13 *It takes at most*

$$2^{2^{9n\log(n+1)}}$$

steps to decide a sentence ϕ with quantifier-depth n in D_p.

Proof From the lemmas it follows that for deciding ϕ we need to check

$$\prod_{k=0}^{n-1} 2m_k^2$$

quantifier free formulas. Hence the total number of steps of checking is at most

$$\prod_{k=0}^{n-1} 2m_k^2 \leqslant \prod_{k=0}^{n-1} 2^n \left(2^{2^{4(n+k+1)\log(n+1)}}\right)^{2n}$$
$$\leqslant 2^{\left(\sum 2^{4(n+k+1)\log(n+1)}\right)2n+n} \leqslant 2^{2^{8n\log(n+1)+n+1}} \leqslant 2^{2^{9n\log(n+1)}}. \qquad \square$$

§ 3. The game $\mathbf{FR}_n(Q_p, Q_p)$ and the upper bound for $\mathbf{Th}(Q_p)$

§ 3.1 The main definitions

We consider Q_p itself as a Q_p-module. There is no danger of confusion to use the same symbol Q_p to denote the module Q_p and the ring Q_p. The zero element of the group Q_p is θ while that of the ring Q_p is written as 0. We also need to define the norm $\| x \|$ for every x in the group Q_p, the coefficient set V_n which is a subset of the ring Q_p, and the function $H(n,k,m): N^3 \to N$. The same symbols are used for most of these concepts as those in Section 2, but they are to be understood as defined in this section.

Definition 3.1 The norm of an element $x \in Q_p$ is

$$\| x \| = \min_{r/s=x} \{\mathrm{Max}\{|r|, |s|\}\}.$$

Definition 3.2 The coefficient sets are

$$V_i = \left\{ \frac{r}{s} \in Q_p : |r|, |s| \leqslant 2^{2^{4i}} \right\} \quad \text{for } i=0,1,\cdots,n.$$

Definition 3.3 For any $\bar{a}_k, \bar{b}_k \in (Q_p)^k$ the equivalence relation $\bar{a}_k E_{n,k} \bar{b}_k$ is true if and only if for every $\bar{v}_k \in (V_n)^k$ the following two conditions are true in Q_p.

(i) $\sum_{i=1}^{k} v_i a_i = \theta \leftrightarrow \sum_{i=1}^{k} v_i b_i = \theta$,

(ii) $p^w \Big| \sum_{i=1}^{k} v_i a_i \leftrightarrow p^w \Big| \sum_{i=1}^{k} v_i b_i$

for all w with $0 \leqslant w \leqslant [\log_p 2^{2^{4n}}]$. Here $n|a$ means that there is a $b \in Q_p$ such that $nb=a$.

The definition of the equivalence relation $E_{n,k}$ is different from that in Section 2. There was only one condition in Definition 2.9 but we need two here. The second condition is needed for the induction steps. We did not give an analogous condition in Section 2 because it is trivial in the group D_p.

Definition 3.4 The bounding function for the game $\mathrm{FR}_n(Q_p, Q_p)$ is

$$H(n,k,m) = (k+1)(m 2^{2^{4n}})^{k+1}.$$

§ 3.2 The congruence relation in Q_p

Before going to the Ehrenfeucht game we need some lemmas about both group and ring Q_p. We extend the concept of congruence modulo p^n to Q_p.

In this section the expression $r/s \in Q_p$ means that r, s are integers, and coprime.

Definition 3.5 Two elements $x, y \in Q_p$ are congruent modulo p^u,

$u \geqslant 0$ written $x \equiv y \pmod{p^u}$, if and only if their difference lies in the ideal $p^u Q_p$. If $x = a/b, y = c/b, p \nmid b, d$, then $x \equiv y \pmod{p^u}$ if and only if $ad \equiv bc \pmod{p^u}$ in Z.

This relation is indeed a congruence on Q_p.

Definition 3.6 Let $x = (p^e a/b), e \geqslant 0, p \nmid a, b$. Then the ideal $(x) = (p^e)$. Thus x is a unit (there exists a y such that $xy = 1$) if $e = 0$, i. e., $x = a/b, p \nmid a, b$. The notation $(a/b, p) = 1$ means $p \nmid a/b$, i. e., $p \nmid x$.

Lemma 3.7 *Suppose that $\bar{a}_k, \bar{b}_k$ are two vectors on $(O_p)^k$. Then the following two conditions are equivalent.*

(i) $\sum_{i=1}^k r_i a_i \equiv \theta \pmod{p^n} \leftrightarrow \sum_{i=1}^k r_i b_i \equiv \theta \pmod{p^n}$. *holds in Q_p for all $\bar{r}_k \in Q_p^k$.*

(ii) *There exists an λ with $(\lambda, p) = 1$ such that $\bar{b}_k \equiv \lambda \bar{a}_k \pmod{p^n}$.*

Proof It is immediate that (ii) implies (i). For the converse, assume (i). Note that $Q_p / p^n Q_p$ is a Q_p-module, cyclic of order p^n.

For each $i (0 \leqslant i \leqslant k)$, let r_i be arbitrary and all other $r_j = 0$. Then

$$r_i a_i \equiv \theta \leftrightarrow r_i b_i \equiv \theta \pmod{p^n}.$$

This implies that a_i and b_i have the same order. Since the group $Q_p / p^n Q_p$ is isomorphic to $C_{p,n}$, this implies that $b_i \equiv \lambda_i a_i$ for some unit $\lambda_i \in Q_p$.

Factor a_i into two parts $a_i \equiv c_i p^{l(i)}$ for $i = 1, 2, \cdots, k$. Choose an a_i such that $l(i)$ reaches the minimum and write $\lambda_i = \lambda$. Let $j \neq i$, then the order of a_j is less than that of a_i. We can find an m such that $a_j \equiv m a_i \pmod{p^n}$. By (i) $b_j \equiv m b_i \pmod{p^n}$. Whence

$$b_j \equiv m \lambda_j a_i \equiv m b_i \equiv m \lambda_i a_i \pmod{p^n}.$$

Hence $m a_i (\lambda_i - \lambda_j) \equiv \theta \pmod{p^n}$. This implies that $m\lambda_i \equiv m\lambda_j \pmod{p^{n-l(i)}}$. Therefore

$$\lambda a_j \equiv \lambda_i a_j \equiv \lambda_i m a_i \equiv \lambda_j m a_i \equiv m \lambda_j a_i \equiv m b_i \equiv b_j \pmod{p^n}. \quad \square$$

§ 3.3 The game $\mathrm{FR}_n(Q_p, Q_p)$

Lemma 3.8 *For any $\bar{a}_k, \bar{b}_k \in (Q_p)^k$ satisfying $\bar{a}_k E_{n+1} \bar{b}_k$ with norms $\| b_1 \|, \| b_2 \|, \cdots, \| b_k \| \leqslant m$ and any $a_{k+1} \in Q_p$ there is a $b_{k+1} \in Q_p$ with norm $\| b_{k+1} \| \leqslant H(n, k, m)$ such that $\bar{a}_{k+1} E_{n,k+1} \bar{b}_{k+1}$.*

Proof Define a set

$$A=\left\{\sum_{i=1}^{k} v_i a_i : v_i \in V_n, i=1,2,\cdots,k\right\}.$$

Our proof is divided into two cases.

Case 1 There is a $v_{k+1} \in V_n - \{0\}$ such that

$$v_{k+1}a_{k+1} = -\sum_{i=1}^{k} v_i a_i \in A.$$

Let $v_i = r_i/s_i, i=1,2,\cdots,k+1, (s_i,p)=1$. From the definition of V_n we know that $|r_i|, |s_i| < 2^{2^{4n}}$. Suppose that $r_{k+1}=p^u r, (r,p)=1$. We choose a suitable $v_{k+1}a_{k+1} \in A$ so that u reaches the minimum. Denote s_{k+1} by s. From above we know that

$$v_{k+1}a_{k+1} = -\sum_{i=1}^{k} v_i a_i,$$

$$\frac{rp^u}{s}a_{k+1} = -\sum_{i=1}^{k}\frac{r_i}{s_i}a_i,$$

$$p^u a_{k+1} = -\sum_{i=1}^{k}\frac{sr_i}{rs_i}a_i, \quad (rs_i,p)=1. \tag{14}$$

Let b_{k+1} be the element satisfying

$$p^u b_{k+1} = -\sum_{i=1}^{k}\frac{sr_i}{rs_i}b_i. \tag{15}$$

The existence of b_{k+1} comes from the condition (ii) of equivalence.

We claim that for any $u_i \in V_n, i=1,2,\cdots,k+1$,

$$\sum_{i=1}^{k+1} u_i a_i = \theta \to \sum_{i=1}^{k+1} u_i b_i = \theta.$$

Let $u_i = x_i/y_i$ where $x_i, y_i \in Z$. $(y_i,p)=1$ and $|x_i|, |y_i| \leqslant 2^{2^{4n}}$ for $i=1, 2,\cdots,k+1$. We factor x_{k+1} into two parts

$$x_{k+1}=p^w x, \quad y_{k+1}=y \quad \text{and} \quad (x,p)=1.$$

From the minimality of u we know that $w \geqslant u$. Hence we derive the following identities.

$$-u_{k+1}a_{k+1} = \sum_{i=1}^{k} u_i a_i, \tag{16}$$

$$\frac{p^{w-u}x}{y}(-p^u a_{k+1}) = \sum_{i=1}^{k}\frac{x_i}{y_i}a_i, \quad \frac{p^{w-u}x}{y}\left(\sum_{i=1}^{k}\frac{sr_i}{rs_i}a_i\right) = \sum_{i=1}^{k}\frac{x_i}{y_i}a_i,$$

$$\sum_{i=1}^{k}\frac{p^{w-u}xsr_i}{yrs_i} - \frac{x_i}{y_i}a_i = \theta. \quad \sum_{i=1}^{k}\frac{p^{w-u}xsr_i - yrs_ix_i}{yrs_iy_i}a_i = \theta. \tag{17}$$

For $i=1,2,\cdots,k$ we estimate the coefficients of ⑰.

$$|yrs_iy_i|\leqslant(2^{2^{4n}})^4<2^{2^{4(n+1)}},$$

$$|p^{w-u}xsr_iy_i-x_iyrs_i|\leqslant 2(2^{2^{4n}})^5\leqslant 2^{2^{4n}5+1}\leqslant 2^{2^{4n}6}<2^{2^{4(n+1)}}.$$

We also know that $(yrs_iy_i,p)=1$. Hence

$$\frac{p^{w-u}xsr_iy_i-yrs_ix_i}{yrs_iy_i}\in V_{n+1}$$

for $i=1,2,\cdots,k$. Therefore from $\bar{a}_kE_{n+1,k}\bar{b}_k$ we have

$$\sum_{i=1}^{k}\frac{p^{w-u}xsr_iy_i-yrs_ix_i}{yrs_iy_i}b_i=\theta.$$

Deriving upward from ⑰ to ⑯ and changing b_i into a_i we obtain

$$\sum_{i=1}^{k+1}u_ib_i=\theta. \qquad ⑱$$

Conversely, suppose we have

$$\sum_{i=1}^{k+1}u_ib_i=\theta,$$

where

$$u_i=\frac{x_i}{y_i}\in V_n\quad\text{for } i=1,2,\cdots,k+1.$$

Again let

$$u_{k+1}=\frac{x_{k+1}}{y_{k+1}}=\frac{p^wx}{y}\quad\text{and}\quad(x,p)=1.$$

We need to prove that

$$\sum_{i=1}^{k+1}u_ia_i=\theta.$$

We claim first that $w\geqslant u$ because $w<u$ leads to a contradiction.

From (2′) we have

$$-\frac{p^wx}{y}b_{k+1}=\sum_{i=1}^{k}\frac{x_i}{y_i}b_i,\quad p^wb_{k+1}=-\sum_{i=1}^{k}\frac{yx_i}{xy_i}b_i.$$

From the definition of b_{k+1} we have

$$p^ub_{k+1}=-\sum_{i=1}^{k}\frac{sr_i}{rs_i}b_i.$$

Hence

$$p^{u-w}\left(-\sum_{i=1}^{k}\frac{yx_i}{xy_i}b_i\right)=-\sum_{i=1}^{k}\frac{sr_i}{rs_i}b_i. \qquad ⑲$$

Transpose ⑲ to one side and estimate the coefficients as we did for ⑰. Hence from $\bar{b}_k E_{n+1,k}\bar{a}_k$ we can transform ⑲ back to

$$p^{u-w}\left(-\sum_{i=1}^{k}\frac{yx_i}{xy_i}a_i\right)=-\sum_{i=1}^{k}\frac{sr_i}{rs_i}a_i.$$

From ⑭ we have

$$p^u a_{k+1}=-\sum_{i=1}^{k}\frac{sr_i}{rs_i}a_i.$$

The right hand side $-\sum_{i=1}^{k}(sr_i/rs_i)a_i$ is p^{u-w} times both $-\sum_{i=1}^{k}(yx_i/xy_i)a_i$ and $p^w a_{k+1}$.

Therefore we have

$$p^w a_{k+1}=-\frac{y}{x}\cdot\sum_{i=1}^{k}\frac{x_i}{y_i}a_i \quad \text{and} \quad \frac{p^w x}{y}a_{k+1}=\sum_{i=1}^{k}\frac{x_i}{y_i}a_i. \qquad ⑳$$

In formula ⑳ $x_i/y_i\in u_i\in V_n$ is from ⑱. This gives

$$|p^w x|<|p^u x|\leqslant 2^{2^{4n}} \quad \text{and} \quad |y|<2^{2^{4n}}.$$

All coefficients of ⑳ are in V_n, but $w<u$. This contradicts the minimality of u. Therefore we have $w\geqslant u$.

By a method similar to that of the first half we can prove that

$$\sum_{i=1}^{k+1}u_i b_i=\theta\rightarrow\sum_{i=1}^{k+1}u_i a_i=\theta.$$

Now we need to check the second condition of equivalence of Definition 3. 4. This is satisfied because, from ⑭ and ⑮, $p^u\sum_{i=1}^{k+1}u_i a_i$ and $p^u\cdot\sum_{i=1}^{k+1}u_i b_i$ can be written as the linear combinations of $a_1,a_2,\cdots,a_k$ and $b_1,b_2,\cdots,b_k$ with the same set of coefficients respectively. By a method similar to that above we can prove that these coefficients are in V_{k+1}. From $\bar{a}_k E_{n+1,k}\bar{b}_k$ we know that

$$p^w \mid p^u\sum_{i=1}^{k+1}u_i a_i\leftrightarrow p^w \mid p^u\cdot\sum_{i=1}^{k+1}u_i b_i$$

for any $u_1,u_2,\cdots,u_{k+1}\in V_n$ and $0\leqslant w\leqslant\log_p 2^{2^{4(n+1)}}$. Hence

$$p^w\left|\sum_{i=1}^{k+1}u_i a_i\leftrightarrow p^w\right|\sum_{i=1}^{k+1}u_i b_i.$$

For any $u_1,u_2,\cdots,u_{k+1}\in V_n$ and $0\leqslant w\leqslant\log_p 2^{2^{4n}}$. Therefore we obtain

$\bar{a}_{k+1}E_{n,k+1}\bar{b}_{k+1}$.

Now we need to check that the norm of b_{k+1} satisfies the requirements. Write

$$b_i=\frac{e_i}{f_i} \quad \text{where} \quad |e_i|,|f_i|\leqslant m.$$

Put these back into ⑮, to obtain

$$p^u b_{k+1}=\sum_{i=1}^{k}\frac{sr_i}{rs_i}b_i=\sum_{i=1}^{k}\frac{sr_ie_i}{rs_if_i}.$$

We can choose b_{k+1} so that

$$b_{k+1}=\sum_{i=1}^{k}\frac{sr_ie_i}{rs_if_i}=\frac{\sum_{i=1}^{k}ss_1f_1\cdots s_{i-1}f_{i-1}r_ie_is_{i+1}f_{i+1}\cdots s_kf_k}{s_0f_0\cdots s_kf_krp^u}=\frac{\sum}{\prod}.$$

Hence

$$\|b_{k+1}\|\leqslant \mathrm{Max}\left\{\left|\sum\right|,|\Pi|\right\}.$$

Recall the upper bounds of the entries,

$$|rp^u|,|s|,|r_i|,|s_i|\leqslant 2^{2^{4n}} \quad \text{and} \quad |e_i|,|f_i|\leqslant m.$$

Hence we have

$$\left|\sum\right|\leqslant k(2^{2^{4n}})^{k+1}m^{k+1}=k(m2^{2^{4n}})^{k+1},$$

$$|\Pi|\leqslant(m2^{2^{4n}})^{k+1}.$$

Therefore

$$\|b_{k+1}\|\leqslant k(m2^{2^{4n}})^{k+1}\leqslant H(n,k,m).$$

Case 2 There is no $v_{k+1}\in V_n-\{0\}$ such that $v_{k+1}a_{k+1}\in A$. By a calculation similar to that in Case 1 we find that

$$v_{k+1}b_{k+1}\in B=\left\{\sum_{i=1}^{k}v_ib_i:v_i\in V_n\right\}$$

only if $\|b_{k+1}\|\leqslant(k+1)(m2^{2^{4n}})^{k+2}$. Hence if b_{k+1} is chosen to satisfy

$$\|b_{k+1}\|>(k+1)(m2^{2^{4n}})^{k+2}.$$

then $v_{k+1}b_{k+1}\notin B$ for all $v_{k+1}\in V_n-\{0\}$. For this b_{k+1} we can easily check that the condition (i) of equivalence

$$\sum_{i=1}^{k+1}v_ia_i=\theta\leftrightarrow\sum_{i=1}^{k+1}v_ib_i=\theta$$

is true because $v_{k+1}=0$.

For the condition (ii) of equivalence we need to prove that

$$p^w \left| \sum_{i=1}^{k+1} v_i a_i \leftrightarrow p^w \right| \sum_{i=1}^{k+1} v_i b_i$$

for all $v_i \in V_n$ and $0 \leqslant w \leqslant \log_p 2^{2^{4n}}$.

From Definition 3. 4 and hypothesis we know that for any sequence $(r_1, r_2, \cdots, r_k) \in V_{n+1}^k$ we have

$$p^u \left| \sum_{i=1}^{k} r_i a_i \leftrightarrow p^u \right| \sum_{i=1}^{k} r_i b_i$$

where $0 \leqslant u \leqslant \log_p 2^{2^{4(n+1)}}$. According to Lemma 3. 7 there is a λ with $(\lambda, p)=1$ such that

$$(b_1, b_2, \cdots, b_k) \equiv \lambda (a_1, a_2, \cdots, a_k) (\bmod\ p^u).$$

Let $t=[\log_p 2^{2^{4n}}]$. We have

$$(b_1, b_2, \cdots, b_k) \equiv \lambda (a_1, a_2, \cdots, a_k) (\bmod\ p^t).$$

Let $\lambda = c/d$ and $a_{k+1}=x/y$. Choose natural numbers x_1, y_1 such that

$$x_1 \equiv cx\ (\bmod\ p^t), \quad y_1 \equiv dy\ (\bmod\ p^t) \quad \text{and} \quad |x_1|, |y_1| \leqslant p^t.$$

Choose a natural number l such that

$$k(m2^{2^{4n}})^{k+1} \cdot 2^{2^{4n}} < x_1 + lp^t \leqslant (k+1)(m2^{2^{4n}})^{k+1}.$$

This is possible because

$$(k+1)(m2^{2^{4n}})^{k+1} - k(m2^{2^{4n}})^{k+1} \geqslant m2^{2^{4n}} \geqslant p^t.$$

Hence if $(x_1+lp^t)/y_1 = x_2/y_2$ where x_2/y_2 is a reduced fraction we still have

$$|x_2| > k(m2^{2^{4n}})^{k+1}.$$

Therefore

$$k(m2^{2^{4n}})(k+1) < \operatorname{Max}\{|x_2|, |y_2|\} \leqslant (k+1)(m2^{2^{4n}})^{k+1}.$$

Let $b_{k+1}=x_2/y_2$. Then we obtain

$$b_{k+1} \equiv \frac{x_2}{y_2} \equiv \frac{x_1+lp^t}{y_1} \equiv \frac{x_1}{y_1} \equiv \frac{cx}{dy} \equiv \lambda a_{k+1} (\bmod\ p^t),$$

$$b_{k+1} \in B \quad \text{and} \quad \| b_{k+1} \| < (k+1)(m2^{2^{4n}})^{k+1} \leqslant H(n,k,m).$$

From $(b_1, b_2, \cdots, b_{k+1}) = (a_1, a_2, \cdots, a_{k+1}) (\bmod\ p^t)$ we know that for any $v_1, v_2, \cdots, v_{k+1} \in V_n$ and any $0 \leqslant w \leqslant t$ the following is true

$$p^w \left| \sum_{i=1}^{k+1} v_i b_i \leftrightarrow p^w \right| \sum_{i=1}^{k+1} v_i \lambda a_i \leftrightarrow p^w \left| \lambda \sum_{i=1}^{k+1} v_i a_i \leftrightarrow p^w \right| \sum_{i=1}^{k+1} v_i a_i.$$

Hence $\bar{a}_{k+1} \bar{E}_{n,k+1} \bar{b}_{k+1}$. □

Theorem 3.9 *It takes at most*

$$2^{2^{8n \log(n+1)}}$$

steps to decide a sentence ϕ *with quantifier-depth* n *in* Q_p.

Proof Similar to the proof of Theorem 2.13. □

§ 4. The reduction of $C_{p,l}$ with large p^l

Let A be a group of type $C_{p,l}$ with $p^l \geqslant 2^{2^{4n-2}}$. Take another group B of type $C_{p,m}$ with also $p^m \geqslant 2^{2^{4n-2}}$. Let l, m and n be fixed throughout this section. We will prove that $A \equiv_n B$. The proof uses the Ehrenfeucht game $G_n(A, B)$. We also need the definitions of the coefficient sets and the equivalence relation but we do not use the definition of norm in this chapter.

Definition 4.1 Choose and keep fixed two sequences of elements in A, B respectively $g_0 = \theta, h_0 = \tau$, where θ and τ are the identities of A and B, and

$$\frac{\theta}{p^i} = g_i \quad \text{for } i = 1, 2, \cdots, l \quad \text{and} \quad \frac{\tau}{p^j} \quad \text{for } j = 1, 2, \cdots, m,$$

such that $g = g_l$ and $h = h_m$ are the generators of A and B.

Definition 4.2 The coefficient sets are

$$V_i = \left\{ \frac{r}{s} : \| r \|, \| s \| \leqslant 2^{2^{4i}} \right\} \quad \text{for } i = 1, 2, \cdots, n+1.$$

Definition 4.3 Let $\bar{a}_k \in A^k$ and $\bar{b} \in B^k$. The equivalence relation $\bar{a}_k E_{v,k} \bar{b}_k$ holds if and only if for $v_0, v_1, \cdots, v_k \in V_v$ and $t = [\log_p 2^{2^{4v}}]$ the following two conditions are satisfied:

(i) $\sum_{i=0}^{k} v_i a_i = \theta \leftrightarrow \sum_{i=0}^{k} v_i b_i = \tau$, where $a_0 = \theta / p^t$ and $b_p = \tau / p^t$;

(ii) $p^u \mid \sum_{i=0}^{k} v_i a_i \leftrightarrow p^u \mid \sum_{i=0}^{k} v_i b_i$ for any $0 \leqslant u \leqslant t$.

Lemma 4.4 *If* v *is less than* n, *then the two terms* $v_0 a_0$ *and* $v_0 b_0$ *are unnecessary for the condition* (ii).

Proof From Definition 4.2 we have

$$t=[\log 2^{2^{4v}}]\leqslant[\log 2^{2^{4n-1}}] \quad \text{and} \quad p^t\leqslant 2^{2^{4(n-1)}}.$$

Hence $a_0=\theta/p^t=p^{l-t}g$ and $b_0=\tau/p^t=p^{m-t}h$ where

$$p^{l-t}=\frac{p^l}{p^t}\geqslant 2^{2^{4n-2}}/2^{2^{4(n-1)}}\geqslant 2^{2^{4(n-1)}}\geqslant p^t.$$

Similarly we have $p^{m-t}\geqslant p^t$. Hence $p^u\mid p^t\mid a_0$ and b_0. □

Lemma 4.5 *The following two conditions are equivalent.*

(i) *For any* $v_0, v_1, \cdots, v_k \in V_v$ *and* $0\leqslant u\leqslant t$ *where* $v<n$ *and* $t=[\log_p 2^{2^{4v}}]$ *we have*

$$p^u\left|\sum_{i=1}^{k} v_i a_i \leftrightarrow p^u\right|\sum_{i=1}^{k} v_i b_i.$$

(ii) *Change* a_i, b_i *to the forms* $a_i=X_i g, b_i=Y_i h$ *where* X_i, Y_i *are integers. Then there exists* $\lambda\in\mathbf{Z}$ *with* $(\lambda,p)=1$ *such that* $\bar{Y}_k=\lambda\bar{X}_k (\mathrm{mod}\ p^t)$.

Proof Similar to the proof of 3.7. □

Lemma 4.6 *There exists a winning strategy for player* Ⅱ *in the Ehrenfeucht game* $G_n(A,B)$.

Proof The game plays as follows.

Step 0. $\varnothing E_{n,0}\varnothing$ is true by Definition 4.3.

Step $k+1$. Suppose that at the end of step k the two players have chosen $\bar{a}_k$ and $\bar{b}_k$ satisfying $\bar{a}_k E_{n-k,k}\bar{b}_k$. Now for every $a_{k+1}\in A$ we want to find a $b_{k+1}\in B$ such that $\bar{a}_{k+1}E_{n-k-1,k+1}\bar{b}_{k+1}$.

Case 1 There are $v_0, v_1, \cdots, v_{k+1}\in V_{n-k-1}$ such that

$$\sum_{i=0}^{k+1} v_i a_i=\theta \quad \text{and} \quad v_{k+1}\neq 0.$$

For checking the condition (i) of Definition 4.3 we can use the same method as we used in the proof of Lemma 2.11 except for the existence of b_{k+1} which was trivial in Lemma 2.11 but now is derived from the induction hypothesis and the condition (ii) of Definition 4.3 satisfied at step k. The condition (ii) can be derived by Lemma 4.5 using a method similar to the proof of Lemma 3.8, Case 2.

Case 2 For any $v_0, v_1, \cdots, v_{k+1}\in v_{k+1}\in V_{n-k-1}$ with $v_{k+1}\neq 0$ we have $\sum_{i=0}^{k+1} v_i a_i\neq\theta$. Suppose that

$$a_i = X_i g \quad \text{for } i=0,1,\cdots,k+1 \quad \text{and}$$
$$b_i = Y_i h \quad \text{for } i=0,1,\cdots,k.$$

From Lemma 4.5 we know that there is a λ with $(\lambda, p)=1$ such that

$$\bar{b}_k = \lambda \bar{a}_k (\bmod\ p^t).$$

Consider the congruence relation

$$Y \equiv \lambda X_{k+1} (\bmod\ p^t)$$

and the inequality

$$\sum_{i=0}^{k} v_i b_i + v_{k+1} Y h \neq \tau.$$

We want to find a common solution for these two conditions.

Let $(rp^u)/s = v_{k+1}$ for $(r,p)=(s,p)=1$. We have

$$p^u Y h = \frac{s}{r} \cdot \sum_{i=0}^{k} v_i (-b_i). \qquad ㉑$$

In this formula $b_0, b_1, \cdots, b_k$ are fixed but $v_0, v_1, \cdots, v_{k+1}$ can be any numbers in V_{n-k-1}. The total number of possible elements for the left hand side is at most

$$[2^{2^{4(n-k-1)}}]^{k+2}.$$

Recall that

$$|v_{n-k-1}| \leqslant 2 \cdot 2^{2^{4(n-k-1)}} \quad \text{and} \quad p^t \leqslant 2^{2^{4n-4}}.$$

From Lemma 2.5 the total number of possible Y's satisfying ㉑ is at most

$$[2^{2^{4(n-k-1)}}]^{k+2} \leqslant 2^{2^{4n-3}}.$$

The total number of elements in A (or, similarly, in B where we use p^m) which are of the form Yh with $Y \equiv \lambda X_{k+1} (\bmod\ p^t)$ is

$$p^l / p^t \geqslant 2^{2^{4n-3}}.$$

Hence we can find a Y_{k+1} such that

$$Y_{k+1} \equiv \lambda X_{k+1} (\bmod\ p^t) \quad \text{and} \quad \sum_{i=0}^{k+1} v_i b_i \neq \tau$$

for any $v_0, v_1, \cdots, v_{k+1} \in V_{n-k-1}$ where $b_{k+1} = Y_{k+1} h$. Therefore from Lemma 4.5 we know that

$$p^u \left| \sum_{i=0}^{k+1} v_i a_i \leftrightarrow p^u \right| \sum_{i=0}^{k+1} v_i b_i$$

for any $v_0, v_1, \cdots, v_{k+1} \in V_{n-k-1}$ and $0 \leqslant u \leqslant t$. Hence there in a winning strategy for player Ⅱ in the Ehrenfeucht game $G_n(A,B)$. □

Theorem 4.7 *It takes at most $2^{2^{4n}}$ steps to decide a sentence ϕ with quantifier-aeptn n in $C_{p,w}$.*

Proof From Lemma 4.6 we know that $C_{p,w}$ is n-equivalent to any group $C_{p,l}$ with $2^{2^{4n-2}} \leqslant p^l$. We can choose a suitable p^l such that $p^l \leqslant 2^{2^{4n-1}}$. Hence the total number of steps needed for checking is at most $[2^{2^{4n-1}}]^n \leqslant 2^{2^{4n}}$. □

§ 5. The reduction of D_p Q_p, Q and $C_{p,w}$ with large p

Let A be a group of type D_p, Q_p or $C_{p,w}$ with $p \geqslant 2^{2^{4n}}$ or the group Q. Take the first prime number $q \geqslant 2^{2^{4n-2}}$. From [6] we know that $q \leqslant 2^{2^{4n}}$. Let $B = C_{q,1}$. We will prove in this chapter that there is a winning strategy for player Ⅱ in the Ehrenfeucht game $G_n(A,B)$.

Definition 5.1 The coefficient sets V_v are the same as in Section 4

$$V_v = \left\{ \frac{r}{s} \in Q_p : |r|, |s| \leqslant 2^{2^{4v}} \right\} \quad \text{for } 0 \leqslant v \leqslant n+1.$$

Lemma 5.2 *Let A, B be defined as above and θ, τ be the zero elements of A, B respectively. Then for any $v \in V_v$ with $v < n$, $a \in A$ and $b \in B$ we have the following properties.*

(i) *va (or similarly vb) exists and is unique.*

(ii) *$va = \theta$ (or $vb = \tau$) implies $v = 0$ or $a = \theta$ (or $b = \tau$).*

(iii) *$v \in V_v$ implies $1/v \in V_v$.*

Proof They are easy to check because $(r,p) = (s,p) = (r,q) = (s,q) = 1$ is true for any $r/s = v \in V_v$. □

Definition 5.3 Suppose that $\bar{a}_k \in A^k$ and $\bar{b}_k \in B^k$. The equivalence relation $\bar{a}_k E_{n-k,k} \bar{b}_k$ holds if and only if for any $v_1, v_2, \cdots, v_k \in V_{n-k}$ we have

$$\sum_{i=1}^{k} v_i a_i = \theta \leftrightarrow \sum_{i=1}^{k} v_i b_i = \tau.$$

Now we give our main theorem of this section. The proof of this theorem is similar to the proof of the corresponding theorems in the preceding sections.

Theorem 5.4 *For the groups A, B given at the beginning of this section there is a winning strategy for player* II *in the Ehrenfeucht game* $G_n(A,B)$.

Proof The game $G_n(A,B)$ plays as follows.

Step 0. $\varnothing E_{n,0}\varnothing$.

Setp $k+1$. Suppose that we have defined up to $\bar{a}_k, \bar{b}_k$ at the end of step k satisfying $\bar{a}_k E_{n-k,k}\bar{b}_k$. We now define $b_{k+1}\in B$ for any $a_{k+1}\in A$.

Case 1 Suppose that there are $v_1, v_2, \cdots, v_{k+1}\in V_{n-k-1}$ with $v_{k+1}\neq 0$ such that

$$\sum_{i=1}^{k+1} v_i a_i = \theta.$$

Hence we have

$$a_{k+1} = \sum_{i=1}^{k}\left(-\frac{v_i}{v_{k+1}}a_i\right).$$

Define

$$b_{k+1} = \sum_{i=1}^{k+1}\left(-\frac{v_i}{v_{k+1}}b_i\right).$$

For any $u_1, u_2, \cdots, u_{k+1}\in V_{n-k-1}$ if $\sum_{i=1}^{k+1} u_i a_i = \theta$, then

$$\sum_{i=1}^{k}\frac{u_i v_{k+1} - u_{k+1}v_i}{v_{k+1}}a_i = \theta.$$

Similarly to the proof of Lemma 2.11, Case 1 we can estimate the coefficients of this identity and prove that

$$\frac{u_i v_{k+1} - u_{k+1}v_i}{v_{k+1}}\in V_{n-k}.$$

Hence from $\bar{a}_k E_{n-k,k}\bar{b}_k$ we obtain

$$\sum_{i=1}^{k}\frac{u_i v_{k+1} - u_{k+1}v_i}{v_{k+1}}b_i = \tau.$$

This can be reduced to

$$\sum_{i=1}^{k+1} u_i b_i = \tau.$$

Similarly we can choose $a_{k+1}\in A$ for $b_{k+1}\in B$.

Case 2 For any $v_1, v_2, \cdots, v_{k+1}\in V_{n-k-1}$ with $v_{k+1}\neq 0$ we have

$$\sum_{i=1}^{k+1} v_i a_i \neq \theta.$$

The total number of possible sums of the form

$$\sum_{i=1}^{k}\left(-\frac{v_i}{v_{k+1}}\right)b_i$$

for different choices of $v_1, v_2, \cdots, v_{k+1} \in V_{n-k-1}$ is at most

$$[2^{2^{4(n-k-1)}}]^{2(k+1)} \leqslant 2^{2^{(4(n-k-1)+1+\log(k+1))}} \leqslant 2^{2^{4(n-k-1)+1+k}} \leqslant 2^{2^{4(n-1)+1}} = 2^{2^{4n-3}}$$

and the number of elements of B is

$$q \geqslant 2^{2^{4n-2}} > 2^{2^{4n-3}}.$$

Hence we can find a $b_{k+1} \in B$ such that

$$b_{k+1} \neq \sum_{i=1}^{k}\left(-\frac{v_i}{v_{k+1}}\right)b_i.$$

Therefore for any $v_1, v_2, \cdots, v_{k+1} \in V_{n-k-1}$

$$\sum_{i=1}^{k} v_i a_i = \theta \leftrightarrow \sum_{i=1}^{k+1} v_i a_i = \theta \quad \text{and} \quad v_{k+1} = 0$$

$$\leftrightarrow \sum_{i=1}^{k} v_i b_i = \tau \quad \text{and} \quad v_{k+1} = 0 \leftrightarrow \sum_{i=1}^{k+1} v_i b_i = \tau.$$

Similarly we can find an $a_{k+1} \in A$ for given $b_{k+1} \in B$. Hence player Ⅱ has a winning strategy in this game. □

§ 6. The reduction of $D_p + \cdots + D_p$ and $C_{p,w} + \cdots + C_{p,w}$ with more than n summands

We give a lemma about the solution of the equation $px = a$ in the direct sum $D_p + \cdots + D_p$ or $C_{p,w} + \cdots + C_{p,w}$ with m summands.

Lemma 6.1 *For every element $a \in A = D_p + \cdots + D_p$ with m summands the equation $px = a$ has p^m distinct solutions when m is finite, and infinitely many solutions when m is infinite.* (A similar result is obtained for $C_{p,w} \dot{+} \cdots + C_{p,w}$ provided that the equation has at least one solution.)

Proof Let $a = (a_1, a_2, \cdots, a_m)$. For any a_i there are exactly p solutions for the equation $px_i = a_i$ for $1 \leqslant i \leqslant m$ and $x = (x_1, x_2, \cdots, x_m)$ forms a solution of $px = a$. □

A similar result is obtained for the subgroups of the above types of groups.

Lemma 6. 2 *Let A_k be a subgroup of $A=D_p+\cdots+D_p$ with m summands where A_k is generated by k elements and k is finite. For any element $a\in A_k$ the equation $px=a$ has at most p^k solutions in A_k.*

Proof A_k is finite. By the fundamental theorem of finitely generated Abelian groups we know that A_k can be decomposed into a direct sum with at most k summands. Then as in the proof of Lemma 6. 1 we know that $px=a$ has at most p^k solutions in A_k. □

Now we give the main theorem of this section.

Theorem 6. 3 *Let $A=D_p+\cdots+D_p$, be a direct sum with m summands where $m\geqslant n$ is finite or infinite. Let $B=D_p+\cdots+D_p$ be a direct sum of D_p's with exactly n summands. Then there is a winning strategy for player* Ⅱ *in the Ehrenfeucht game $G_n(A,B)$.*

Proof *Step* 1. For a given $a_1\in A$ let p^t be the order of a_1. Choose any element $b_1\in B$ of order p^t. It is easy to see that $f(a_1)=b_1$ can be extended to an isomorphism between $A_1=\langle a_1\rangle$ and $B_1=\langle b_1\rangle$.

Similarly we can choose a_1 for a given b_1.

Step $k+1$. Suppose at the end of the kth step we have defined $(a_1, a_2,\cdots,a_k)$ and $(b_1,b_2,\cdots,b_k)$ such that

$$f(a_1)=b_1,b_2,\cdots,f(a_k)=b_k$$

can be extended to an isomorphism between $A_k=\langle a_1,a_2,\cdots,a_k\rangle$ and $B_k=\langle b_1,b_2,\cdots,b_k\rangle$. For a given $a_{k+1}\in A$ if the order of a_{k+1} is p^t, then $p^ta_{k+1}=\theta\in A_k$. Let p^u be the smallest power such that $p^ua_{k+1}\in A_k$. For every element $x\in A_{k+1}=\langle A_k,a_{k+1}\rangle$, x can be decomposed as

$$x=g+la_{k+1}$$

where $g\in A_k$ and $l=0,1,\cdots,$ or p^u-1. We claim that this decomposition is uniquely determined by x. For if

$$x=g_1+l_1a_{k+1}=g_2+l_2a_{k+1}$$

where $0\leqslant l_2\leqslant l_1\leqslant p^u-1$, then

$$(l_1-l_2)a_{k+1}=g_2-g_1\in A_k,$$

but $0\leqslant l_1-l_2<p^u$. Hence $l_1-l_2=0$ and $g_2-g_1=0$.

Consider the isomorphism $f: A_k \to B_k$. Let $b=f(a)\in B_k$ where $a=p^u a_{k+1}$. If $u=0$, then let $b_{k+1}=b$ and f itself is the isomorphism extended between A_{k+1} and B_{k+1}. If $u\neq 0$, then from Lemmas 6.1 and 6.2 the equation $px=b$ has m distinct solutions in B but at most $k(k<n\leqslant m)$ of them are in B_k. Hence we can choose $c_1\in B-B_k$ such that

$$pc_1=b.$$

Choose consecutively $c_2, c_3, \cdots, c_u$ such that

$$pc_i=c_{i-1}$$

for $i=3,4,\cdots,u$. None of the c_i's is in B_k. Otherwise we would have $c_1\in B_k$, a contradiction. Let

$$b_{k+1}=c_u \quad \text{and} \quad B_{k+1}=\langle B_k, b_{k+1}\rangle.$$

We claim that u is the smallest natural number satisfying $p^u b_{k+1}\in B_k$. For, if $p^w b_{k+1}\in B_k$ and $0\leqslant w\leqslant u$, then some of the c_i's would be in B_k. This contradicts the choices of the c_i's. As above we know that for any $y\in B_{k+1}$, y can be decomposed uniquely as

$$y=h+lb_{k+1}$$

where $h\in B_k$ and $l=0,1,\cdots$, or p^u-1. Hence

$$b_1=f(a_1), b_2=f(a_2), \cdots, b_{k+1}=f(a_{k+1})$$

can be extended to an isomorphism between A_{k+1} and B_{k+1}.

Similarly we can choose $a_{k+1}\in A$ for a given $b_{k+1}\in B$.

The winning strategy for player Ⅱ is to keep choosing elements to match the isomorphism. □

Theorem 6.4 *Let $A=C_{p,w}+\cdots+C_{p,w}$ be a direct sum of $C_{p,w}$ with m summands and $B=C_{p,w}+\cdots+C_{p,w}$ a direct sum of $C_{p,w}$ with n summands where $m\geqslant n$, then there is a winning strategy for player Ⅱ in the Ehrenfeucht game $G_n(A,B)$.*

§ 7. The reduction of $Q_p+\cdots+Q_p$ with more than $2^{2^{4n}}$ summands

§ 7.1 The main definitions

In this section we are going to prove a reduction for the direct sum of group Q_p's with a large number, possibly infinite, of summands. Let G be a direct sum of Q_p's with a fixed p. $G = Q_p^{(1)} + Q_p^{(2)} + \cdots + Q_p^{(m)}$. Every element $a \in G$ can be decomposed as $a = a^{(1)} + a^{(2)} + \cdots + a^{(m)}$ where $a^{(i)} \in Q_p^{(i)}$. $a^{(i)}$ is called the ith coordinate or component of a.

In Section 3 we defined a congruence relation modulo p^t among the elements of Q_p. Now we prove that there are integers in every congruence class.

Lemma 7.1 *Every coset in $Q_p / p^t Q_p$ contains an integer.*

Proof $Q_p / p^t Q_p$ is isomorphic to $C_{p,t}$ as Q_p module. □

Come back to the direct product $G = Q_p^{(1)} + Q_p^{(2)} + \cdots + Q_p^{(m)}$. Let $a_1, a_2, \cdots, a_k$ be a sequence of elements of G. Each a_j is an m-vector of elements of Q_p.

$$a_j = (a_j^{(1)}, a_j^{(2)}, \cdots, a_j^{(m)}) \quad \text{for } 1 \leqslant j \leqslant k.$$

For a fixed i the sequence of elements $(a_1^{(i)}, a_2^{(i)}, \cdots, a_k^{(i)})$ is called a column or a k-column. From Lemma 7.1 we know that there are integers $0 \leqslant r_1, r_2, \cdots, r_k < p^t$ such that

$$(a_1^{(i)}, a_2^{(i)}, \cdots, a_k^{(i)}) \equiv (r_1, r_2, \cdots, r_k) \pmod{p^t}.$$

For developing a winning strategy for player Ⅱ in the Ehrenfeucht game we also need to define a norm, an equivalence relation and a bounding function as in the earlier chapters but now the equivalence relation is a little more complicated than before.

Definition 7.2 For $a = (a^{(1)}, a^{(2)}, \cdots, a^{(m)}) \in G = Q_p^{(1)} + Q_p^{(2)} + \cdots + Q_p^{(m)}$, the norm of a in G is defined as

$$\| a \| = \mathrm{Max}_i \{ \| a^{(i)} \|, \text{the norm in } Q_p^{(i)} \}$$

Definition 7.3 The coefficient sets are

$$V_v = \left\{ \frac{r}{s} \in Q_p : |r|, |s| \leqslant 2^{2^{4v}} \right\} \quad \text{for } 0 \leqslant v \leqslant n.$$

Definition 7.4 Suppose that A and B are two direct sums of Q_p's with λ and μ summands respectively, that $\bar{a}_k \in A^k$ and that $\bar{b}_k \in B^k$. Then $\bar{a}_k E_{n-k,k} \bar{b}_k$ holds if and only if the following two conditions are satisfied.

(i) $\sum_{i=1}^k v_i a_i = \theta \leftrightarrow \sum_{i=1}^k v_i b_i = \tau$ for all $v_1, v_2, \cdots, v_k \in V_{n-k}$.

Before stating condition (ii) we need some preparations.

Let $a_i=(a_i^{(1)},a_i^{(2)},\cdots,a_i^{(\lambda)})$ and $b_i=(b_i^{(1)},b_i^{(2)},\cdots,b_i^{(\mu)})$ for $i=1,2,\cdots,k$. The columns composed by the elements of a_i's and b_i's are

$$C=\{(a_1^{(j)},a_2^{(j)},\cdots,a_k^{(j)}):\text{for } j=1,2,\cdots,\lambda\} \quad \text{and}$$
$$D=\{(b_1^{(j)},b_2^{(j)},\cdots,b_k^{(j)}):\text{for } j=1,2,\cdots,\mu\}.$$

Let $\Delta=(r_1,r_2,\cdots,r_k)$ where $1\leqslant r_1,r_2,\cdots,r_k<p^t$ and $t=[\log_p 2^{2^{4(n-k)}}]$. And let

$$M_\Delta=\{j\in[1,\lambda]:(a_1^{(j)},a_2^{(j)},\cdots,a_k^{(j)})\equiv\Delta(\operatorname{mod} p^\tau)\} \quad \text{and}$$
$$N_\Delta=\{j\in[1,\mu]:(a_1^{(j)},a_2^{(j)},\cdots,b_k^{(j)})\equiv\Delta(\operatorname{mod} p^\tau)\}.$$

Then condition (ii) is as follows:

(ii) If one of $|M_\Delta|$ and $|N_\Delta|$ is smaller than $2^{2^{4(n-k)}}$, then $|M_\Delta|=|N_\Delta|$. And otherwise both of them are greater than or equal to $2^{2^{4(n-k)}}$.

Definition 7.5 The bounding function is

$$H(n-k,k,m)=(k+1)(m2^{2^{4(n-k)}})^{k+1}.$$

This is the same function as we used in Section 3.

Now we are ready for a new Ehrenfeucht game.

§7.2 The Ehrenfeucht game $G(A,B)$

Theorem 7.6 *Let A, B be direct sums of Q_p's with λ, μ summands respectively where $\lambda,\mu\geqslant 2^{2^{4n}}$ and μ may be infinite. Then there is a winning strategy for player* Ⅱ *in the Ehrenfeucht game $G_n(A,B)$.*

Proof Suppose that $\bar{a}_k\in A^k$, $\bar{b}_k\in B^k$ are two vectors satisfying $\bar{a}_k E_{n-k,k}\bar{b}_k$ where $\|b_i\|\leqslant m$ for $i=1,2,\cdots,k$. Then for every $a_{k+1}\in A$ there is a $b_{k+1}\in B$ such that

$$\bar{a}_{k+1}E_{n-k-1,k+1}\bar{b}_{k+1} \quad \text{and} \quad \|b_{k+1}\|\leqslant H(n-k,k,m).$$

The proof of the existence of b_{k+1} is divided into two cases. We define a set C first. Let

$$C=\left\{\sum_{i=1}^{k} v_i a_i : v_i\in V_{n-k-1}\right\}.$$

Case 1 There is a $v_{k+1}\in V_{n-k-1}$, $v_{k+1}\neq 0$ such that $v_{k+1}a_{k+1}\in C$. From induction hypothesis we know that for any $\Delta=(r_1,r_2,\cdots,r_k)$ where $0\leqslant r_i<p^T$ and

$$T=[\log_p 2^{2^{4(n-k)}}].$$

If there is a column $(a_1^{(j)}, a_2^{(j)}, \cdots, a_k^{(j)}) \equiv \Delta (\bmod p^T)$, then there is at least one column $(b_1^{(j')}, b_2^{(j')}, \cdots, b_k^{(j')}) \equiv \Delta (\bmod p^T)$.

It is convenient to insert soma lemmas into the long proof of the theorem.

Lemma 7.7 *Under the conditions above we have*

$$p^u \left| \sum_{i=1}^{k} v_i a_i \leftrightarrow p^u \right| \sum_{i=1}^{k} v_i b_i$$

for any $v_1, v_2, \cdots, v_k \in V_{n-k}$ *and all* $0 \leqslant u \leqslant T$.

Proof We need only to show one direction, that is if $p^u \nmid \sum_{i=1}^{k} v_i a_i$, then $p^u \nmid \sum_{i=1}^{k} v_i a_i^{(j)}$ for some j. From condition (ii) of equivalence E we can find a suitable j' such that

$$(b_1^{(j')}, b_2^{(j')}, \cdots, b_k^{(j')}) \equiv (a_1^{(j)}, a_2^{(j)}, \cdots, a_k^{(j)}) (\bmod p^T).$$

Hence

$$\sum_{i=1}^{k} v_i a_i^{(j)} \equiv \sum_{i=1}^{k} v_i b_i^{(j')} (\bmod p^T).$$

Therefore

$$p^u \nmid \sum_{i=1}^{k} v_i b_i^{(j')} \quad \text{and then} \quad p^u \nmid \sum_{i=1}^{k} v_i b_i.$$

Go back to the proof of Theorem 7.6. Let

$$v_{k+1} = \frac{r p^u}{s} \quad \text{where} \quad (r, p) = (s, p) = 1,$$

$$0 \leqslant u < p^t \quad \text{and} \quad t = [\log_p 2^{2^{4(n-k-1)}}].$$

Choose $v_{k+1} a_{k+1} \in C - \{\theta\}$ such that u is as small as possible. From

$$a_{k+1} = \frac{1}{p^u} \cdot \sum_{i=1}^{k} \left(-\frac{v_i s}{r}\right) a_i$$

and Lemma 7.7 we know that

$$b_{k+1} = \frac{1}{p^u} \cdot \sum_{i=1}^{k} \left(-\frac{v_i s}{r}\right) b_i$$

exists. The checking of the condition (i) of $\bar{a}_{k+1} E_{n-k-1} \bar{b}_{k+1}$ is similar to the corresponding part of the proof of Lemma 3.8. To check condition (ii) of the equivalence we suppose that

$$(a_1^{(j)}, a_2^{(j)}, \cdots, a_k^{(j)}) \equiv (b_1^{(j')}, b_2^{(j')}, \cdots, b_k^{(j')}) \equiv \Delta \equiv (r_1, r_2, \cdots, r_k) (\bmod p^T).$$

Hence we can find an r_{k+1} such that

$$\sum_{i=1}^{k}\left(-\frac{v_i s}{r}\right)a_i \equiv \sum_{i=1}^{k}\left(-\frac{v_i s}{r}\right)b_i \equiv \sum_{i=1}^{k}\left(-\frac{v_i s}{r}\right)r_i \equiv r_{k+1} (\text{mod } p^T).$$

From the property of congruence modulo p^T we know that

$$\frac{1}{p^u}\cdot\sum_{i=1}^{k}\left(-\frac{v_i s}{r}\right)a_i \not\equiv \frac{1}{p^u}\cdot\sum_{i=1}^{k}\left(-\frac{v_i s}{r}\right)b_i (\text{mod } p^T),$$

only if $r_{k+1}\equiv 0$ (mod p^T). But in this case we also have

$$\frac{1}{p^u}\cdot\sum_{i=1}^{k}\left(-\frac{v_i s}{r}\right)a_i \equiv \frac{1}{p^u}\cdot\sum_{i=1}^{k}\left(-\frac{v_i s}{r}\right)b_i \equiv 0 (\text{mod } p^t),$$

because $u+t<T$. Therefore

$$(a_1^{(j)},a_2^{(j)},\cdots,a_{k+1}^{(j)})\equiv(b_1^{(j')},b_2^{(j')},\cdots,b_{k+1}^{(j')})\equiv$$
$$\delta\equiv (r_1,r_2,\cdots,r_{k+1})(\text{mod } p^t).$$

Hence for δ we can define the set M_δ, N_δ as in Definition 7.4 and we have $M_\delta=M_\Delta$ and $N_\delta=N_\Delta$. Therefore

$$|M_\delta|=|N_\delta| \quad \text{for } |M_\delta| \text{ or } |N_\delta|<2^{2^{4(n-k-1)}}.$$

The condition (ii) is satisfied.

As in the proof of Lemma 3.8 we can estimate the norm of b_{k+1} as

$$\| b_{k+1}^{(j)} \| \leqslant k(m2^{2^{4(n-k-1)}})^{k+1}.$$

Hence

$$\| b_{k+1} \| \leqslant k(m2^{2^{4(n-k-1)}})^{k+1}\leqslant H(n-k,k,m).$$

Case 2 For every $v\in V_{n-k-1}-\{0\}$ we have $va_{k+1}\notin C$. Define a set D in B.

$$D=\left\{\sum_{i=1}^{k}v_i b_i : v_1,v_2,\cdots,v_k\in V_{n-k-1}\right\}.$$

As in the calculation of Lemma 3.8 we know that for $v\in V_{n-k-1}-\{0\}$ and $vb_{k+1}\in D$ the norm of b_{k+1} is at most $m2^{2^{4(n-k-1)}}$. Conversely, if we can find some b_{k+1} with a norm greater than the above number, then $vb_{k+1}\notin D$ for all v.

From the induction hypothesis for fixed $\Delta=(r_1,r_2,\cdots,r_k)\in[0,p^T-1]^k$ the numbers $|M_\Delta|$, $|N_\Delta|$ satisfy

$$|M_\Delta|=|N_\Delta|<2^{2^{4(n-k)}} \quad \text{or both} \quad |M_\Delta|,|N_\Delta|\geqslant 2^{2^{4(n-k)}}.$$

A similar property is also true when we change T into t. Because, if we take all the sequences $\Delta=\Delta_0,\Delta_1,\cdots,\Delta_e$ which are equivalent to each other modulo p^t but not p^T, then

$$\overline{M}=\{i:(a_1^{(i)},a_2^{(i)},\cdots,a_k^{(i)})\equiv\Delta\ (\mathrm{mod}\ p^t)\}=\bigcup_{0\leqslant i\leqslant e}M_{\Delta_i}.$$

$$\overline{N}=\{j:(b_1^{(i)},b_2^{(i)},\cdots,b_k^{(i)})\equiv\Delta\ (\mathrm{mod}\ p')\}=\bigcup_{0\leqslant j\leqslant e}N_{\Delta_j}.$$

If for some i, $|M_{\Delta_i}|\geqslant 2^{2^{4(n-k)}}$ then $|N_{\Delta_j}|$ is at least $2^{2^{4(n-k)}}$. Hence both $|\overline{M}|,|\overline{N}|\geqslant 2^{2^{4(n-k)}}$. If for all i, $|M_{\Delta_i}|<2^{2^{4(n-k)}}$, then for all i, $|M_{\Delta_i}|=|N_{\Delta_i}|$. Hence we also have

$$|\overline{M}|=|\overline{N}|\quad \text{or both}\quad |\overline{M}|,|\overline{N}|\geqslant 2^{2^{4(n-k)}}.$$

For a given r_{k+1} let

$$(r_1,r_2,\cdots,r_k,r_{k+1})=(\Delta,r_{k+1})=\delta.$$

Consider the set of $(k+1)$-columns

$$M_\delta=\{i:(a_1^{(i)},a_2^{(i)},\cdots,a_k^{(i)},a_{k+1}^{(i)})\equiv\delta\ (\mathrm{mod}\ p^t)\}.$$

We define b_{k+1} according to the following steps.

(a) If $|M_\delta|<2^{2^{4(n-k-1)}}$, then choose a subset N_δ of $\overline{N}$ such that $|N_\delta|=|M_\delta|$ and assign the coordinate $b_{k+1}^{(j)}=r_{k+1}$ for all $j\in N_\delta$.

(b) If $|M_\delta|\geqslant 2^{2^{4(n-k-1)}}$, then choose a subset N_δ of $\overline{N}$ such that $|N_\delta|=2^{2^{4(n-k-1)}}$ and assign the coordinate $b_{k+1}^{(j)}=r_{k+1}$ for all $j\in N_\delta$.

(c) Do above (a) and (b) for all possible δ's with the same Δ.

(d) If there are still some of the coordinates of $b_{k+1}^{(j)}$ $(j\in\overline{N})$ undefined we assign $b_{k+1}^{(j)}=r_{k+1}$ for all these j's where r_{k+1} is the smallest number with $|M_\delta|\geqslant 2^{2^{4(n-k-1)}}$.

(e) Do the above (a)—(c) for all possible Δ's. Change one of the $b_{k+1}^{(j)}$ to $b_{k+1}^{(j)}+Lp^t$ so that

$$k(m2^{2^{k-1}})^{k+1}<b_{k+1}^{(j)}+Lp^t\leqslant(k+1)(m2^{2^{4(n-k-1)}})^{k+1}.$$

We claim that this procedure is effective. For according to (a) and (b) we need to assign r_{k+1} as a coordinate $b_{k+1}^{(j)}$ for all $j\in N_\delta$ and

$$N_\delta\cap N_{\delta'}=\varnothing\quad \text{for } \delta\neq\delta'.$$

This can be done because for a fixed Δ the number of δ's with the same initial segment Δ(i. e. ,$\delta=(\Delta, r_{k+1})$) is p^t. For each δ, $|N_\delta|$ is $< 2^{2^{4(n-k-1)}}$. Hence

$$|U_{\delta=(\Delta, r_{k+1})} N_\delta| \leqslant p^t \cdot 2^{2^{4(n-k-1)}} < 2^{2^{4(n-k)}}.$$

If $|\overline{M}_\Delta| \geqslant 2^{2^{4(n-k)}}$, then $|\overline{N}_\Delta| \geqslant 2^{2^{4(n-k)}}$. There is enough room for our assignments. If $|\overline{M}_\Delta| < 2^{2^{4(n-k)}}$, then $|\overline{N}_\Delta| = |\overline{M}_\Delta|$ we also have enough room for our assignments.

We need to use (d) only if for some δ we have $|M_\delta| > |N_\delta|$. In this case both of them are greater than $2^{2^{4(n-k-1)}}$. Hence we can find the smallest r_{k+1} and assign all the other coordinates as r_{k+1}.

The change in (e) is to increase the norm of b_{k+1} while keeping the congruence. This is possible because

$$(k+1)(m2^{2^{4(n-k-1)}})^{k+1} - k(m2^{2^{4(n-k-1)}})^{k+1} \geqslant p^t.$$

From the assignments (a) to (e) we have defined completely an element $b_{k+1} = (b_{k+1}^{(1)}, b_{k+1}^{(2)}, \cdots, b_{k+1}^{(l)})$. We need to check that $\overline{b}_{k+1}$ satisfies the requirements.

From (e) there is a j such that

$$\| b_{k+1}^{(j)} \| > k(m2^{2^{4(n-k-1)}})^{k+1}.$$

hence

$$\| b_{k+1} \| > k(m2^{2^{4(n-k-1)}})^{k+1}.$$

Therefore

$$\sum_{i=1}^{k+1} v_i a_i = \theta \leftrightarrow \sum_{i=1}^{k} v_i a_i = \theta \text{ and } v_{k+1} = 0$$

Consider $a_i = (a_i^{(1)}, a_i^{(2)}, \cdots, a_i^{(l)})$ and $b_i = (b_i^{(1)}, b_i^{(2)}, \cdots, b_i^{(m)})$ for $i = 1, 2, \cdots, k+1$. Take $\delta = (r_1, r_2, \cdots, r_{k+1})$ and some form the sets

$$M'_\delta = \{i \in [1, l]: (a_1^{(i)}, a_2^{(i)}, \cdots, a_{k+1}^{(i)}) \equiv \delta \pmod{p^t}\}$$

and

$$N'_\delta = \{j \in [1, m]: (b_1^{(j)}, b_2^{(j)}, \cdots, b_{k+1}^{(j)}) \equiv \delta \pmod{p^t}\}.$$

From (a) to (e) we know that $M'_\delta = M_\delta$ and $N'_\delta = N_\delta$. Hence

$$|M'_\delta| = |N'_\delta| \quad \text{when } |M'_\delta| \text{ or } |N'_\delta| \text{ is} < 2^{2^{4(n-k-1)}}.$$

This proves that the condition (ii) of Definition 7. 4 is true.

From the assignment (a) to (e) we know that for every $1 \leqslant j \leqslant m$

the norm $\| b_{k+1}^{(j)} \| \leqslant p^t$ or $\| b_{k+1}^{(j)} \| \leqslant (k+1)(m2^{2^{4(n-k-1)}})^{k+1}$. Hence $\| b_{k+1} \| \leqslant H(n-k,k,m)$.

The induction is now completed and the winning strategy for player Ⅱ in the Ehrenfeucht game $G_n(A,B)$ exists. □

§ 8. The upper bound for a group in SG_2

In Section 1 we proved that $\mathrm{Th}(AG)=\mathrm{Th}(SG_1)$ where SG_1 consists of all finite direct sums of groups with the following types

$$D_p, Q_p \quad \text{and} \quad C_{p,w}$$

where $p=2,3,5,\cdots$ and $w=0,1,2,\cdots$ The class SG_1 is still too big for us because it has infinitely many members. From Sections 4 to 7 we have proved a series of reduction theorems. Now for a given group $A \in SG_1$ we can use these theorems to find a group B with better structure and satisfying the same set of sentences with quantifier-depth at most n. The reduction steps are as follows.

(1) Change all the summands of A of types D_p, Q_p and $C_{p,w}$ with $p \geqslant 2^{2^{4n}}$ to suitable summands of type $C_{q,1}$ with $q \leqslant 2^{2^{4n}}$ and keep fixed the other summands to form a group A_1. From Theorems 5.4, 1.4 we know that $A \equiv_n A_1$.

(2) Change all the summands of A_1 of type $C_{p,w}$ with $p^w > 2^{2^{4n}}$ to suitable summands of type $C_{p,u}$ with $p^u \leqslant 2^{2^{4n}}$ and keep fixed the other summands to form a group A_2. From Theorems 4.7, 1.4 we know that

$$A_1 \equiv_n A_2.$$

(3) If the number of summands of A_2 which are isomorphic to a fixed D_p is more than n, then change this part to a direct sum of the same D_p with exactly n summands and keep fixed the other summands to form a group A_3. From Theorem 6.4 we know that $A_2 \equiv_n A_3$.

(4) If the number of summands of A_3 which are isomorphic to a fixed $C_{p,w}$ is more than n, then change this part to a direct sum of the same $C_{p,w}$ with exactly n summands and keep fixed the other summands

to form a group A_4. From Theorems 6.4, 1.4 we know that $A_3 \equiv_n A_4$.

(5) If the number of summands of A_4 which are isomorphic to a fixed Q_p is more than $2^{2^{4n}}$, then change this part to a direct sum of the same Q_p with exactly $2^{2^{4n}}$ summands and keep fixed the other summands to form a group B. From Theorems 7.6, 1.4 we know that $A_4 \equiv_n B$.

We summarize this result into Definition 8.1 and Theorem 8.2.

Definition 8.1 SG_2 is a subset of SG_1 and consists of the direct sums of groups of the following types

$$D_p, Q_p \quad \text{and} \quad C_{p,w}$$

where p, w and the numbers of isomorphic summands in every $A \in SG_2$ satisfy the following:

(i) $p \leqslant 2^{2^{4n}}$.

(ii) $p^w \leqslant 2^{2^{4n}}$.

(iii) The number of summands isomorphic to a fixed group of type D_p or $C_{p,w}$ is at most n.

(iv) The number of summands isomorphic to a fixed group of type Q_p is at most $2^{2^{4n}}$.

Theorem 8.2 $\mathrm{Th}_n(AG) = \mathrm{Th}_n(SG_2)$.

From now on we can simply discuss the groups of SG_2. In Sections 2 and 3 we have defined norms, equivalence relations and bounding functions for groups of types D_p and Q_p. To unify our results we define some trivial cases of norm, equivalence relation and bounding function for groups of type $C_{p,w}$.

Definition 8.3 (i) The norm of $a \in C_{p,w}$ is defined as $\| \theta \| = 0$, $| a \| = 1$ for $a \neq 0$.

(ii) The equivalence relation $\bar{a}_k E_{n,k} \bar{b}_k$ holds in $C_{p,w}$ if and only if $\bar{a}_k = \bar{b}_k$.

(iii) The bounding function for $C_{p,w}$ is $H(n,k,m) = 1$.

Now we have definitions for all three types of groups D_p, Q_p and $C_{p,w}$. Consider a group $A \in SG_1$:

$$A = A^{(1)} + A^{(2)} + \cdots + A^{(r)}$$

where $A^{(j)}$ is one of the three types. We assume that $\| a^{(j)} \|$ takes the same value as it does in $A^{(j)}$, $\bar{a}_k^{(j)} E_{n,k} \bar{b}_k^{(j)}$ is true if and only if it is true in $A^{(j)}$ and $H^{(j)}(n,k,m)$ takes the same value as it does for group $A^{(j)}$ according to their types.

Definition 8.4 Let $A=A^{(1)}+A^{(2)}+\cdots+A^{(r)}$ be a group of SG_1.

(i) For any $a=(a^{(1)},a^{(2)},\cdots,a^{(r)})\in A$ the norm of a is

$$\| a \| = \max_{1\leqslant j\leqslant r}\{ \| a^{(j)} \| \}.$$

(ii) For any two columns $a_i=(a_i^{(1)},a_i^{(2)},\cdots,a_i^{(r)})$, $b_i=(b_i^{(1)},b_i^{(2)},\cdots,b_i^{(r)})\in A$ for $i=1,2,\cdots,k$ the equivalence relation $\bar{a}_k E_{n,k} \bar{b}_k$ is true if and only if $\bar{a}_k^{(j)} E_{n,k} \bar{b}_k^{(j)}$ is true for all $j=1,2,\cdots,r$.

(iii) The bounding function of A is

$$H(n,k,m)=\max_{1\leqslant j\leqslant r}\{H^{(j)}(n,k,m)\}.$$

Now we can prove the existence of a winning strategy for player Ⅱ in the game $\mathrm{FR}_n(A,A)$ with bounding function $H(n,k,m)$ for every group $A\in SG_1$.

Theorem 8.5 *For any $\bar{a}_k, \bar{b}_k\in A^k$ satisfying $\bar{a}_k E_{n,k} \bar{b}_k$ and $\| b_i \|\leqslant m$ for $i=1,2,\cdots,k$ and for every $a_{k+1}\in A$ there is a $b_{k+1}\in A$ such that*

$$\bar{a}_{k+1} E_{n-1,k+1} \bar{b}_{k+1} \quad \text{and} \quad \| b_{k+1} \| \leqslant H(n,k,m).$$

Proof Let $A=A^{(1)}+A^{(2)}+\cdots+A^{(r)}$ where r is finite. Consider $a_i=(a_i^{(1)},a_i^{(2)},\cdots,a_i^{(r)})$ and $b_i=(b_i^{(1)},b_i^{(2)},\cdots,b_i^{(r)})$ for $i=1,2,\cdots,k$. From $\bar{a}_k E_{n,k} \bar{b}_k$ and $\| b_i \|\leqslant m$ we know that $\bar{a}_k^{(j)} E_{n,k} \bar{b}_k^{(j)}$ and $\| b_i^{(j)} \| m$ for all $i=1,2,\cdots,k$ and $j=1,2,\cdots,r$.

If $A^{(j)}$ is of type D_p or Q_p, then from Lemma 2.11 or Lemma 3.8 we know that there is a $b_{k+1}^{(j)}\in A^{(j)}$ such that

$$\bar{a}_{k+1}^{(j)} E_{n-1,k+1} \bar{b}_{k+1}^{(j)} \quad \text{and} \quad \| b_{k+1}^{(j)} \| \leqslant H^{(j)}(n,k,m)\leqslant H(n,k,m).$$

If $a^{(j)}$ is of type $C_{p,w}$, then $a_i^{(j)}=b_i^{(j)}$ for all $i=1,2,\cdots,k$. Let $b_{k+1}^{(j)}=a_{k+1}^{(j)}$. Then we have

$$\bar{a}_{k+1}^{(j)}=\bar{b}_{k+1}^{(j)}, \quad \bar{a}_{k+1}^{(j)} E_{n-1,k+1} \bar{b}_{k+1}^{(j)} \quad \text{and}$$
$$\| b_{k+1}^{(j)} \| \leqslant 1 = H^{(j)}(n,k,m)\leqslant H(n,k,m).$$

Therefore b_{k+1} is found and satisfies

$$\bar{a}_{k+1} E_{n-1,k+1} \bar{b}_{k+1} \quad \text{and} \quad \| b_{k+1} \| \leqslant H(n,k,m). \qquad \square$$

Theorem 8.6 *It takes at most $2^{2^{2^{5n}}}$ steps to decide a sentence ϕ with* *quantifier-depth n in a group* $A\in SG_2$.

Proof We have shown in Theorem 8. 5 that for any group $G\in SG_2$ there is an equivalence relation $E_{n,k}$ for the game $\mathrm{FR}_n(G,G)$ with bounding function $H(n,k,m)$ as defined in Definition 8. 4. From Theorem 1. 3 we know that any first-order sentence ϕ with quantifier-depth less than, or equal to n can be decided in $\prod_{i=1}^{n} F(M_i)$ steps where $M_0=1, M_{i+1}=H(n-i,k,M_i)$ for $i=1,2,\cdots,n-1$ and $F(M)$ is the number of elements in G with norms greater than M.

From Definitions 2. 10, 3. 4 and 8. 4 we have

$$H(n-k,k,M)\leqslant 2[(k+1)(M2^{2^{4n}})^{k+2}]^2.$$

From Lemma 2. 12 we know that

$$M_n\leqslant 2^{2^{4(n+k)\log(n+1)}}\leqslant 2^{2^{8n\log(n+1)}}.$$

Now we need to estimate the value of function $f(M_n)$. Since

$$F(M_1)\leqslant\cdots\leqslant F(M_n)$$

we can use $F(M_n)$ as the upper bound of the other factors.

According to the definition a group $G\in SG_2$ may have the following types of summands.

(i) At most n summands are isomorphic to a fixed D_p for $p\leqslant 2^{2^{4n}}$.

(ii) At most n summands are isomorphic to a fixed $C_{p,w}$ for $p^w\leqslant 2^{2^{4n}}$.

(iii) At most $2^{2^{4n}}$ summands are isomorphic to a fixed Q_n for $p\leqslant 2^{2^{4n}}$.

(iv) At most $2^{2^{4n}}$ non-isomorphic summands of each of the types D_p, $C_{p,w}$ or Q_p.

Hence the total number of summands of G is at most

$$n\cdot(2^{2^{4n}}2^{2^{4n}})+2^{2^{4n}}\cdot 2^{2^{4n}}\leqslant 2^{2^{4n+2}}.$$

Therefore the number of different elements of G with norm less than $2^{2^{8n\log(n+1)}}$ is

$$[2^{2^{8n\log(n+1)}}]^{2^{2^{4n+2}}}=2^{2^{(8n\log(n+1)+2^{4n+2})}}.$$

Finally the number of steps needed for checking the sentence ϕ is at most

$$[2^{2^{(8n\log(n+1)+2^{4n+2})}}]^n\leqslant 2^{2^{2^{5n}}}.\quad\square$$

§ 9. The upper bound for *AG* and *FG*

We have shown in Section 8 that $\mathrm{Th}_n(AG)=\mathrm{Th}_n(SG_2)$ and for any sentence ϕ with quantifier-depth n the number of steps needed for checking ϕ in a group $G\in SG_2$ is at most $2^{2^{2^{5n}}}$. Now we want to know how many nonisomorphic groups there are in SG_2.

Theorem 9.1 (Ⅰ) $|SG_2|\leqslant 2^{2^{2^{5n}}}$.

(Ⅱ) *It takes at most* $2^{2^{2^{6n}}}$ *steps to decide a sentence* ϕ *of quantifier-depth* n *for all of the groups of* SG_2.

Proof (Ⅰ) From the proof of Theorem 8.6 we know that every group $G\in SG_2$ is a direct sum of the following items:

(i) At most n summands are isomorphic to a fixed D_p for $p\geqslant 2^{2^{4n}}$.

(ii) At most n summands are isomorphic to a fixed $C_{p,w}$ for $p^w\leqslant 2^{2^{4n}}$.

(iii) At most $2^{2^{4n}}$ summands are isomorphic to a fixed Q_p for $p\leqslant 2^{2^{4n}}$.

(iv) At most $2^{2^{4n}}$ nonisomorphic summands of each of the types D_p, $C_{p,w}$ or Q_p.

Each of the above items might or might not appear in G. Hence the number of possibilities for G is at most $2^{2^{2^{5n}}}$.

(Ⅱ) The total number of steps for checking ϕ for all groups in SG_2 is at most $2^{2^{2^{5n}}}\cdot 2^{2^{2^{5n}}}\leqslant 2^{2^{2^{6n}}}$. □

In the above discussion we consider only the big steps. For converting a big step into Turing steps we have to estimate how many Turing steps are needed for a big step. It is more convenient to give our theorem with respect to the size of a sentence.

Theorem 9.2 *It takes* $2^{2^{2^{cn}}}$ *Turing steps with a suitable constant* c *to decide a sentence* ϕ *with size* n *for all of the groups in* SG_2.

Proof (outline) We need to store the elements of $2^{2^{2^{5n}}}$ different groups each of which consists of at most $2^{2^{2^{5n}}}$ elements. Hence the space needed for storage is at most $2^{2^{dn}}$ Turing space units with a suitable constant d. The multiplication of two elements $x, y \in G \in SG_2$ can be done in at most $2^{2^{2^{cn}}} \cdot 2^{2^{2^{cn}}}$ steps. Hence the total space needed is $2^{2^{dn}}$ Turing space units and the number of steps is $[2^{2^{2^{cn}}}]^{2n} \leqslant 2^{2^{2^{c'n}}}$ where c' is a constant. □

Theorem 9.3 *It takes* $2^{2^{2^{cn}}}$ *Turing steps and* $2^{2^{dn}}$ *Turing space units to decide a sentence* ϕ *with size less than or equal to* n *for* AG *and* FG.

Proof Because $\mathrm{Th}_n(AG) = \mathrm{Th}_n(SG_2)$ and $\mathrm{Th}_n(FG) = \mathrm{Th}_n(FG \cap SG_2)$, the discussion of Theorem 9.2 is good for both AG and FG. In the latter case we only need to take out all of the summands of type Q_p. □

§ 10. The upper bound for Z^*

In Section 9 we gave an upper bound for the computational complexity of the theories of AG and FG. Now we want to get the same result for the theory of the countably direct sum Z^*.

Theorem 10.1 *Let* $G = \prod_{i=1}^{\infty} Q_{p_i}$. *Then* $\mathrm{Th}(G) = \mathrm{Th}(Z)$.

Proof From [8] we only need to check the basic sentences.

(i) Z does not have elements of order p^k. Hence in Z we have $R^{(i)}[p,k,0]$ and $\neg R^{(i)}[p,k,m]$ for $i=1,3$ and $m \geqslant 1$. This is also true in G.

(ii) Z has exactly one strongly independent element with respect to each p^k. Hence in Z we have $R^{(2)}[p,k,0]$, $R^{(2)}[p,k,1]$ and $\neg R^{(2)}[p,k,m]$ for $m \geqslant 2$. This is also true in G.

(iii) Z does not have nontrivial elements of finite order. Hence in Z we have $\neg K[n]$ for all $n \neq 0$. And this is also true in G.

Therefore G and Z both have the same set of true basic sentences. We obtain $\mathrm{Th}(G) = \mathrm{Th}(Z)$. □

Similarly we obtain the following theorem.

Theorem 10.2 *Let* G *be the same group as in Theorem* 10.1. *Then*

$$\mathrm{Th}(G^*) = \mathrm{Th}(Z^*).$$

We can not simply use Theorem 5.4 to reduce the summands of G' with large p because we can only reduce the summands with one p at a time but there are summands with infinitely many p's to be taken out. We intend to prove a theorem to reduce all of the summands with large p's at once.

Theorem 10. 3 *Let p_k be the first prime greater than or equal $2^{2^{4n-2}}$ such that*

$$G^* = \sum_{i=1}^{\infty} Q_{p_i}^* = \sum_{i=1}^{k} Q_{p_i}^* + \sum_{i=k+1}^{\infty} Q_{p_i}^* \quad \text{and} \quad H^* = \sum_{i=1}^{k} Q_{p_i}^* + C_{p_{k+1},1}.$$

Then $\mathrm{Th}_n(G^*)=\mathrm{Th}_n(H^*)$.

Proof Similar to the proof of Theorem 5. 4. □

Now we can use the reduction theorem in Sections 6 and 7.

Theorem 10. 4 *Let p_k, H^* be as in Theorem 10. 3 and*

$$H = \sum_{i=1}^{k} \sum_{j=1}^{r} Q_{p_i} + C_{p_{k+1},1}.$$

Then $\mathrm{Th}_n(H)=\mathrm{Th}_n(H^*)$ *where* $r \leqslant 2^{2^{4n}}$ *by Theorem 7. 6.*

Theorem 10. 5 *It takes at most $2^{2^{2^{cn}}}$ Turing steps and $2^{2^{dn}}$ Turing space units to decide a sentence ϕ with size n in $\mathrm{Th}_n(H)=\mathrm{Th}_n(Z^*)$ where c and d are suitable constants.*

Proof Group H has at most $[2^{2^{4n}}]^{2^{2^{4n}}} = 2^{2^{4n}+2^{4n}}$ elements. Hence the number of big steps needed for checking ϕ is at most $[2^{2^{4n}+2^{4n}}]^n \leqslant 2^{2^{2^{5n}}}$.

As in the proof of Theorem 9. 2 we know that the Turing time and the Turing space units needed for this checking are at most $2^{2^{2^{cn}}}$ and $2^{2^{dn}}$ respectively. □

Acknowledgments

This paper is the main part of my doctoral dissertation at The University of Michigan. It is my pleasure to thank my thesis advisors Professors Roger C. Lyndon and Yuri Gurevich. I also want to thank Professor Andreas Blass and Dr. Jane Kister for many useful suggestions.

References

[1] Ehrenfeucht A. An application of games to the completeness problem for formalized theories. Fund. Math. ,1961,49:129—141.

[2] Eklof P and Fischer E. The elementary theory of Abelian groups. Ann. Math.

Logic,1972,4:115—171.

[3] Ferrante J and Rackoff C. The Computational Complexity of Logical Theories. Springer,Berlin,1979.

[4] Fischer M J and Rabin M O. Super exponential complexity of Presburger arithmetic,Complexity of real computational processes. SIAM-AMS Proceedings,1974,7:27—41.

[5] Fuchs L. Abelian Groups. Pergamon Press,New York,3rd ed. ,1960.

[6] Nagura J. On the interval containing at least one prime number. Proc. Japan Acad. ,1952,28(4):177—181.

[7] Rackoff C. On the complexity of the theories of weak direct products:A preliminary report. Proc. 6th ACM Symp:On the Theory of Computing ACM, New York,1976,149—160.

[8] Szmielew W. Elementary properties of Abelian groups. Fund. Math. ,1955, 41:203—271.

名古屋商科大学论集，1991，36(1)：191～204

实数加法的正式子的计算复杂性

The Computational Complexity of Positive Formulas in Real Addition

Abstract Categories and Subject Descriptors: F. 2 [Analysis of algorithms and problem complexity]: Computational complexity. F. 4[Mathematical logic and formal languages]: Decision problem of logic systems.

Keywords Computational complexity, Polynomial time computation, Exponential time computation.

§ 1. Introduction

The computational complexity of the first order theory of real numbers with addition was studied by many authors. For the lower bounds Fischer and Rabin in [2] prove the following theorem:

Theorem(Fischer and Rabin). There is a rational constant $c>0$ such that if M is a Turing machine which accepts $Th(R,+)$, then M runs for at least $c^{|s|}$ steps when started on input s, for infinitely many sentences.

This is up to now the best result on this line. For the upper bounds instead of dealing with $Th(R,+)$ people usually discuss $Th(R,+,>)$ because '$>$' gives more convenience in working with formulas. We do not know whether upper bounds get worst because of introducing the relation '$>$'. The best upper bound is given by Ferrante and Rackoff in [1]. Their theorem is as follows.

Theorem (Ferrante and Rackoff). There is a rational constant $d>0$

and a Turing machine M such that M accepts $Th(R,+,>)$ and requires at most $d^{|s|}$ storage units in computing a formula s.

A new direction is to narrow the set of formulas hoping that we can get better upper and lower bounds which are closer to the limits for practical implementations. In our paper we work on positive formulas of $Th(R,+)$. We will prove that for this class of formulas we can have a better algorithm to solve the decision problem. Our algorithm requires $2^{|s|}$ time for a formula s. With a smaller class of formulas we can reduce the complexity substantially.

Acknowledgement. I would like to thank Professor Yuri Gurevich and Professor Marek Suchenek for many useful suggestions.

§ 2. The upper bound of the computation

All formulas in this paper are supposed to be positive. A positive formula is formed by means of the following symbols:

(1) logical symbols $\vee$, $\wedge$, $\forall$, $\exists$, (,),

(2) variables symbols x^*, y^*, z^*, …

(3) operation symbol $+$, and

(4) predicate symbol $=$.

If m_i, $i=1,2,\cdots,n+1$ are integers then formula $\sum_{i=1}^{n} m_i x_i = m_{n+1}$ is an atomic formula.

If a_i, $i=1,2,\cdots,n+1$ are real numbers then formula $\sum_{i=1}^{n} a_i x_i = a_{n+1}$ is equivalent to an atomic formula $\sum_{i=1}^{n} m_i x_i = m_{n+1}$.

The size of the latter formula could grow as a square of the previous one. we will discuss this part later.

If Ψ_i, $i=1,2,\cdots,n$ are conjunctive formulas then formula $\bigvee_{i=1}^{n} \Psi_i$ is called a disjunctive-conjunctive formula.

Our decision procedure will look like this. We have a formula Φ with length n. In general it is equivalent to a formula $(QX)(\vee \wedge \Psi_{ij})$

where (QX) is the set of all quantifiers and Ψ_{ij} is an atomic formula. The length of the latter could grow to 2^{cn}. After that we move the quantifiers in across the symbol $\vee$. In general this is not true but in our case we can prove that this kind of moves will not change the validity of the formula. This part is accomplished in Theorem 1 and Theorem 2. In Theorem 3 we discuss the case when a formula is conjunctive with quantifiers. This kind of formulas can be decided in a procedure with a polynomial number of real operations. In Theorem 4 we sum up the decision procedure. Theorem 5 will give a lower bound of the computation. Finally in Theorem 6 we improve our result to having an NP decision algorithm for the real positive formulas requiring a polynomial number of operations.

We discuss the conjunctive formulas first. Theorem 1 asserts that if a conjunctive formula $\Phi(x)$ is true for two different numbers, then it is true for all real numbers. This theorem allows us to generalize a formula from 2 values to all values.

Theorem 1 Let

$$\Psi(x_1, x_2, \cdots, x_n) = \bigwedge_{j=1}^{m} \left(\sum_{k=1}^{n} a_{jk} x_k = b_j\right)$$

be a conjunctive formula. If formula $(Q_2 x_2)\cdots(Q_n x_n)\ \Psi$ is true when $x_1 = c$ and $x_1 = d$ where $c \neq d$, then formula $(\forall x_1)\ (Q_2 x_2)\cdots(Q_n x_n)\Phi$ is true.

Proof Divide $(Q_2 x_2)\cdots(Q_n X_n)$ into blocks:

$$\forall x_2 \cdots x_{n_1}\ \exists x_{n_1+1} \cdots x_{n_2}\ \forall x_{n_2+1} \cdots x_{n_3}\ \exists x_{n_3+1} \cdots x_{n_4} \cdots$$

Substitute c and d into the formula. From the Skolem Function Theorem we know that for the free variable sequence $x_2, \cdots, x_{n_1}, x_{n_{2+1}}, \cdots, x_{n_3}, \cdots$ we can find $x_{n_1+1}, \cdots, x_{n_2}, x_{n_3+1}, \cdots, x_{n_4}, \cdots$ where

$x_{n_1+1} = X_{n_1+1}(x_2, \cdots, x_{n_1}), \cdots$ and

$x_{n_3+1} = X_{n_3+1}(x_2, \cdots, x_{n_1}, x_{n_2+1}, \cdots, x_{n_3}), \cdots$

such that they are the functions of the previous variables. The new variable list together with c satisfy the formula $\Psi(c, x_2, \cdots, x_n)$.

The same is true for $x_1 = d$. In this case we change the variable list to $y_2, y_3, \cdots, y_n$. Therefore we have the following equations.

$$(1)\begin{cases}\bigwedge_{j=1}^{m}(a_{j1}c+\sum_{k=2}^{n}a_{jk}x_k=b_j),\\ \bigwedge_{j=1}^{m}(a_{j1}d+\sum_{k=2}^{n}a_{jk}y_k=b_j).\end{cases}$$

where

$$x_2,\cdots,x_{n_1},x_{n_2+1},\cdots,x_{n_3},\cdots\text{and } y_2,\cdots,y_{n_1},y_{n_2+1},\cdots,y_{n_3},\cdots$$

are free variables

$$x_{n_1+1},\cdots,x_{n_2},x_{n_3+1},\cdots,x_{n_4},\cdots\text{and } y_{n_1+1},\cdots,y_{n_2},y_{n_3+1},\cdots,y_{n_4},\cdots$$

are functions of the previous free variables.

$x_{n_1+1}=X_{n_1+1}(x_2,x_3,\cdots,x_{n_1}),\cdots,$

$y_{n_1+1}=X_{n_1+1}(y_2,y_3,\cdots,y_{n_1}),\cdots,$

$x_{n_3+1}=X_{n_3+1}(x_2,x_3,\cdots,y_{n_1},x_{n_2+1},\cdots,x_{n_3}),\cdots,$

$y_{n_3+1}=X_{n_3+1}(y_2,y_3,\cdots,y_{n_1},y_{n_2+1},\cdots,y_{n_3}),\cdots,$

Now we want to prove that formula $(\forall x_1)(Q_2x_2)\cdots(Q_nx_n)\Phi$ is true.

Let $z_2,z_3,\cdots,z_{n_1},z_{n_2+1},\cdots,z_{n_3},\cdots$ be a sequence of free variables. We define the sequences of x's and y's accordingly.

The equation system

$$(2)\begin{cases}\lambda c+\mu d=z_1,\\ \lambda+\mu=1.\end{cases}$$

has a uniqud solution (λ,μ) because $c\neq d$. Let

$$x_i=y_i=z_i,\text{for } i=2,3,\cdots,n_1,n_2+1,\cdots,n_3,\cdots$$

by the above we know that we can find

$$x_i,y_i,\text{for } i=n_1+1,\cdots,n_2,n_3+1,\cdots,n_4,\cdots$$

such that they are functions of the previous variables satisfying (1). Multiplying by λ, and μ separately:

$$(3)\begin{cases}\bigwedge_{j=1}^{m}(a_{j1}\lambda c+\sum_{k=2}^{n}a_{jk}\lambda x_k=\lambda b_j),\\ \bigwedge_{j=1}^{m}(a_{j1}\mu d+\sum_{k=2}^{n}a_{jk}\mu y_k=\mu b_j),\end{cases}$$

Combining the two equations together we get the following:

$$\bigwedge_{j=1}^{m}(a_{j_1}(\lambda c+\mu d)+\sum_{k=2}^{n}a_{jk}(\lambda x_k+\mu y_k)=(\lambda+\mu)b_j).$$

From (2) we can simplify the formula.

$\lambda c+\mu d=z_1$, and

$\lambda x_k+\mu y_k=(\lambda+\mu)z_k=z_k$, for $k=2,3,\cdots,n_1,n_2+1,\cdots,n_3,\cdots$

Let $z_k=\lambda x+\mu y$, for $k=n_1+1,\cdots,n_2,n_3+1,\cdots,n_4,\cdots$ These z's are the functions of the previous z's on the z-sequence. The formula becomes:

$$\bigwedge_{j=1}^{m}\left(\sum_{k=1}^{n}a_{jk}z_k=b_j\right).$$

Therefore $(Q_2x_2)(Q_3x_3)\cdots(Q_nx_n)\Psi$ is true. Our theorem is proved.

In general the symbol '$\exists$' can move across the symbol '$\vee$' without changing the validity of the formula but the symbol '$\forall$' can not move across the symbol '$\vee$'. We will prove in the following theorem that this kind of moves is allowed in the case of positive formulas of real numbers in disjunctive-conjunctive forms.

Theorem 2 Let

$$\Phi(x_1,x_2,\cdots,x_n)=\bigvee_{i=1}^{l}\left(\bigwedge_{j=1}^{m}\left(\sum_{k=1}^{n}a_{ijk}x_k=d_{ij}\right)\right)$$

be a disjunctive-conjunctive normal formula. Then $(Q_1x_1)(Q_2x_2)\cdots(Q_nx_n)\Phi$ is true if and only if there is an i such that

$$\Psi_i(x_1,x_2,\cdots,x_n)=\bigwedge_{j=1}^{m}\left(\sum_{k=1}^{n}a_{ijk}x_k=b_{ij}\right).$$

and $(Q_1x_1)(Q_2x_2)\cdots(Q_nx_n)\Psi_i$ is true.

Proof It is easy to check that theorem is true when $n=1$. Assuming that the theorem is true for the case of n we now prove the case of $n+1$. "If case". Suppose that $(Q_1x_1)(Q_2x_2)\cdots(Q_nx_n)\Psi_i$ is true. Let $(Q_1x_1)(Q_2x_2)\cdots(Q_nx_n)$ be grouped into sections.

$$\forall x_1\cdots x_{n_1}\ \exists x_{n_1+1}\cdots x_{n_2}\ \forall x_{n_2+1}\cdots x_{n_3}\ \exists x_{n_3+1}\cdots x_{n_4}\cdots$$

From the Skolern Function Theorem we know that for the free variable sequence $x_1,x_2,\cdots,x_{n_1},x_{n_2+1},\cdots,x_{n_3},\cdots$ we can find $x_{n_1+1},x_{n_2+1},\cdots,x_{n_2},x_{n_3+1},\cdots,x_{n_4},\cdots$, where

$$x_{n_1+1}=X_{n_1+1}(x_1,x_2,\cdots,x_{n_1}),\cdots$$

and

$$x_{n_3+1}=X_{n_3+1}(x_1,x_2,\cdots,x_{n_1},x_{n_2+1},x_{n_3}),\cdots$$

such that they are the functions of the previous variables. The new variable list satisfies the formula $\Psi_i(x_1,x_2,\cdots,x_n)$.

The choosing functions $X_{n_1+1},X_{n_1+2},\cdots,X_{n_1+3},\cdots$ work for formula Φ also. Hence$(Q_1x_1)(Q_2x_2)\cdots(Q_nx_n)\Phi$ is true.

"Only if case". Suppose that $(Q_1x_1)(Q_2x_2)\cdots(Q_nx_n)\Phi$ is true.

(1)Let the first quantifier be $(\exists x_1)$. Then there is a $c \smile R$ for the free variable sequence $x_2, x_3, \cdots, x_{n_1}, x_{n_2+1}, \cdots, x_{n_3}, \cdots$ we can find x_{n_1+1}, $x_{n_1+2}, \cdots, x_{n_2}, x_{n_3+1}, \cdots, x_{n_4}, \cdots$, where

$x_{n_1+1} = X_{n_1+1}(x_1, x_2, \cdots, x_{n_1}), \cdots$ and

$x_{n_3+1} = X_{n_3+1}(x_1, x_2, \cdots, x_{n_1}, x_{n_2+1}, x_{n_2+2}, \cdots, x_{n_3}), \cdots$

such that they are the functions of the previous variables. The new variable list satisfies the formula $\Phi i(c, x_2, x_3, \cdots, x_n)$. Therefore we have

$$\bigvee_{j=1}^{l} \left(\bigwedge_{j=1}^{m} \left(a_{ij1}c + \sum_{k=2}^{n} a_{ijk} x_k = b_{ij} \right) \right).$$

Move $a_{ij1}c$ to the other side and give a name as Φ' to the new formula we get

$$\Phi' = \bigvee_{i=1}^{l} \left(\bigwedge_{j=1}^{m} \left(\sum_{k=2}^{n} a_{ijk} x_k = b_{ij} + a_{ija} c \right) \right).$$

This implies that $(Q_2 x_2)(Q_3 x_3)\cdots(Q_n x_n)\Phi'$ is true. By induction $(Q_2 x_2)$ $(Q_3 x_3)\cdots(Q_n x_n)\Psi'_i$ is true, where

$$\Psi'_i = \bigwedge_{j=1}^{m} \left(\sum_{k=2}^{n} a_{ijk} x_k = b_{ij} + a_{ij1} c \right).$$

is one of the sub-formulas of Φ'. Moving the term $a_{ij1}c$ back to the left hand side and name the new formula as Ψ_i we get.

$$\Psi_i = \bigwedge_{j=1}^{m} \left(a_{ij1} c + \sum_{k=2}^{n} a_{ijk} x_k = b_{ij} \right).$$

and $(Q_2 x_2)(Q_3 x_3)\cdots(Q_n x_n)\Psi_i$ is true. Therefore $(\exists x_1)(Q_2 x_n)(Q_3 x_n)\cdots$ $(Q_n x_n)\Psi_i$ is true. This is what we want to prove.

(2) Let the first quantifier be $(\forall x_1)$ and assume that $(Q_1 x_1)$ $(Q_2 x_2)(Q_3 x_3)\cdots(Q_n x_n)\Phi(x_1, x_2, \cdots, x_n)$ is true

$$\Phi(x_1, x_2, \cdots, x_n) = \bigvee_{i=1}^{l} \left(\bigwedge_{j=1}^{m} \left(\sum_{k=2}^{n} a_{ijk} x_k = b_{ij} \right) \right).$$

Substitute $c = 1, 2, \cdots, l+1$ *for* x_1 into the formula we know that

$$(Q_2 x_2)(Q_3 x_3)\cdots(Q_n x_n)\Phi(c, x_2, x_3, \cdots, x_n)$$

for $c = 1, 2, \cdots, l+1$. By induction $(Q_2 x_2)(Q_3 x_3)\cdots(Q_n x_n)\cdots\Psi_{ic}(c, x_2, x_3, \cdots, x_n)$ is true for one of the sub-formulas

$$\Psi_{ic}(x_1, x_2, \cdots, x_n) = \bigwedge_{j=1}^{m} \left(\sum_{k=1}^{n} a_{icjk} x_k = b_{icj} \right).$$

$i_1, i_2, \cdots, i_{l+1}$ can not be all different. For example if $i_u = i_v = i$, then sub-formula $(Q_2 x_2)(Q_3 x_3)\cdots(Q_n x_n)\Psi_i(x_1, x_2, \cdots, x_n)$ is true for $x_1 = u$ and

$x_1=v$ where $u\neq v$. By Theorem 1 we know that $(\forall x_1)(Q_2x_2)(Q_3x_3)\cdots(Q_nx_n)\Psi_i(x_1,x_2,\cdots,x_n)$ is true. Our theorem is proved.

Gaussian procedure gives a solution to a linear equation system consuming a polynomial number of steps of real operations. We extend his procedure to give a decision algorithm for the positive conjunctive formulas with quantifiers.

Theorem 3 Formula

$$(Q_1x_1)(Q_2x_2)\cdots(Q_nx_n)\bigwedge_{j=1}^{m}\left(\sum_{k=1}^{n}a_{jk}x_k=b_j\right).$$

can be decided in a polynomial many steps of operations.

Proof We apply the quantifier elimination procedure.

(1) Let the last quantifier be $\exists x_n$. The formula

$$\Psi=\begin{cases}a_{11}x_1+a_{12}x_2+\cdots+a_{1n}x_n+b_1\wedge\\a_{21}x_1+a_{22}x_2+\cdots+a_{2n}x_n=b_2\wedge\\\cdots\wedge\\a_{m1}x_1+a_{m2}x_2+\cdots+a_{mn}x_n=b_m.\end{cases}$$

If all $a_{1n},a_{2n},\cdots,a_{mn}=0$, Then we simply take off the $(\exists x_n)$. Now suppose that $a_{1m}\neq 0$. Using Gaussian procedure we can eliminate the x_n terms except for the one on the first row. We obtain

$$\Psi'=\begin{cases}c_{11}x_1+c_{12}x_2+\cdots+c_{1n}x_n=d_1\wedge\\c_{21}x_1+c_{22}x_2+\cdots+c_{2,n-1}x_{n-1}=d_2\wedge\\\cdots\wedge\\c_{m1}x_1+c_{m2}x_2+\cdots+c_{m,n-1}x_{n-1}=d_m.\end{cases}$$

$(\exists x_n)\Psi'$ is equivalent to

$$\Psi''=\begin{cases}e_{21}x_1+e_{2,n-1}x_{n-1}=f_2\wedge\\\cdots\wedge\\e_{m1}x_1+e_{m2}x_2+\cdots+e_{m,n-1}x_{n-1}=f_m.\end{cases}$$

The quantifier $\exists x_n$ is now eliminated.

(2) Let the last quantifier be $\forall x_n$. Similarly we have a formula

$$\Psi=\begin{cases}a_{11}x_1+a_{1n}x_n=b_1\wedge\\a_{21}x_1+a_{22}x_2+\cdots+a_{2n}x_n=b_2\wedge\\\cdots\wedge\\a_{m1}x_1+a_{m2}x_2+\cdots+a_{mn}x_n=b_m.\end{cases}$$

If all $a_{1n}, a_{2n}, \cdots, a_{mn}=0$, Then we simply take off the $(\forall x_n)$. Now suppose that $a_{1m}\neq 0$. For any $x_1, x_2, \cdots, x_{n-1}$ there is no way to let Ψ be true for all x_n. Therefore

$$(Q_1x_1)(Q_2x_2)\cdots(Q_nx_n)\Psi$$

is not true.

The above procedures (1) and (2) consume only a polynomial number of steps of real operations.

Joining the above theorems together we give our main theorem of this paper which provides an algorithm to the decision problem of positive formulas of $Th^+(R,+)$.

Theorem 4 A positive formula in $Th^+(R,+)$ of length n can be decided in time 2^{cn}.

Proof Let $(QX)\Phi$ be our original formula. $|(QX)\Phi|=n$. Move the quantifiers to the front and expand the formula to a disjunctive-conjunctive normal form. The formula will be equivalent to $(QX)\Phi'=(Q+x+)\cdots(Q_nx_n)\Phi$ where

$$\Phi'(x_1,x_2,\cdots,x_n)=\bigvee_{i=1}^{l}\Big(\bigwedge_{j=1}^{m}\Big(\sum_{k=1}^{n}a_{ijk}x_k=b_{ij}\Big)\Big)$$

We can simply write it as $\Phi'=\vee\wedge\Sigma$. The length of Φ' is at most 2^{cn}.

By Theorem 2 the decision of formula $(QX)(\vee\wedge\Sigma)$ is equivalent to the decision of $\vee(QX)(\wedge\Sigma)$. The last formula can be decided one by one. If one of the formulas $(QX)\wedge\Sigma$ is true, then the original formula is true. Otherwise the original formula is not true.

By Theorem 3 formula $(QX)\wedge\Sigma$ can be decided in a polynomial time of n. Therefore $(QX)\Phi$ can be decided in time 2^{cn}.

§ 3. The lower bound

In the case of all formulas of the theory of real numbers $Th(R,+)$ the upper bound and the lower bound are 2^{cn} space and 2^{dn} time. People agree that they match each other because we rarely find the same limit for both the upper bound and the lower bound in this kind of computations. In our case for the theory of $Th^+(R,+)$ the upper bound is reduced to 2^{cn} time. It is not likely to get an lower bound of 2^{dn} time because they are so close. We will prove that the decision procedure for the positive formulas in real numbers is NP hard. This will act as a lower bound.

Theorem 5 The decision problem for positive formulas of real addition is NP hard.

Proof From [3], p. 324 we only need to prove that the decision problem of the positive formulas is at least as hard as the satisfiability problem. Again from [3], p. 330, Theorem 13. 1 we only need to find a p-time reduction for any instance of L_{3SAT} to an instance of real positive formula. A boolean formula Φ in 3-CNF can be written in the following form:

$$\Phi(A_1, A_2, \cdots, A_n) = \bigwedge_{i=1}^{l} \left(\bigvee_{j=1}^{3} \varepsilon_{ij} A_{ij} \right)$$

Where ε_{ij} can be positive or negative and A_{ij} can be any variable on the sequence $A_1, A_2, \cdots, A_n$. Define δ_{ij} as follows.

$$\delta_{ij} = \begin{cases} 0, & if\ \varepsilon_{ij}\ is\ positive. \\ 1, & if\ \varepsilon_{ij}\ is\ negative. \end{cases}$$

Let $x_1, x_2, \cdots, x_n$ be a sequence of real variables. We construct a positive formula Ψ in the real with addition.

$$\Psi = (\exists x_1 x_2 \cdots x_n) \left(\bigwedge_{i=1}^{l} \left(\bigvee_{j=1}^{3} \delta_{ij} = x_{ij} \right) \right)$$

It is easy to check that Ψ is true if and only if Φ is in L_{3SAT}. Therefore the decision problem of L_{3SAT} can be reduced polynomially to the decision problem of positive formulas in the real numbers with addition. $Th^+(R, +)$ is now NP hard.

§ 4. Step counts and operation counts

We all know that the Gaussian procedure for solving linear equation system in real numbers consumes $|s|^{\frac{3}{2}}$ steps of real operations, where $|s|$ is the length of the formula. Can we improve our result on upper bound of the computation on positive formulas in counting the number of operations? The following theorem will give an upper bound for the number of operations in the decision procedure of a positive formula.

Theorem 6 There is an NP procedure for deciding a positive formula requiring a polynomial number of operations.

Proof Let n be the length of the formula. As the first step we move all the quantifiers to the front part of the formula. If we try to expand the formula into a disjunctive-conjunctive form the length of the formula could increase to 2^{cn}. This is what we don't want to do. We will try not to shift around but find out all possible formulas of the form:

$$\bigwedge_{i=1}^{k} \sum_{j=1}^{l} a_{ij} x_j = b_i$$

which can appear in the disjunctive-conjunctive expansion of the original formula as a coniuctive term.

First of all reduce all unitary parenthesis except for the once with only an atomic formula in it. A parenthesis is called a $\wedge$-parenthesis if the last operation in it is a $\wedge$. A parenthesis is called a $\vee$-parenthesis if the last operation in it is a $\vee$. By using consecutive $\wedge$'s in a parenthesis or consecutive $\vee$'s in a parenthesis we reduce our formula. A parenthesis P within another parenthesis Q is called an immediate parenthesis of P if there is no parenthesis R such that P is in the scope of R and R is in the scope of Q. A parenthesis is reduced if it is a $\wedge$-parenthesis and all its immediate parentheses are not $\vee$-parentheses or if it is a $\vee$-parenthesis and all its immediate parentheses are not $\vee$-parentheses. A formula is completely reduced if all its sub-formulas are reduced.

Now we start issuing subscripts to all parentheses of the formula. If the outermost parenthesis is a $\wedge$-parenthesis its subscript is $\wedge 1$ if the outermost parenthesis is a $\vee$-parenthesis its subscript is $\vee 1$. Let $P=(Q_1 \wedge Q_2 \wedge \cdots \wedge Q_k)$ be a parenthesis with k immediate parentheses and the subscript of P is X. Then the subscript for Q_i is $X \vee i$ for $i=1,2,\cdots,k$. Let $P=(Q_1 \vee Q_2 \vee \cdots \vee Q_k)$ be a parenthesis with k immediate parentheses and the subscript of P is X. Then the subscript for Q_i is $X \wedge i$ for $i=1,2,\cdots,k$.

As an example the following is a set of subscript of a formula.

$(((\cdots)_{\wedge 1 \vee 1 \wedge 1} \vee (\cdots)_{\wedge 1 \vee 1 \wedge 2})_{\wedge 1 \vee 1} \wedge ((\cdots)_{\wedge 1 \vee 2 \wedge 1} \vee (\cdots)_{\wedge 1 \vee 2 \wedge 2})_{\wedge 1 \vee 2} \wedge$

$((\cdots)_{\wedge 1 \vee 3 \wedge 1} \vee (\cdots)_{\wedge 1 \vee 3 \wedge 2})_{\wedge 1 \vee 3})_{\wedge 1}$

The above procedure expand the length of the formula polynomially. In this way every sub-parenthesis has a subscript. Especially the atomic subfornmlas have their subscripts.

Our NP procedure to find a conjunctive product of atomic formulas is to choose a sequence SQ of atomic formulas with their subscripts in the order from left to right as in the formula. This takes NP time. Whenever the sequence is chosen the procedure starts checking the intersection of the sequence and the original sub-parentheses of the formula.

1)If the intersection of the sequence SQ with a $\wedge$-parenthesis P is not empty, then the intersection of SQ with every immediate parenthesis of P is not empty.

2)If the intersection of the sequence SQ with a $\vee$-parenthesis Q is not empty, then the intersection of SQ with one and only one immediate parenthesis of Q is not empty.

We can prove by induction that if the intersection of a parenthesis P with the sequence SQ is not empty, then the intersection forms a maximum conjunctive sub-formula of the disjunctive-conjunctive expansion of parenthesis P. Because for the atomic formulas our argument is trivially true. 1) and 2) insure the induction step to get through. By Theorem 3 we know that the decision procedure for a conjunctive sub-formula with quantifiers requires only polynomial many steps of operations. Therefore the entire procedure requires NP steps of operations.

Reference

[1]Ferrante J and Rackoff C. A decision procedure for the first order theory of real addition with order. SIAM Journal on Computing, 1975, 4: 69—76.

[2] Fischer M L. and Rabin M O. Super-exponential complexity of Presburger arithmetic, Complexity of computation (R. Karp, editor). SIAM—AMS Proceedings, American Mathematical Society, Providence, Rhode Island, 1974, 7: 27—42.

[3]Hopcroft J E and Ullman J D. Introduction to automata theory languages and computation. Addison-Wesley, 1979.

The Journal of Symbolic Logic, 1992, 57(1): 118～130

有限系统上的函数与泛函数①

Functions and Functionals on Finite Systems

§0. Introduction

The global function on finite systems is a new concept defined by Gurevich in [1] and discussed in [2] and [3]. In the last ten years this concept has become more and more useful in computer science and logic. Gurevich also pointed out the importance of global functionals on finite systems. In this paper we will give a brief introduction to the concepts of global functions and global functionals on finite systems.

In studying the natural number system $N=\langle N,+,0\rangle$ we often refer to its functions and functionals. There are a lot of books and papers in this area. Kleene in [5] gave a detailed introduction to the recursive functions of N. The functionals of N are normally very difficult to compute because here we need to tell the machine what the input function is, which is not very easy to do. In developing the theory of finite systems the functions and functionals are also very useful. For computing the functionals in finite systems we can take the entire graph of a function as the input, which is not possible in N. We will discuss recursive

① Received April 25, 1989; revised September 10, 1990, February 7, 1991, and March 5, 1991.

functions and functionals for finite systems. The definitions of recursive functions are very similar to the case in N, but we will have a very different situation. In N the number of elements is infinite. The number of all possible functions from N to N is the continuum. In a finite system the number of all possible functions is finite. It seems that there is no necessity to define the global functions. The need to develop such global functions arises because we discuss the same function in different finite systems. It is not very clear when we say that "+" is "a function" on different finite systems; for example, "+" in a system of one element acts very different from the case when it is in a system of three elements. That is why we call it a global function. A global function φ on finite systems acts as a function in each system. We can also consider a global function as $\varphi=\{f_1, f_2, \cdots, f_i, \cdots\}$, where f_i is a function in a finite system with i elements. In a system S of s elements the number of all possible partial functions from S to S is s^{s+1}, because we allow undefined values. For a global function φ we do not set any restrictions on what function it can act as in different finite systems. Therefore the number of all possible global functions on all finite systems is also the continuum.

We use the standard notation of LOGSPACE, PTIME and PSPACE for the computational complexity classes. The reader can find the definitions in [4]. We also use the standard notions for complexity class EXPTIME and EXP, where EXPTIME: $=\bigcup_{c>0}$ DTIME$[2^{cn}]$ and EXP: $=\bigcup_{c>0}$ DTIME$[2^{c^n}]$.

In [1] Gurevich examined the global functions on finite systems and gave the following very interesting results: (1) A global function is LOGSPACE computable if and only if it is primitive recursive. (2) A global function is PTIME computable if and only if it is recursive. Here the formulas in the definitions of recursive (global) functions are almost the same as those we have seen for N (like the formulas in [5]), but the meanings are different. In N we discuss only one system, where the recursion goes infinitely; but in the present case we deal with infinitely many systems and in each system the recursion goes finitely. In this paper we give a brief review of Gurevich's definitions and apply the diagonalization procedure to find lower bounds for the computation on global functions.

There are many ways to define global functionals. We say that a functional of a system is a function from the set of all (partial) functions into the set of all (partial) functions of the same system. A global functional acts as a functional in each finite system. For example, we define a functional $F(u,v,w)=(u_1,v_1)$. Here we need to answer the following three questions. 1. Are the arities and the co-arity of the global functional fixed? Yes, in the example; the arity of F is 3 and co-arity of F is 2. 2. Are the arities of the functions u, v, and w as arguments fixed? In general this is unnecessary, but the functional is also considered as a partial function. So, for some inputs F might not produce any output functions at all. 3. Do we define our global functional as a function from the set of all global functions into (or onto) the set of all global functions? We say "yes" because a global functional acts as a functional in each finite system, which in turn defines a function from global functions to global functions. The input and output functions are normally given by their graphs. That is how we compute the length of the input and the output. In §7 we will generalize Gurevich's results to prove that a global functional is primitive recursive if and only if it is PSPACE computable and a global functional is recursive if and only if it is EXPTIME computable.

§1. Preparations

Our discussion is based on the *second order logic* of finite systems. We have a series of finite system S_i of $s=i$ elements (for $i=1,2,\cdots$) with a linear order on each system S_i. A global function f acts as a function on each system S, but the arity (and co-arity) of f cannot change when we discuss different systems S_i and S_i. Let FN be the set of all functions from S to S. A functional is a function of FN to FN.

The following symbols are used.

$a, \cdots, e$ are general variables.

f, g, h and Greek letters are variables for functions on S.

F, G, H and upper case names are variables or constants for functionals on S.

$i, \cdots, s$ are variables or constants for natural numbers.

$t, \cdots, z$ are variables for elements of S.

For convenience, we refer to arities (and co-arities) l, m and n throughout, and use overbars, etc., for vectors of these quantities; viz., $\bar{a} := (a_1, a_2, \cdots, a_l)$, $\tilde{a} := (a_1, a_2, \cdots, a_m)$ and $\hat{a} := (a_1, a_2, \cdots, a_n)$.

A finite system S is a set with s elements $0 < 1 < \cdots < e$, where 0 is the first element and $e = s - 1$ is the last element of S.

For the global functions we give the following symbols and explanations. A function F of arity l and co-arity n is defined if for every $\bar{x} = (x_1, x_2, \cdots, x_l)$ there is at most one $\hat{y} = (y_1, y_2, \cdots, y_n)$ such that $\hat{y} = F(\bar{x})$. We allow the arity of functions to be zero. A lunction of arity 0 and co-arity m is called a vector of length m. A vector of length 1 is identified with the corresponding element of S: $(x) = x$. It is convenient to consider elements as functions.

We give the following examples:

1. The function "+" (addition) with an overflowing signal is a global function. "+" is a partial function.

2. The function " * " (multiplication) is also a global function. It is also a partial function.

3. Some of the global functions may not be very natural. For example, we can define the following global function φ.

3. 1. φ acts like "+" in a system with an odd number of elements.

3. 2. φ acts like " * " in a system with an even number of elements.

§ 2. Primitive recursive functions

Consider the following equations and equation systems (Ⅰ)~(Ⅴ).

Each of them defines a function φ on S, where l and m are positive integers, i is an integer such that $1\leqslant i\leqslant l$, x is an element of S, $\overline{x}=(x_1, x_2, \cdots, x_l)$, and $\Psi, \chi_1, \chi_2, \cdots, \chi_m$ and χ are given functions of the indicated arities.

(Ⅰ) The successor function $\varphi(\overline{x})=\sigma(\overline{x})$, where

$$\sigma(\overline{x})=(0,\cdots,0,\sigma(x_i),x_{i+1},\cdots,x_l);$$

here $\overline{x}=(e,\cdots,e,x_i,x_{i+1},\cdots,x_l)$, x_i being the first component of $\overline{x}$ which is different from e.

(Note that the system S is considered numerically; $\sigma(x)=x+1$ and $\sigma(\overline{x})$ can be written as $\overline{x}+1$; e does not have a successor, so $\sigma(\overline{e})$ is undefined.)

(Ⅱ) The constant functions $\varphi(\overline{x})=0$ or $\varphi(\overline{x})=e$.

(Ⅲ) The projection functions $\varphi(\overline{x})=(x_{i1},x_{i2},\cdots,x_{in})$ for every sequence $x_{i1},x_{i2},\cdots,x_{in}$ satisfying $1\leqslant x_{ij}\leqslant l$.

(Ⅳ) The composition of functions

$$\varphi(\overline{x})=\psi(\chi_1(\overline{x}),\chi_2(\overline{x}),\cdots,\chi_m(\overline{x}))$$

where χ_i is a function of arity l and co-arity r_i, $r=r_1+r_2+\cdots+r_m$, and ψ is a function of arity r and co-arity n.

(Ⅴ) Primitive recursion:

(Ⅴa) $\varphi(\overline{x},\widetilde{0})=\psi(\overline{x})$.

(Ⅴb) $\varphi(\overline{x},\sigma(\overline{y}))=\chi(\varphi(\overline{x},\overline{y}),\overline{x},\overline{y})$, where ψ is a function of arity l and co-arity n and χ is a function of arity $I+m+n$ and co-arity n. (Ⅴa) and (Ⅴb) define a function φ of arity $I+m$ and co-arity n.

A function φ is called an *initial function* if φ satisfies equation (Ⅰ), or (Ⅱ) for a particular l, or (Ⅲ) for a particular l and a sequence $x_{i1},x_{i2},\cdots,x_{in}$.

A function φ is called an *immediate dependent* of other function if φ satisfies equation (Ⅳ) for a particular l and m with ψ and $\chi_1,\chi_2,\cdots,\chi_m$ as the other functions, or equations (Ⅴa) and (Ⅴb) for a particular l with ψ and χ as the other functions.

A function φ is *primitive recursive* if there is a finite sequence φ_1,

$\varphi_2, \cdots, \varphi_k (k \geqslant 1)$ of functions such that each function of the sequence is either an initial function or an immediate dependent of preceding functions of the sequence, and the last function φ_k is the function φ.

A primitive recursive function may be a partial function in finite systems. A *primitive recursive predicate* is a total primitive recursive function with values in $\{0,1\}$ only.

We also need some lemmas which will give more primitive recursive functions. Their proofs are routine.

Lemma 1 *The bounded sum* $\sum_{y<z} \psi(\bar{x}, y)$ *and bounded product* $\prod_{y<z} \psi(\bar{x}, y)$ *are primitive recursive in* ψ. *If the sum or the product gets too large, the function may be undefined for some values.*

Lemma 2 *A predicate* u *obtained by substituting functions* χ_1, $\chi_2, \cdots, \chi_m$ *for the respective variables of a predicate* $v(x_1, x_2, \cdots, x_m)$ *is primitive recursive in* $\chi_1, \chi_2, \cdots, \chi_m$ *and* v.

Lemma 3 *The predicate* $\neg v(\bar{x})$ *is primitive recursive in* v. *The predicates* $v(\bar{x}) \vee w(\bar{x})$, $v(\bar{x}) \wedge w(\bar{x})$, $v(\bar{x}) \rightarrow w(\bar{x})$ *and* $v(\bar{x}) \leftrightarrow w(\bar{x})$ *are primitive reeursive in* v *and* w.

§ 3. The characterization by formulas

It is easy to see that every first order formula without function symbols can be computed in LOGSPACE. However, because of the undefined values resulting from overflows in our finite systems we cannot prove the converse of [5, Theorem I, p. 241]. To overcome this difficulty, and allow formulas to simulate the LOGSPACE computation, we introduce the symbols for functions into the formulas. Suppose from now on that the function symbols $f, g, h, \cdots$ can be written into a formula as function constants. We give a format to write the formulas.

The following 16 symbols are used:

$$\wedge, \vee, \neg, \forall, \exists, =, 0, 1, e, {}_{0,1}, x, f, F, \sigma, |, (,).$$

They are called *basic symbols*.

The elements of S are coded by strings over $\{0,1\}$ in the standard manner, while the small symbols $_0$ and $_1$ encode natural numbers used in subscripts. Let l represent a natural number.

x and xl are the variable symbols.

f and fl are the function constants.

σ is a function constant expressing the successor of an element in S.

| is used to separate the arguments.

From the definition of a primitive recursive function we know that if f is primitive recursive, then there is a sequence of functions f_1, $f_2, \cdots, f_k$ such that each f_i is either an initial function or an immediate consequence of the previous ones, and f_k is f. Now we want to show how to write f as a computable formula. For each f_i, $1 \leqslant i \leqslant k$, we define φ_i according to the following cases.

(Ⅰ) If f_i is the successor function, let φ_i be the formula

$$(\forall x) f_i(x) = \sigma(x).$$

(Ⅱ) If f_i is a constant function, let φ_i be the formula

$$(\forall \bar{x}) f_i(\bar{x}) = 0 \text{ or } (\forall \bar{x}) f_i(\bar{x}) = e.$$

(Ⅲ) If f_i is a projection function, let φ_i be the formula

$$(\forall x_1, \forall x_2, \cdots, x_l) f_i(x_1, x_2, \cdots, x_l) = (x_{j1}, x_{j2}, \cdots, x_{jr}).$$

(Ⅳ) If f_i is derived by composition scheme let φ_i be the formula

$$(\forall x) f_i(\bar{x}) = g(h_1(\bar{x}), h_2(\bar{x}), \cdots, h_m(\bar{x})),$$

where $h_1, h_2, \cdots, h_m$ and G are functions from $f_1, f_2, \cdots, f_{i-1}$.

(Ⅴ) *If* f_i is derived by the primitive recursion scheme, let φ_i be the formula

$$(\forall \bar{x}, \tilde{y})(f_i(\bar{x}, \tilde{0}) = g(\bar{x}).\ \wedge.\ f_i(\bar{x}, \tilde{y}+1) = h(f_i(\bar{x}, \tilde{y}), \bar{x}, \tilde{y})).$$

(Ⅵ) Finally we define the formula φ:

$$\varphi :\equiv: \varphi_1.\ \wedge.\ \varphi_2.\ \wedge.\ \cdots.\ \wedge.\ \varphi_k.\ \wedge.\ \hat{z} = f_k(\bar{x}, \tilde{y}).$$

Theorem 1 (Gurevich) *A function can be computed in* LOGSPACE *if and only if it can be represented by a formula using scheme* (Ⅵ).

We do not repeat the proof here (see [2]), but we do explain how we compute the functions. The reader can refer to the proofs in Gurevich's papers. In the computation of a function $\tilde{y} = f(\bar{x})$ with arity l

and co-arity m we assume that it is done by a multitape offline Turing machine. The numbers of input, output and working tapes are l, m and n. The length of the input is $\ulcorner \log x_1 \urcorner + \ulcorner \log x_2 \urcorner + \cdots + \ulcorner \log x_l \urcorner$, and the lengths of the output and working tapes are also $\ulcorner \log s \urcorner$, where s is the number of elements of the finite system S being considered. In the proof, to check that all primitive global functions are computable in LOGSPACE we go through the recursion schema and prove that they produce LOGSPACE functions from LOGSPACE functions. Conversely, to check that a LOGSPACE computable function is primitive recursive we make use of the recursions and prove that it can simulate LOGSPACE computable Turing machines.

§ 4. Diagonalization for LOGSPACE computability

Gurevich's theorem shows that a function u is LOGSPACE computable if and only if it can be defined by formula (Ⅵ) in § 3. If we write an element $x \in S$ as in the positional notation with 16 symbols and interpret each digit as a basic symbol, then it becomes a formula φ. We say that x is the Gödel number of φ, and write $x = \ulcorner \varphi \urcorner$. We know that well-formed formulas can be decided in LOGSPACE. Therefore, by Theorem 1 the well-formed formulas can be defined using function constants and induction. We now define an informal predicate $\varphi(y, z)$ as follows:

"$\varphi(y, z)$ is true if and only if the following three conditions are true:

(1) y is the Gödel number of a formula $\psi(x)$,

(2) z is an element of S,

(3) (a) ψ is not a good formula, or (b) $\psi(z)$ does not exist, or (c) $\psi(z)$ is not true."

One would expect that the function $\psi(y, z)$ cannot be computed in LOGSPACE. This expectation is confirmed in the following theorem.

Theorem 2 *$\varphi(y, z)$ cannot be computed in* LOGSPACE.

Proof Suppose on the contrary that $\varphi(y, z)$ is computed in

LOGSPACE. From Theorem 1 we know that it can be computed by a formula $\chi(y,z)$. Let γ be the predicate defined by $\gamma(y)\leftrightarrow\chi(y,y)$; γ is also computed in LOGSPACE. Now consider

(4) $$\gamma(\ulcorner\gamma(x)\urcorner).$$

If it is true, then

(5) $$\chi(\ulcorner\gamma(x)\urcorner,\ulcorner\gamma(x)\urcorner)$$

is true. By the meaning of χ we know that

(6) $$\varphi(\ulcorner\gamma(x)\urcorner,\ulcorner\gamma(x)\urcorner)$$

is true. From the above conditions (1), (2) and (3) we know that either

(a) $\gamma(x)$ is not a good formula, or

(b) $\gamma(\ulcorner\gamma(x)\urcorner)$ does not exist, or

(c) $\gamma(\ulcorner\gamma(x)\urcorner)$ is not true.

Since conditions (a) and (b) do not hold for γ, (c) contradicts the truth of (4). We get a contradiction. Conversely, suppose that (4) is not true. Then (c) is not true. From (1)(2) and (3) the informal predicate $\varphi(\ulcorner\gamma(x)\urcorner,\ulcorner\gamma(x)\urcorner)$ is true. We then derive (6)(5) and(4) consecutively, obtaining another contradiction. Hence χ does not exist.

The explanation of the above theorem is that there is no LOGSPACE algorithm for computing all formulas with length $\leqslant \log s$.

For if there were such an algorithm AL, then the Gödel number y of a formula ψ of length $\leqslant c * \log s$ would be in S. Here $y=\ulcorner\psi\urcorner$ and $c=1/\log_2 16$, where 16 is the number of symbols in constructing the formulas of finite systems. Therefore AL would compute the informal predicate $\varphi(y,z)$ in Theorem 2. This contradicts the theorem.

Theorem 3 *The function $\varphi(x,y)$ in Theorem 2 can be computed in $\log^2 s$ space.*

Proof Decode y into a formula ψ. Then compute $\psi(x)$. There are at most $l * \log s$ different variables (including bounded variables) in ψ. A space of $l * \log^2 s$ units to store the values for the variables is sufficient. By induction we know that a formula of scheme (Ⅵ) with length k can be computed in a storage of k tapes with $\log s$ units on each tape.

We do not claim that this is the best upper bound. There should be some improvements in this respect.

§ 5. Formulas for recursive functions

The above sections discussed primitive recursive functions. Now we turn to the recursive functions. We give a new scheme which simplifies the formulas for writing recursive functions in [1]. It requires information about the values of f on three consecutive $\bar{x}$'s for the previous $\bar{y}$ in computing a new value of f.

(Ⅶ. a) $f(\bar{x},\tilde{0})=g(\bar{x})$,

(Ⅶ. b) $f(\bar{x},\bar{y}+1)=h(f(\bar{x}-1,\bar{y}),f(\bar{x},\bar{y}),f(\bar{x}+1,\bar{y}),\bar{x},\bar{y})$

where the $\bar{x}-1$ and $\bar{x}+1(\bar{x}\neq\bar{0}$ or $\bar{e})$ represent the predecessor and successor functions. The remaining cases are computed by

$$f(\bar{0},\bar{y}+1)=h(f(\bar{0},\bar{y}),f(\bar{0},\bar{y}),f(\bar{0}+1,\bar{y}),\bar{x},\bar{y})$$

and

$$f(\bar{e},\bar{y}+1)=h(f(\bar{e}-1,\bar{y}),f(\bar{e},\bar{y}),f(\bar{e},\bar{y}),\bar{x},\bar{y}).$$

We also need a scheme to write the formulas for recursive functions.

(Ⅷ) This scheme is similar to scheme (Ⅵ). In addition we are allowed to use recursion scheme (Ⅶ) to define any f_i.

Now we show that this representation is equivalent to the one given by Gurevich.

Theorem 4 (1) (Gurevich). *A function can be computed in* PTIME *if and only if it is recursive.*

(2) *A function is recursive if and only if it can be represented by a formula using scheme* (Ⅷ).

Proof *If.* Suppose that g and h are recursive. By [3, Theorem 2] we know that they can be computed by PTIME Turing machines. For computing $h(f(\bar{x}-1,\bar{y}),f(\bar{x},\bar{y}),f(\bar{x}+1,\bar{y}),\bar{x},\bar{y})$ in polynomial time, here is a direct method: simply write the values of $f(\bar{x},\bar{y})$ on some tapes over all $\bar{x}\in S^l$ and look them up as needed. Since l is fixed, the time $O(s^{l+1})$ this takes is polynomial in s. The new Turing machine M goes back and forth to work on 3 consecutive numbers $f(\bar{x}-1,\tilde{y})$, $f(\bar{x},\tilde{y})$, $f(\bar{x}+1,\tilde{y})$ and the numbers $\bar{x}$ and $\tilde{y}$ to compute the value of $f(\bar{x},\tilde{y}+1)$. M also consumes time s^c for a suitable constant c.

Only if. Suppose that the function f can be computed by a PTIME Turing machine. (Ⅷ) can be used to simulate the moves of the machine, and the other schema give the desired function.

§ 6. Diagonalization for PTIME

Similarly to the proof of Theorem 2 we can prove a lower bound for the PTIME computations. By Theorem 4 a function f is PTIME computable if and only if it can be defined by an equation system. Using the same Gödel numbering as in § 3, we can also define the Gödel number $\ulcorner\varphi(x)\urcorner$ of an equation system $\varphi(x)$, which in turn defines a function on $\{0,1\}$. Again we define an informal predicate $\varphi(y,z)$ as follows:

"$\varphi(y,z)$ is true if and only if the following three conditions are true:

(1) y is the Gödel number of an equation system $\psi(x)$,

(2) z is an element of S,

(3) (a) ψ is not in a good form as an equation system, or

(b) $\psi(z)$ does not exist (including the case where two or more values of $\psi(z)$ can be produced by the equation systems), or

(c) $\psi(z)$ is not true."

One would expect that the function $\varphi(y,z)$ cannot be computed in PTIME. The expectation is also confirmed.

Theorem 5 *$\varphi(y,z)$ cannot be computed in* PTIME.

Proof Suppose on the contrary that $\varphi(y,z)$ is computed in PTIME. From Theorem 4 we know that it can be computed by an equation system $\chi(y,z)$. Let γ be the predicate defined by $\gamma(\gamma)\leftrightarrow\chi(y,y)$; γ is also computed in PTIME. Now consider

(4) $$\gamma(\ulcorner\gamma(x)\urcorner).$$

If it is true then

(5) $$\chi(\ulcorner\gamma(x)\urcorner,\ulcorner\gamma(x)\urcorner)$$

is true. By, the meaning of χ we know that

(6) $$\varphi(\ulcorner\gamma(x)\urcorner,\ulcorner\gamma(x)\urcorner)$$

is true. Consider the conditions in the definition of $\varphi(y,z)$ where (1) $y=\ulcorner\gamma(x)\urcorner$ is a Gödel number, (2) $z=y$ is an element of S, and (3) we know that either

(a) $\gamma(\ulcorner\gamma(x)\urcorner)$ is not a good formula, or

(b) $\gamma(\ulcorner\gamma(x)\urcorner)$ does not exist, or

(c) $\gamma(\ulcorner\gamma(x)\urcorner)$ is not true.

Since conditions (a) and (b) do not hold for γ, (c) contradicts the truth of (4). We get a contradiction. Conversely, suppose that (4) is not true. Then (c) is not true. From (1) (2) and (3) the informal predicate $\varphi(\ulcorner\gamma(x)\urcorner,\ulcorner\gamma(x)\urcorner)$ is true. We then derive (6) (5) and (4) consecutively, obtaining another contradiction. Hence χ does not exist.

The explanation of the above theorem is that there is no PTIME algorithm for computing all equation systems with length $\leqslant \log s$. The reason is similar to the explanation for the LOGSPACE computation in § 4.

Theorem 6 *The function $\varphi(x,y)$ in Theorem 5 can be computed in time $s^{l*\log s}$.*

Proof Let ψ be the formula decoded from y. The decoding procedure will use the remainder function, so the time consumed is $O(\log s)$. There are at most $l * \log s$ variables in ψ. In our case $l=1/\log 16$, which is a constant. The variables are divided into groups associated with each f_i. To compute $f_k(x)$ we store all possible values of each $f_i(\overline{x})$ for $\overline{x}$ one by one whenever they are available. The machine will scan from old values to new values. If no more new values can be produced, then the machine comes to a halt. Finally it outputs the value of $f_k(x)$. The time consumed is at most $s^{l*\log s}$.

§ 7. Definition of functionals

We will introduce global functionals of fixed arities. They are global functional constants. F, G, H and upper case symbols will represent the global functionals. We switch to using u, v, w and lower case symbols to represent the function variables. A functional $\hat{v}=F(\bar{u})$ will take input functions $\bar{u}=(u_1, u_2, \cdots, u_l)$ and produce output functions $\hat{v}=(v_1, v_2, \cdots, v_n)$. In the computation $\bar{u}$ and $\hat{v}$ will be represented by their graphs.

A function $y=u(x)$ cannot be recognized by a Turing machine unless it is printed out with machine symbols. From [4] we know that if a functional is computable at all, then it is computable by a Turing machine which uses only 3 symbols $\{0,1,b\}$, where blank symbols b appear only outside the used part of the tape. The global functional constants $v=\text{PRINTOUT}(u)$ and $u=\text{ORIGINAL}(v)$ are discussed here. More functionals are introduced later.

Let u be a function of arity l and co-arity m, and let $s=|S|$.

An element $x\in S$ is printed as

$$\text{PRINTOUT}(x)=0^{x-1}10^{s-x}.$$

A vector $t=(x_1, x_2, \cdots, x_m)$ is printed as

$$\begin{aligned}&\text{PRINTOUT}(t)\\&=\text{PRINTOUT}(x_1)\text{PRINTOUT}(x_2)\cdots\text{PRINTOUT}(x_m)0^{s(s-m)}.\end{aligned}$$

An undefined value is printed as 0^s.

Up to now the tape of a vector occupied s^2 positions.

A function u of arity l and co-arity m is printed in blocks of s^2 positions, as follows.

The first block equals $\text{PRINTOUT}(\bar{y})$, where $\bar{y}=u(\bar{0})$. The second block equals $\text{PRINTOUT}(\bar{y})$, where $\bar{y}=u(\bar{0}+1)$. And so on up to $y=u(\bar{e})$.

If $\bar{x}=(x_1, x_2, \cdots, x_l)$ and $\bar{y}=u(\bar{x})$, then the symbols of $\text{PRINTOUT}(\bar{y})$ are printed in a block numbered

$$x_1+x_2 * s+\cdots+x_l * s^{l-1}.$$

Therefore the length of $\text{PRINTOUT}(u)$ is s^{l+2}. The above description is formally written in the following definition.

Definition For any function u with arity l and co-arity m the print function $v=\text{PRINTOUT}(u)$ is defined as a function of arity $l+2$ and co-arity 1 such that if $(y_1, y_2, \cdots, y_m)=u(x_1, x_2, \cdots, x_l)$, then $v(y_i, i, x_1, \cdots, x_l)=1$ and $v(z, j, x_1, \cdots, x_l)=0$, where $z\neq y_j$ or $j>m$.

Definition. $u=\text{ORIGINAL}(v)$ if and only if $v=\text{PRINTOUT}(u)$.

§ 8. Primitive recursive functionals

Now we start discussing the relation between global functionals and PSPACE computations. Gurevich in [3] proves that a function f is LOGSPACE computable if and only if it is primitive recursive. We adopt his procedure to prove that a functional is PSPACE computable if and only if it is primitive recursive. Before doing this we need some definitions and lemmas. We code a number of length s^c as a function u with arity c, co-arity 1 and values on $\{0,1,2\}$.

DEFINITION GROUP FL. (FL. 0) NUMBER(u) :↔: (Arity(u)= c. ∧. Co-arity(u)=1. ∧. $(\forall \bar{x})(u(\bar{x})=0.\ \vee.\ u(\bar{x})=1.\ \vee.\ u(\bar{x})=2)$).

A NUMBER is similar to an element of S where "0" and "1" are understood as symbols to form the digits and "2" represents the blank. The number of digits of a NUMBER could be as high as s^c, where c is the arity of this particular NUMBER.

(0. 1) nul is a NUMBER satisfying $(\forall \bar{x})\mathrm{nul}(\bar{x})=0$.

(0. 2) end is a NUMBER satisfying $(\forall \bar{x})\mathrm{end}(\bar{x})=1$.

(FL. 1. 1) There is a functional constant NULL for each arity l and co-arity n such that $\mathrm{NULL}(\bar{u})=\hat{n}\hat{u}\hat{l}$.

(1. 2) There is a functional constant END for each arity and co-arity such that $\mathrm{END}(\bar{u})=\hat{e}\hat{n}\hat{d}$.

The functional NEXT acts as the successor functional in the NUMBER's of arity c. The NEXT functional is defined for only NUMBER's, so it is a partial functional. "<" is the dictionary order of vectors or arity c.

(FL. 2. 1) NUMBER(u). ∧. $u\neq$end. ∧. NUMBER(v). ∧. arity(u)= arity(v) :→:

$(v=\mathrm{NEXT}(u)$:↔: $(\exists \bar{x})((\forall \bar{y})(\bar{y}<\bar{x}$:→: $u(\bar{y})=1.\ \wedge.\ v(\bar{y})=0)$

$.\ \wedge.\ u(\bar{x})=0.\ \wedge.\ v(\bar{x})=1.\ \wedge.\ (\forall \bar{z})(\bar{z}>\bar{x}$:→:$u(\bar{z})=v(\bar{z}))))$.

When the arity of u is 1 we define the successor functional as

(FL. 2. 2) SUCC(u)=NEXT(u).

If the arity of $\bar{u}$ is $p\neq 1$, we define the successor functional as

$$\mathrm{SUCC}(u_1,u_2,\cdots,u_l)=(\mathrm{nul},\cdots,\mathrm{nul},\mathrm{NEXT}(u_i),u_{i+1},\cdots,u_l),$$

where u_i is the first NUMBER≠end on the list $u_1,u_2,\cdots,u_l$.

If $\bar{u}=\overline{\mathrm{end}}$, then SUCC($\bar{u}$) is undefined.

(FL. 3) Projective fundtionals. For every $l\geqslant 1$ and $1\leqslant i_1,i_2,\cdots,i_n\leqslant l$ a separate projective functional takes $(u_1,u_2,\cdots,u_l)$ to $(u_{i1},u_{i2},\cdots,u_{in})$:

$$\mathrm{PROJ}(u_1,u_2,\cdots,u_l)=(u_{i1},u_{i2},\cdots,u_{in}).$$

(FL. 4) Composition functionals:

$$F(\bar{u})=G(H_1(\bar{u}),H_2(\bar{u}),\cdots,H_l(\bar{u})),$$

where each $H_i(\bar{u})$ is a vector of functionals, and $H_1(\bar{u}),H_2(\bar{u}),\cdots,H_l(\bar{u})$ concatenate together to form a long vector of functions.

(FL. 5) Primitive recursion.

(5. 1) $F(\bar{u},\widetilde{\overline{\mathrm{nul}}}):=G(\bar{u})$.

(5. 2) $F(\bar{u},\mathrm{SUCC}(\tilde{v})):=H(F(\bar{u},\tilde{v}),\bar{u},\tilde{v})$.

(FL. 6) General recursion.

(6. 1) $F(\bar{u},\widetilde{\overline{\mathrm{nul}}}):=G(\bar{u})$.

(6. 2) When $\bar{u}\neq\overline{\mathrm{nul}}$ or $\overline{\mathrm{end}}$, we define

$$F(\bar{u},\mathrm{SUCC}(\tilde{v})):=$$
$$H(F(\mathrm{PRED}(\bar{u}),\tilde{v}),F(\bar{u},\tilde{v}),F(\mathrm{SUCC}(\bar{u}),\tilde{v}),\bar{u},\tilde{v}).$$

(6. 3) When $\bar{u}=\overline{\mathrm{nul}}$, the formula is

$$F(\bar{u},\mathrm{SUCC}(\tilde{v})):=H(F(\bar{u},\tilde{v}),F(\bar{u},\tilde{v}),F(\mathrm{SUCC}(\bar{u}),\tilde{v}),\bar{u},\tilde{v}).$$

(6. 4) When $\bar{u}=\overline{\mathrm{end}}$, the formula is

$$F(\bar{u},\mathrm{SUCC}(\tilde{v})):=$$
$$H(F(\mathrm{PRED}(\bar{u}),\tilde{v}),F(\bar{u},\tilde{v}),F(\bar{u},\tilde{v}),\bar{u},\tilde{v}),$$

where PRED($\tilde{v}$) and SUCC($\tilde{v}$) are the predecessor and the successor of $\tilde{v}$ respectively.

A primitive recursive functional can be obtained by using (FL. 1)～(FL. 5), and a recursive functional can be obtained by using all schema (FL. 1)～(FL. 6). Similarly to the cases in discussing the functions, we know that all usual functionals like addition, multiplication, exponentiation, factorial, predecessor, positive difference, maximum, minimum, …can be defined in this system with only (FL. 1)～(FL. 5). We can also give definitions for primitive recursive relations (relationships) and predicates. We write down some lemmas here. They are similar to Lemmas 1～3.

Lemma 4 *Any Boolean combination of primitive recursive predicates is primitive recursive.*

Lemma 5 *A concatenation* $(F_1(\bar{u}_1), F_2(\bar{u}_2), \cdots, F_m(\bar{u}_m))$ *of primitive recursive functionals is a primitive recursive functional.*

Lemma 6 *Suppose that global functionals* $F_1, F_2, \cdots, F_m$ *are defined by a simultaneous primitive recursion*

$$F_i(\bar{u}, \widetilde{\text{nul}}) := G_i(\bar{u}),$$

$$F_i(\bar{u}, \text{SUCC}(\tilde{v})) := H_i(F_1(\bar{u}, \tilde{v}), \cdots, F_m(\bar{u}, \tilde{v}), \bar{u}, \tilde{v}).$$

If the global funtionals G_i *and* H_i *are primitive recursive, so are* F_1, $F_2, \cdots, F_m$.

Theorem 7 *A global functional is primitive recursive if and only if it can be computed by a* PSPACE *Turing machine.*

Proof *Only if.* We need to set up a PSPACE Turing machine to compute the primitive recursive functionals.

(1) (2) and (3). It is easy to set up PSPACE Turing machines to compute the SUCC functional, constant functionals and the projection functionals.

(4) A PSPACE Turing machine computing the composition functional

$$F(\bar{u}) = G(H_1(\bar{u}), H_2(\bar{u}), \cdots, H_m(\bar{u}))$$

is also easy provided that G and H_1, H_2, $\cdots$, H_m are computed in PSPACE.

(5) The primitive recursion step:

(5.1) $F(\bar{u}, \widetilde{\text{nul}}) := G(\bar{u})$. Provided that $G(\bar{u})$ is computed in PSPACE, then so is $F(\bar{u}, \widetilde{\text{nul}})$. Suppose that $\bar{u} = (u_1, u_2, \cdots, u_l)$ and $\tilde{v} = (v_1, v_2, \cdots, v_m)$. The value of each u_i or v_j occupies a tape. Here the u_i's and v_j's are functions, so they are stored by their graphs. The storage for this part is counted as input. The values of $F(\bar{u}, \tilde{v})$ are also stored on one group of tapes. Consider the following computation:

(5.2) $F(\bar{u},\mathrm{SUCC}(\tilde{v})):=H(\bar{u},\tilde{v},F(\bar{u},\tilde{v}))$.

At this moment we have already had the values of $\bar{u},\tilde{u}$ and $F(\bar{u},\tilde{v})$. For computing H we need a fixed amount of storage, which is 2^c for a suitable constant c. The resulting value of $H(\bar{u},\tilde{v},F(\bar{u},\tilde{v}))$ is stored in another group of tapes. The tapes for storing $F(\bar{u},\tilde{v})$ are now erased for future computations.

If. Suppose that a functional F is computed by a Turing machine M with a storage of s^m units. From [4, Theorem 12.10] we know that the computation takes time $\leqslant\exp(s^m)$. We set up our NUMBER's to simulate the moves of the tapes where 0, 1 and 2 represent the tape symbols 0, 1 and b respectively. The arity of each NUMBER is m. The time $\tilde{t}$ runs from $\widetilde{\mathrm{nul}}$ to $\widetilde{\mathrm{end}}$. The head positions on the tapes are represented as functionals: $v_1=\mathrm{HEAD}_1(\tilde{t}),\mathrm{HEAD}_2(\tilde{t}),\cdots,v_m=\mathrm{HEAD}_m(\tilde{t})$.

The contents of the tapes are represented by the functionals $w_1=\mathrm{SYMB}_1(\tilde{v},\tilde{t}),w_2=\mathrm{SYMB}_2(\tilde{v},\tilde{t}),\cdots,w_m=\mathrm{SYMB}_m(\tilde{v},\tilde{t})$. Therefore the moves of the machine M can be written as a simultaneous recursion of $\mathrm{HEAD}_i,\mathrm{SYMB}_i(i=1,2,\cdots,m)$ and STAT for the head positions, scanning symbols and the state of the machine. The functional F computed by M is now a primitive recursive functional.

§ 9. Recursive functionals

We also obtain a similar result about the recursive functionals.

Theorem 8 *A global Junctional is recursive if and only if it can be computed by an* $\exp(s,s^c)$ *time bounded Turing machine. Since*

$$s^{s^c}=2^{s^{(c+\log\log_2 s/\log_2 s)}}<2^{s^{c+1}}\ (\textit{provided } s\geqslant 2),$$

we are actually referring to EXP *time.*

Proof *Only if*. The checking of the schema (FL. 1)～(FL. 5) is relatively easy. We only show that scheme (FL. 6) defines an $\exp(s,s^c)$ computable function F whenever the functions G and H are $\exp(s,s^a)$ computable for a suitable constant $a>0$. This is true because at each round the function vector $\tilde{v}$ runs through $(\exp(s,s^c))^m=\exp(s,ms^c)$ different functions. From $F(\bar{u},\bar{v})$ to $F(\bar{u},\mathrm{SUCC}(\bar{v}))$ we need to compute $\exp(s,s^b)$ steps for another suitable constant $b>0$. Therefore the total time consumed is $\exp(s,s^d)$ for a suitable constant $d>0$.

If. Suppose that F is computed by a Turing machine M using time $\exp(s,s^c)$. M uses p working tapes. At the beginning the input functions are printed on these tapes. After the computation the output functions are also printed on the working tapes. We make use of two functions u and v of arity c and co-arity 1. The function u indicates the position on the tapes, and the function v is for timing. The symbols on each tape are now a function $\mathrm{SYMB}_i(u,v)$. Using schema (FL. 1)～(FL. 6), we can describe the moves of M. And the output functions will be picked up when the function STAT(end) is a final state.

§ 10. Conclusion

1. We have explained the concepts of global functions, primitive recursive global functions and recursive global functions in finite systems and compared them with similar concepts in the natural number system and quoted two of Gurevich's theorems. The words used here differ a little from the text:

Theorem 1 (Gurevich) *A function can be computed in* LOGSPACE *if and only if it is primitive recursive.*

Theorem 4 (Gurevich) *A function can be computed in* PTIME *if and only if it is recursive; we give one more scheme to define such functions.*

2. After characterizing the primitive global functions and recursive global functions it is very natural to ask what kinds of global functions are not computable in LOGSPACE or PTIME, and what results will be obtained if we apply the diagonalization technique to the global functions. These questions are answered in Theorems 2, 3, 5 and 6.

In Theorems 2 and 3 we define a global function which is not computable in LOGSPACE but is computable in $(\mathrm{LOG})^2$ space.

In Theorems 5 and 6 we define a global function which is not computable in PTIME but is computable in time $s^{l*\log s}$.

3. We define primitive recursive global functionals and recursive global functionals in detail and give two theorems to characterize their computational complexity.

Theorem 7 *A global functional is primitive recursive if and only if it can be computed by a* PSPACE *Turing machine.*

Theorem 8 *A global functional is recursive if and only if it can be computed by an* EXPTIME *bounded Turing machine.*

Acknowledgements

I want to thank Professor Yuri Gurevich for suggesting that I work on primitive recursive functionals and recursive functions. I also want to thank Professors R. F. Roggio and J. F. Porter for arranging for me to stay at the beautiful University of Mississippi for the years 1988—1989, where the first version of this work was done. The revised version was done at Nagoya University of Commerce and Business Administration, where Professor J. A. Nord and Mrs. B. E. Lafaye gave a lot of help. The author also gives special thanks to the referee for very many valuable comments and suggestions.

References

[1] Gurevich Y. Algebras of feasible functions. Proceedings of the 24th IEEE Symposium on Foundations of Computer Science, 1983, 210—214.

[2] Gurevich Y. Toward logic tailored for computational complexity. Computation and proof theory(M. Richter et al. , editors), Lecture Notes in Mathematics, vol. 1104, Springer-Verlag, Berlin, 1984, 175—216.

[3] Gurevich Y. Logic and the challenge of computer science. Trends in theoretical computer science(E. Börger, editor), Computer Science Press, Rockville, Maryland, 1988, 1—57.

[4] Hopcroft J E and Ullman J D. Introduction to automata theory, languages, and computation. Addison-Wesley, Reading, Massachusetts, 1979.

[5] Kleene S C. Introduction to metamathematicso. Van Nostrand, New York, 1952.

南京大学学报数学半年刊，1998，15(5)：100～107

数论中的多项式时间可计算算法①

P-Time Algorithms in Number Theory

Abstract The modular equation $x^k \equiv a \pmod{p}$ with a primitive root is studied in this paper. The existing results at this direction are not polynomials time computable because they involve calculating the index table which consumes exponential time. Gauss studied. some of the equations and gave partial results. Since then very few improvements were discovered. In this paper we first give p-time algorithms in solving the equation $x^2 \equiv a \pmod{p}$ and then generalize this result to solve the equation $x^k \equiv a \pmod{p}$. There are no restrictions on our input, the time consumption of our algorithm is at most $c \log^4 p$.

Keywords **PTIME computation, EXPTIME computation.**

P-time algorithms are studied very often in some mathematical branches like graph theory, Boolean algebra and geometry. In number theory only a few algorithms are known to be p-time computable. Recently computer scientists turn their attention to number theory and found out some very interesting results. Pratt in [1] finds that to judge that a number is a prime it is sufficient to use a nondeterministic Turing

① Project supported by the National Natural Science Foundation of China numbered 19571009.

machine to compute in polynomial time. Let $PRIME$ represent the membership problem of prime numbers and NP represent the class of problems that can be computed in nondeterministic polynomial time. We write $PRIME \in NP$. Miller (see Pajunen in [2]) sharpens this result to $PRIME \in P$ (P represents the polynomial time computable calss) assuming that the Extended Riemann Hypothesis is true. Of course nobody has proved the Extended Riemann Hypothesis up to now. In our paper we extend the Gaussian procedure for solving some special modular equations to the general cases of the equation $x^2 \equiv a$ (mod p) and generalize this result to solve the equation $x^k \equiv a$ (mod p) provided that the number p is a prime and a primitive root g of p is known which could be found by a p-time procedure with Extended Riemann Hypothesis or by some other sources. Although there is an existing solution for the problem but the procedure consumes exponential time. Our procedure is p-time computable. We first discuss the p-time computability of some number theoretical procedures. They will be given as lemmas. Then we start discussing the modular equation $x^2 \equiv a$ (mod p). Hua in his book [3] gave detailed explanation about the Gaussian procedure in solving some special cases of the equation. Our procedure will be a generalization of this procedure.

§ 1. Some Lemmas

In our paper all numbers are supposed to be written in binary as in input, output and in the computation. The length of the input is a logarithmic function on base 2 of the real value of the number. For example 10110(bin)=22(dec) the length of this numbers is $\lceil \log 22 \rceil = 5$. Now we give the definition of p-time computable problems.

Definition 1 A decision problem or a problem with input and output is p-time computable if and only if there is a deterministic Turing machine M with constants c and d such that for any input of length n the machine M can complete the task in cn^d steps.

It is well known that multiplication and division with a remainder are cn^2 time computable. We give a lemma showing that the Euclidean procedure in finding greatest common divisor is also cn^2 time computable.

Lemma 1 *Let $d=(a,b)$ be the greatest common divisor of a and b. There is a Turing machine M computing d from input a and b. M takes time cn^2 where $n=\lceil \log a \rceil+\lceil \log b \rceil$.*

Proof The machine M uses two working tapes. M subtracts one number or its 2^k multiples from another block by block until one of the tapes becomes zero. The number on the other tape is the output. Machine M runs on every round of the subtraction taking time $c_1 \max\{\log a, \log b\}$ and eliminating at least one digit on the numbers. Therefore M stops in cn^2 steps.

We also need another lemma showing that in p modular operations to compute an exponential of two numbers with values again equal to exponentials of the input length it is possible to use a p-time Turing machine.

Lemma 2 *Let a, d be two numbers with lengths $\leqslant n=\lceil \log p \rceil$. We can find a p-time Turing machine M to compute the number $e \equiv a^d \pmod p$. The time limit is cn^3.*

Proof Let $(d_{n-1}d_{n-2}\cdots d_0)$ be the binary expression of d where $d_i=0$ or 1 for $i=0,1,\cdots,n-1$. We compute $a^2, a^4, a^8, \cdots, a^{2^{n-1}}$ one after another using the results from previous computation in modular field of prime p. Each of them is shortened to within length $\leqslant n$. The computation uses only multiplications and divisions with remainders. Now we compute a^d.

$$a^d \equiv (a^{d_0}) \times (a^2)^{d_1} \times \cdots (a^{2^{n-1}})^{d_{n-1}} \pmod p.$$

Again we use only multiplications and divisions with remainders. The multiplications and divisions consumes time $c_1 n^2$. Therefore computing a^d consumes time cn^3.

§ 2. Modular equation of order 2

To find the solution of modular equation $x^2 \equiv a \pmod p$ is a difficult problem. Only two special cases are realtively simple. The cases of $p \equiv 3 \pmod 4$ and $p \equiv 5 \pmod 8$ were solved by Gauss and written down in details in Hua's [3]. For the general case first we need to compute the factorial $\left(\frac{p-1}{2}\right)!$. Let n denote the length of the input: $n = [\log p]$. To find out the factorial takes $c_1(2^{\frac{n}{2}})^2$ steps. This is not polynomial time computable. Therefore we have to give up the checking with the Legendre symbol. Another way to find out the solvability is to compute $a^{\frac{p-1}{2}}$. From Lemma 2 it is p-time computable. Therefore the Gaussian procedure consumes polynomial time. The problem is that the Gaussian procedure works only for some special cases of the equation. For the general cases we need to find out a primitive root and then to compute the index table. The solution is obtained with the help of the index table. Let again the input length be $n = [\log p]$ and g be a primitive root of p. We assume that $\lceil \log a \rceil, \lceil \log g \rceil \leqslant n$. Therefore they do not give more trouble in computation. Only the index table is the most difficult to compute because it contains at least 2^{n-1} items. So again it is not p-time computable. We will develop a p-time procedure to solve this problem. Our procedure goes as follows.

We know that

$$a^{p-1} \equiv g^{p-1} \equiv 1 \pmod p.$$

Therefore

$$a^{\frac{p-1}{2}} \equiv \pm 1 \pmod p.$$

Case 0 If $a^{\frac{p-1}{2}} \equiv 1 \pmod p$, we write $a^{\frac{p-1}{2}} \equiv g^{\varepsilon_1 \frac{p-1}{2}} \pmod p$ where

$$\varepsilon_1 = 0.$$

Case 1 If $a^{\frac{p-1}{2}} \equiv -1 \pmod p$, we also write

$$a^{\frac{p-1}{2}} \equiv g^{\varepsilon_1 \frac{p-1}{2}} \pmod p \text{ where } \varepsilon_1 = 1.$$

In the next step we work on the equations

$$a^{\frac{p-1}{2}} \equiv g^{\varepsilon_1 \frac{p-1}{2}} \pmod p,$$
$$a^{\frac{p-1}{4}} \equiv \pm g^{\varepsilon_1 \frac{p-1}{4}} \pmod p.$$

Case ε_1 0 If $a^{\frac{p-1}{4}} \equiv g^{\varepsilon_1 \frac{p-1}{4}} \pmod p$, we write

$$a^{\frac{p-1}{4}} \equiv g^{\varepsilon_1 \frac{p-1}{4} + \varepsilon_2 \frac{p-1}{2}} \pmod p \text{ where } \varepsilon_2 = 0.$$

Case ε_1 1 If $a^{\frac{p-1}{4}} \equiv -g^{\varepsilon_1 \frac{p-1}{4}} \pmod p$, we also write

$$a^{\frac{p-1}{4}} \equiv g^{\varepsilon_1 \frac{p-1}{4} + \varepsilon_2 \frac{p-1}{2}} \pmod p \text{ where } \varepsilon_2 = 1.$$

Let 2^l be the largest 2 power factor of $p-1$ and $p = 2^l . q + 1$ where q is and odd number. The above procedure can go on for l times. At the end we obtain:

$$\text{Case } \varepsilon_1 \cdots \varepsilon_l : a^{\frac{p-1}{2^l}} \equiv g^{\left(\varepsilon_1 \frac{p-1}{2^l} + \varepsilon_2 \frac{p-1}{2^{l-1}} + \cdots + \varepsilon_l \frac{p-1}{2}\right)} \pmod p.$$

Since $q = \dfrac{p-1}{2^l}$ is an odd number we discuss it in the following cases

Case A $q = 1$. We have

$$a \equiv g^{\left(\varepsilon_1 + \varepsilon_2 \frac{p-1}{2^{l-1}} + \cdots + \varepsilon_l \frac{p-1}{2}\right)} \pmod p.$$

If $\varepsilon_1 = 0$, then

$$\begin{aligned} x &\equiv \pm g^{\frac{1}{2}\left(\varepsilon_2 \frac{p-1}{2^{l-1}} + \varepsilon_3 \frac{p-1}{2^{l-2}} + \cdots + \varepsilon_l \frac{p-1}{2}\right)} \\ &\equiv \pm g^{\left(\varepsilon_2 \frac{p-1}{2^l} + \varepsilon_3 \frac{p-1}{2^{l-1}} + \cdots + \varepsilon_l \frac{p-1}{4}\right)} \pmod p \end{aligned}$$

are the solutions of modular equation $x^2 \equiv a \pmod p$.

If $\varepsilon_1 = 1$, then there is no solution for the equation because the index of a is

$$\left(\varepsilon_1 + \varepsilon_2 \frac{p-1}{2^{l-1}} + \cdots + \varepsilon_l \frac{p-1}{2}\right)$$

which is an odd number.

Case B $q \neq 1$.

Suppose that $\varepsilon_1 = 0$. Let $\bar{\varepsilon}_i = 1 - \varepsilon_i$ for $i = 2, 3, \cdots, l$. The formula in case $\varepsilon_1 \varepsilon_2 \cdots \varepsilon_l$ becomes

$$a^q \equiv g^{\left(\varepsilon_1 \frac{p-1}{2^l} + \varepsilon_2 \frac{p-1}{2^{l-1}} + \cdots + \varepsilon_l \frac{p-1}{2}\right)} \pmod p.$$

Hence

$$\begin{aligned} & a^q \cdot g^{\left[\left(\bar{\varepsilon}_2 \frac{p-1}{2^{l-1}} + \bar{\varepsilon}_3 \frac{p-1}{2^{l-2}} + \cdots + \bar{\varepsilon}_l \frac{p-1}{2}\right) + \frac{p-1}{2^{l-1}}\right]} \\ \equiv & g^{\left(\varepsilon_2 \frac{p-1}{2^{l-1}} + \varepsilon_3 \frac{p-1}{2^{l-2}} + \cdots + \varepsilon_l \frac{p-1}{2}\right)} \cdot g^{\left[\left(\bar{\varepsilon}_2 \frac{p-1}{2^{l-1}} + \bar{\varepsilon}_3 \frac{p-1}{2^{l-2}} + \cdots + \bar{\varepsilon}_l \frac{p-1}{2}\right) + \frac{p-1}{2^{l-1}}\right]} \end{aligned}$$

$$\equiv g^{\left[(\varepsilon_2+\bar{\varepsilon}_2)\frac{p-1}{2^{l-1}}+(\varepsilon_3+\bar{\varepsilon}_3)\frac{p-1}{2^{l-2}}+\cdots+(\varepsilon_l+\bar{\varepsilon}_l)\frac{p-1}{2}+\frac{p-1}{2^{l-1}}\right]}$$

$$\equiv g^{p-1}\equiv 1 \pmod p.$$

Define

$$x\equiv \pm a^{\frac{p-1+2^l}{2^{l+1}}}\cdot g^{\frac{1}{2}\left[\left(\bar{\varepsilon}_2\frac{p-1}{2^{l-1}}+\bar{\varepsilon}_3\frac{p-1}{2^{l-2}}+\cdots+\bar{\varepsilon}_l\frac{p-1}{2}\right)+\frac{p-1}{2^{p-1}}\right]} \pmod p.$$

Where $p-1\equiv q\cdot 2^l$, $p-1+2^l=(q+1)\cdot 2^l$ is a multiple of 2^{l+1} and the index of g on the other side is also an integer. Hence x is computable.

$$x^2\equiv\left(\pm a^{\frac{p-1+2^l}{2^{l+1}}}\cdot g^{\frac{1}{2}\left[\left(\bar{\varepsilon}_2\frac{p-1}{2^{l-1}}+\bar{\varepsilon}_3\frac{p-1}{2^{l-2}}+\cdots+\bar{\varepsilon}_l\frac{p-1}{2}\right)+\frac{p-1}{2^{l-1}}\right]}\right)^2$$

$$\equiv a\cdot a^q\cdot g^{\left[\left(\bar{\varepsilon}_2\frac{l-1}{2^{l-1}}+\bar{\varepsilon}_3\frac{l-1}{2^{l-2}}+\cdots+\bar{\varepsilon}_l\frac{p-1}{2}\right)+\frac{p-1}{2^{l-1}}\right]}$$

$$\equiv a \pmod p.$$

Suppose that $\varepsilon_1=1$. The formula in case $\varepsilon_1\cdots\varepsilon_l$ becomes

$$a^q\equiv g^{(\varepsilon_1\cdot q+\varepsilon_2\cdot 2\cdot q+\cdots+\varepsilon_l\cdot 2^{l-1}\cdot q)} \pmod p.$$

If the index of a in g is an even number, then the index of a^q is also an even number which is impossible because on the other side of the above formula g has an odd exponential. Therefore the index of a in g is an odd number and the equation $x^2\equiv a \pmod p$ does not have a solution. We come to a conclusion and state our results in the next theorem.

Theorem 1 *The modular equation* $x^2\equiv a \pmod p$ *has solutions if and only if* $\varepsilon_1=0$ *and the solutions are*

$$x\equiv \pm a^{\frac{p-1+2^l}{2^{l+1}}}\cdot g^{\left[\left(\bar{\varepsilon}_2\frac{p-1}{2^{l}}+\bar{\varepsilon}_3\frac{p-1}{2^{l-1}}+\cdots+\bar{\varepsilon}_l\frac{p-1}{4}\right)+\frac{p-1}{2^{l}}\right]} \pmod p.$$

The time consumed in computing x *is a polynomial of* $n=[\log p]$.

Proof We only need to check the time in computing x. In each round of the computation in formula

$$a^{\frac{p-1}{2^i}}\equiv g^{\left(\varepsilon_1\frac{p-1}{2^i}+\varepsilon_2\frac{p-1}{2^{i-1}}+\cdots+\varepsilon_i\frac{p-2}{2}\right)} \pmod p.$$

for $i=1,2,\cdots,l$.

By Lemma 2 it takes time $\leqslant c_1 n^3\cdots$Repeating the procedure l times the time consumed is $\leqslant c_2 n^4$. Similarly in cases A and B all operations are multiplications, divisions or exponentials. They consume another $\leqslant c_3 n^4$ time. Therefore the time consumed in the entire procedure is $\leqslant cn^4$ for a suitable constant c.

§ 3. Modular Equations of Arbitrary Degrees

As in Section 2 modular equations of arbitrary degrees are not considered as p-time computable even with a given primitive root. The traditional solution involved the calculation of an index table which takes naturally exponential time. In this section we will discuss the p-time computability of finding a solution for the modular equation $x^4 \equiv a \pmod{p}$ with a prime number p and its primitive root g where k is arbitrary and $a \not\equiv 1 \pmod{p}$. The procedure will be similar to that in Section 2 but for the general k we need more discussions and a new lemma.

Lemma 3 *If $x^k \equiv a \pmod{p}$ has a solution, then it has $d=(k, p-1)$ many solutions. If x is a solution and g is a primitive root of p. We define $r=g^{\frac{p-1}{d}}$, Then*

$$r, r^2, \cdots, r^d$$

are d different dth roots of 1 and

$$rx, r^2x, \cdots, r^dx$$

are d different roots of a.

Proof r^ix and r^jx are different from each other for $1 \leqslant i \neq j \leqslant d$ because

$\frac{r^ix}{r^jx} \equiv \frac{r^i}{r^j} \equiv r^{i-j} \not\equiv 1$. They are all kth roots of a because

$$(r^ix)^k \equiv (r^j)^k x^k \equiv x^k \equiv a \pmod{p}.$$

Now we come back to the original problem. Let p be a prime and g is its primitive root. We consider modular equation $x^k \equiv a \pmod{p}$ where k is an arbitrary number and $a \not\equiv 1 \pmod{p}$. In the problem k is supposed to be a constant, $n=[\log p]$ is the length of the input and the length of a is less than or equal to n. The factorization of $p-1$ and k on their common factors are as follows:

$$p-1=p_1^{\alpha_1} p_2^{\alpha_2} \cdots p_l^{\alpha_l} \cdot q$$

and

$$k=p_1^{\beta_1} p_2^{\beta_2} \cdots p_l^{\beta_l} \cdot r$$

where $(q,k)=1$ and $(r,p-1)=1$. The greatest common divisor of $p-1$ and k is

$$d=(p-1,k)=p_1^{\gamma_1}p_2^{\gamma_2}\cdots p_l^{\gamma_l}$$

where $\gamma_i=\min\{\alpha_i,\beta_i\}$ for $i=1,2,\cdots,l$.

We first prove that the condition for $x^k\equiv a(\bmod p)$ to have a solution is that $a^{\frac{p-1}{d}}\equiv 1(\bmod p)$. For suppose that $a^{\frac{p-1}{d}}\equiv 1(\bmod p)$. Let the index of a in g be m (i. e. $a\equiv g^m(\bmod p)$), then we have $g^{\frac{m(p-1)}{d}}\equiv 1(\bmod p)$. Therefore $d|n$ and $a\equiv g^{d\varepsilon}(\bmod p)$. From $(k,p-1)=d$ we can find u and v such that $uk+v(p-1)=d$. Hence $a\equiv g^{d\varepsilon}\equiv g^{uk\varepsilon+v(p-1)\varepsilon}\equiv(g^{u\varepsilon})^k$ $(\bmod p)$. $g^{u\varepsilon}$ is a solution for $x^k\equiv a(\bmod p)$. Conversely suppose that $x^k\equiv a$ $(\bmod p)$ has a solution. Let the index of x be m. Then the index of a is km. We have $d|km$ and

$$a^{\frac{p-1}{d}}\equiv a^{\frac{km}{d}(p-1)}\equiv 1\ (\bmod p).$$

The condition $a^{\frac{p-1}{d}}\equiv 1\ (\bmod p)$ can be checked in p-time but to find the number m for $a=g^m(\bmod p)$ is not naturally in p-time. This is why we need the following procedure. But before that we need a lemma for taking p_ith roots on both sides of an equation.

Lemma 4 *If $a^{up_i}\equiv g^{vp_i}(\bmod p)$ and $p_i|(p-1)$, then*

$$a^u\equiv(g^v)\cdot(g^{\frac{p-1}{p_i}})^{\varepsilon}$$

where $1\leqslant\varepsilon\leqslant p_i$.

Proof From $a^{up_i}\equiv g^{vp_i}(\bmod p)$ and $p_i|(p-1)$ we have

$$\frac{a^{up_i}}{g^{vp_i}}\equiv\left(\frac{a^u}{g^v}\right)^{p_i}\equiv 1\ (\bmod p).$$

Hence

$$\frac{a^u}{g^v}\equiv(g^{\frac{p-1}{p_i}})^{\varepsilon}$$

for a suitable ε satisfying $1\leqslant\varepsilon\leqslant p_i$.

Now we start our procedure at the following formula:

$$a^{\frac{p-1}{d}}\equiv a^{\overline{\frac{p-1}{p_1^{\gamma_1}p_2^{\gamma_2}\cdots p_l^{\gamma_l}}}}\equiv g^{p-1}\equiv 1\ (\bmod p).$$

Consider $\gamma_i\leqslant\alpha_i$ for $i=1,2,\cdots,l$. We obtain

$$(a^{\overline{\frac{p-1}{p_1^{\gamma_1+1}p_2^{\gamma_1+2}\cdots p_l^{\gamma_l}}}})^{p_1}\equiv(g^{\frac{p-1}{p_1}})^{p_1}(\bmod p),$$

$$a^{\frac{p-1}{p_1^{\gamma_1+1}p_2^{\gamma_1+2}\cdots p_l^{\gamma_l}}}\equiv g^{\frac{\varepsilon_{1,\gamma_1+1}(p-1)}{p_1}}\ (\mathrm{mod}\ p),$$

$$\cdots$$

$$a^{\frac{p-1}{p_1^{\alpha_1}p_2^{\gamma_2}\cdots p_l^{\gamma_l}}}\equiv g^{\left(\varepsilon_{1,\gamma_1+1}\frac{p-1}{p_1^{\alpha_1-\gamma_1}}+\varepsilon_{1,\gamma_1+2}\frac{p-1}{p_1^{\alpha_1-\gamma_1-1}}+\cdots+\varepsilon_{1,\alpha_1}\frac{p-1}{p_1}\right)}\ (\mathrm{mod}\ p),$$

$$a^{\frac{p-1}{p_1^{\alpha_1}p_2^{\gamma_2+1}\cdots p_l^{\gamma_l}}}\equiv g^{\left(\varepsilon_{1,\gamma_1+1}\frac{p-1}{p_1^{\alpha_1-\gamma_1}p_2}+\varepsilon_{1,\gamma_1+2}\frac{p-1}{p^{\alpha_1-\gamma_1-1}p_2}+\cdots+\varepsilon_{1,\alpha_1}\frac{p-1}{p_1p_2}+\varepsilon_{2,\gamma_2+1}\frac{p-1}{p_2}\right)}\ (\mathrm{mod}\ p),$$

$$\cdots$$

$$a^{\frac{p-1}{p_1^{\alpha_1}p_2^{\alpha_2}\cdots p_l^{\alpha_l}}}\equiv g^{\left(\varepsilon_{1,\gamma_1+1}\frac{p-1}{p_1^{\alpha_1-\gamma_1}\cdots p_l^{\alpha_l-\gamma_l}}+\cdots+\varepsilon_{1,\alpha_1}\frac{p-1}{p_1p_2^{\alpha_2-\gamma_2}\cdots p_l^{\alpha_l-\gamma_l}}+\cdots+\varepsilon_{l,\gamma_l+1}\frac{p-1}{p^{\alpha_l-\gamma_l}}+\cdots+\varepsilon_{l,\alpha_l}\frac{p-1}{p_l}\right)}$$

$$\equiv g_M^{\frac{p-1}{p_1^{\alpha_1-\gamma_1}p_2^{\alpha_2-\gamma_2}\cdots p_l^{\alpha_l-\gamma_l}}}\ (\mathrm{mod}\ p),$$

And

$$M=\varepsilon_{1,\gamma_1+1}+\cdots+\varepsilon_{1,\alpha_1}p_1^{\alpha_1-\gamma_1-1}+\cdots+\varepsilon_{l,\gamma_l+1}p_1^{\alpha_1-\gamma_1}\cdots p_{l-1}^{\alpha_{l-1}-\gamma_{l-1}}+\cdots+$$
$$\varepsilon_{l,\alpha_l}p_1^{\alpha_1-\gamma_1}\cdots p_{l-1}^{\alpha_{l-1}-\gamma_{l-1}}p_l^{\alpha_l-\gamma_l-1}$$

Where $0\leqslant\varepsilon_{ij}\leqslant p_i$ and $j=\gamma_{i+1},\cdots,\alpha_i$ for $i=1,2,\cdots,l$. To compute ε_{ij} at each step we need at most p_i checks. Every check requires a multiplication and a comparison between two numbers. $p_i\leqslant k$ is kept in less than a constant number. Therefore the procedure up to now is p-time computable. The last formula can be written as

$$a^q\equiv g^{qd(\varepsilon_1+\varepsilon_2+\cdots+\varepsilon_l)}\equiv g^{qd\varepsilon}\ (\mathrm{mod}\ p)$$

where

$$\varepsilon_i\equiv\varepsilon_{i,\gamma_{i+1}}p_1^{\alpha_1-\gamma_1}\cdots p^{\alpha_{i-1}i-1-\gamma_{i-1}}+\cdots+\varepsilon_{i,\alpha_{i+1}}p_1^{\alpha_1-\gamma_1}\cdots$$
$$p_{i-1}^{\alpha_{i-1}-\gamma_{i-1}}p_i^{\alpha_i-\gamma_{i-1}}\ (\mathrm{mod}\ p-1)$$

for $i=1,2,\cdots,l$ and $\varepsilon\equiv\varepsilon_1=\cdots+\varepsilon_l(\mathrm{mod}\ p-1)$.

From $(q,k)=1$ we can find u,v satisfying $uq+vk=1$. Hence

$$a\equiv a^{uq+vk}\equiv a^{uq}\cdot a^{vk}\equiv(a^q)^u(a^{vk})\equiv g^{qd\varepsilon u}\cdot a^{vk}\ (\mathrm{mod}\ p).$$

From $(k,p-1)=d$ we can find w,s satisfying $wk+s(p-1)=d$. Hence

$$g^d\equiv g^{wk}\cdot g^{s(p-1)}\equiv g^{wk}\ (\mathrm{mod}\ p).$$

$$a\equiv g^{qd\varepsilon u}\cdot a^{vk}\equiv g^{wkq\varepsilon u}\cdot a^{vk}(\equiv g^{wq\varepsilon u}\cdot a^v)^k\ (\mathrm{mod}\ p).$$

Therefore we obtain our theorem.

Theorem 2 $x\equiv g^{wq\varepsilon u}\cdot a^v$ *is a solution for the modular equation* $x^k\equiv a\ (\mathrm{mod}\ p)$ *and it is p-time computable.*

Proof Checking the time, the input p could be very large but the power k is supposed to be a constant. The length of the input is $n=\lceil \log p \rceil$. The lengths of g and a are less than or equal to n which we do not count as in the length of the input. At the beginning of the procedure we factor $k=p_1^{\beta_1} p_2^{\beta_2} \cdots p_l^{\beta_l} \cdot r$. Since k is a constant the factorization takes at most a constant number many tries. Then we use division with remainders to find out the factors of $p-1=p_1^{\alpha_1} p_2^{\alpha_2} \cdots p_l^{\alpha_l} \cdot q$ where q is not necessary to be a prime. The number of factors for $p-1$ is at most $n=\lceil \log p \rceil$. The factorization of $p-1$ takes time $\leqslant c_1 n^2$. Then we compute d and check the condition $a^{\frac{p-1}{d}} \equiv 1 \pmod p$. The time for this step is $\leqslant c_2 n^3$. The most complicated step is computing ε_{ij}. We first notice that the total number of ε_{ij} is at most $\alpha_1+\alpha_2+\cdots+\alpha_l \leqslant n$. In computing each ε_{ij} we need to find out two exponentials which takes time $\leqslant c_3 n^3$. Then we check to find out the right ε_{ij} satisfying $a^\lambda \equiv g^{\mu\varepsilon_{ij}}$. This part of the procedure takes at most

$$\max_{1\leqslant i\leqslant l}\{p_i\}$$

checks which is less than the constant k. Therefore the total time for deciding all ε_{ij} is $\leqslant c_4 n^4$. In computing $\varepsilon_1, \varepsilon_2, \cdots, \varepsilon_l$ and ε we mainly use multiplication and divisions with remainders in modular p and $p-1$. A time of $c_5 n^3$ would be enough for this part. Finally we compute u, v, w and x. We now use multiplications and Euclidean procedures and computing exponentials. This part takes another $c_6 n^3$ time. Therefore altogether in the entire procedure at most cn^4 time is consumed. The computation is now in p-time.

References

[1] Pratt. V R. Every prime has a succinct certificate. SIAM JC, 1975, 4: 214.

[2] Pajunen S. On two theorems of Lenstra. IPL, 1980, 11: 224.

[3] Hua L K. Introduction to Number Theory. Science Press (Chinese), 1957.

[4] Hopcroft J E. and Ullman J D. Introduction to Automata Theory. Languages and Computation, Addison-Wesley, Reading (Mass.), 1979.

2005 年 6 月. 广州中山大学第 4 届逻辑与认知国际会议.

在计算机科学中去掉无限

Remove Infinity from Computer Science

Abstract We give the following suggestions for computer science:

1. To do '+1' consumes resources. We can not put infinitely many natural numbers together. There is not an infinite set. The natural numbers form a finite set.

2. The length of a decimal is finite. The length of decimal p is finite. The lengths of them depend on the resource in a computer.

3. The number of real numbers is finite. How many of them can be used also depends on the resource in a computer.

4. Functions $\sin x$, $\cos x$ and some other functions are computed by finite procedures.

5. Differentiation and integration are computed by finite procedures.

6. A Turing machine with a tape of infinite length is not the correct model of a computer. A computer is a machine with finitely many states.

Keywords **finite machine, finite procedure, model of a computer, infinity.**

§ 1. Infinite sets and paradoxes

Chinese Zhuangzi said that "One foot of rope can not be eliminated within ten thousand generations if one cuts it in half everyday." Ancient

Greeks also had an argument about"Achilles can not catch up with a turtle". It was like this: Suppose that Achilles is ten times faster than the turtle and the turtle is ten feet ahead of Achilles. In the first period Achilles runs ten feet, but the turtle is still a foot ahead of Achilles; In the second period Achilles runs a foot, but the turtle is still a tenth of a foot ahead of Achilles … and so on. Therefore Achilles can not catch up with the turtle. These concepts and paradoxes were developed continuously in the years there after and a series of theories about infinity were developed and completed.

At first people assume that the number of natural numbers is infinite. Every natural number can produce the number following it by adding 1 to itself. Collecting all natural numbers together can form the set of natural numbers which is an infinite set. As consequences the numbers of all rational numbers, all real numbers are all infinite. Moreover the number of various kinds of sets is also infinite. By a similar reasoning, people assume that there are infinite procedures like taking a limit. Therefore we can define the decimals with infinite length. The differentiation and integration are invented. To put all of these on a sound foundation people start discussing a very impotent problem: Do'infinite sets' truly exist or are they only false images from logic inference? Are there truly infinitely many natural numbers? Is the number of rational numbers or real numbers truly infinite? "What is infinity?"is also debating in the range of philosophy. I suggested that there be no necessity to use infinity in the the computer science [1]. People can say that originally the number of all natural numbers is infinite and this statement is repeated in many books. But I ask that who really sees infinitely many numbers and who can really use infinitely many numbers? It is just on the opposite side. The existence of infinite sets (including the existence of infinitely many numbers) is supposed to be an axiom which is unprovable (see [2] p. 42) by the other axioms.

Principle of infinity (V). There exists at least one infinite set: the set of all natural numbers.

In his book Fraenkel named the axioms as principles and the principle quoted above is called **Principle of infinity**. He provided a set theory system which is called ZF system. He stated his system by using natural language. Although he did not consider using the formalized languages to present his theory but he had made a detailed derivation to the theory.

In [3] Gödel gave a formal system for set theory. The axiom group D has the following axiom:

$$\neg Em(A) \rightarrow (\exists u)[u \in A \wedge Ex(u, A)]$$

It is called the 'axiom of infinity'. As an axiom it can not be proved by the other axioms. This is called the independence problem of an axiom system.

First Difficulty The existence of infinity can not be proved.

Another important concept is the consistency of an axiom system. Mathematicians have noticed this question for a long time. They have constructed a formalized number theory system and tried to prove that the system is consistent. Unfortunately this project is not successful. On the contrary they proved that 'If the number theoretic formal system is consistent, then the consistency of the formalized number theory system' is unprovable (see Kleene [4], p. 210, Theorem 30). Here is the first difficulty for having infinity.

Theorem 30 If the number theoretic formal system is (simply) consistent, then not $\vdash$ Consis; i. e. if the system is consistent, then there is no consistency proof for it by the methods formalizable in the system. (Gödel's second theorem.)

To prove that the formalized number theory system is consistent we need to put it into a formal system which is the formalized number theory system itself. In the theorem 'Consis' represents 'the consistency of the formalized number theory system' and '$\vdash$' represents something is provable by the methods formalizable in the system. Kleene's **Theorem 30** says exactly that '$\vdash$ Consis' is not provable in the system. This tricky theorem is proved in a very strict way. But some people still doubt that a system used to prove the consistency of some other systems ought not to be used to prove the consistency of the system itself. It is like someone tries to prove the innocence of oneself in a lawsuit procedure. This is the second difficulty.

Second Difficulty The consistency of infinity can not be proved.

Because of supposing the existence of infinite sets we get into a lot of trouble. Suppose that we have a very strong power to put any things together we would form'the set of all sets'. Cantor's paradox says that it is wrong to do so. Otherwise you will get into contradiction.

We will briefly introduce the Cantor's paradox and the solution to the problem. First we say that if we can form'the set of all sets'we will get a contradiction. If the set exists let it be M. Therefore M is itself a member of M which gets $M \in M$. Among all member of M some of them satisfy $M \in M$ and the others satisfy $M \notin M$. All the sets satisfying $M \notin M$ form a set A. Now we ask whether $A \notin A$ or not.

If $A \in A$, then by the definition of A, a member of A should satisfy $A \in A$. We obtain $A \notin A$ which contradicts $A \in A$.

On the other hand, if $A \notin A$, then by the definition of A, A should be a member of A. we obtain $A \in A$ which is also impossible.

Therefore we get a contradiction. The set of all sets does not exist. This is the third difficulty. The solution to this difficulty is to call the thing having all sets together a'class'which is not a set any more. Using this method the set theory system provided by Gödel has escaped contradiction temporary so this is the third difficulty:

Third Difficulty A set can not be too large.

If we assume that the set of all natural numbers has infinitely many elements, the sets with the same number of elements are called countable sets. If we assume the existence of countable sets, then there must be uncountable sets. Especially the set of continuum (an uncountable set formed by all real numbers) exists. Now we quote a theorem in [4].

Corollary 2. 1. 4 (Downward Löwenheim-Skolem-Tarski Theorem) Every consistent theory T in $\mathcal{L}$ has a model of power at most $|\mathcal{L}|$ (see [5], p. 66).

From this Corollary because the number of symbols in constructing the set theory is countable there exists a set theory model *S* which is itself countable. Since S is a set theory model there must be a set *C* representing continuum in the model. All real numbers are the elements of set *C*. Therefore *C* has uncountably many elements. But the entire model S has only countably many elements. It seems that we can suppose that there are only countably many real numbers. This is a very strange phenomenon. How can a set be uncountable which is formed by countably many members. One explanation is like this: Inside the model the set *C* is uncountable but it is countable outside. It is like if we embedded a Euclidean plan into a sphere with different measurements. A line on the sphere is of infinite length inside the plan but it looks finite from outside of the plan. With this explanation we have escaped a contradiction, but at the same time we see that the number of real numbers depends on how you look at the system. Therefore the number of real numbers is not really uncountable or we say that it has relativity. So this is the forth difficulty for having infinity.

Forth Difficulty We do not know how many real numbers there are for sure.

Mathematicians are used to assume infinite sets but as computer scientists there will not be any obstacles for us not to use the concept of infinite sets. In my paper [1] I suggest that it is possible for computer science not to use infinite sets because there are only a finite number of numbers in a computer. Using these numbers we can solve all kinds of problems from reality with the computer if they can be solved by infinite procedures. We can also use differentiation and integration but they are all considered as finite procedures. We will discuss this problem in the following sections.

§ 2. The numbers in the computer

There is not a warehouse for numbers in a computer. The numbers are ready for use if you want to use them but they disappear if you do not want to use them. They are not stored in a device but can be called by a program in programming languages. For example in the programming language PASCAL the integers are divided into three categories: short integers, integers (for clearly distinguishing the three categories of integers we call them normal integers) and long integers (in symbols shortint, integer and longint respectively).

Their properties and ranges are described in the following table:

Name	Range	bytes	bits
shortint	−128 to 129	1	8
integer	−32768 to 32767	2	16
longint	−2147483648 to 2147483647	4	32

The above divisions are for saving the resource in running programs. A problem can be solved by using shorter integers will not use the longer integers. In all three categories there are largest numbers. The largest integer of normal integer is 32767 which has a special symbol 'maxint'. Some compilers would give $-32768=32767+1$ and some other compilers would not run the program if there is the expression $32767+1$ in it. It is called overflowing in running the program. The largest number of longint is maxlongint=2147483647. A program can not run or it produces an error if it goes over these limits in the running procedure. Therefore in a PASCAL program we can use at most $2^{32}=4294967296$ integers regularly. You can still use larger integers but you have to define or set up a procedure by yourself. We can not give an exact maximum number of integers that you can use in a computer. It depends on the type of the computer and the storage device connecting to the computer.

The number of real numbers in a computer is also finite. Some programming languages provide regular real numbers from $\pm 10^{-128}$ to $\pm 10^{+128}$ but they have only 7 digits in precision. You can also extend these limitations by using double precision real numbers or if you can use a super computer the precision would increase to about 21 digits. Anyway the total number of real numbers in a computer is finite and the number of digits of a real number is also finite. There is a book that prints from the first page to the last page a decimal with one million digits for the real number $p=3.141\,5\cdots$ but its number of digits is still finite.

§ 3. Computer is a finite machine

People say that the Turing machine is the model of a computer. Actually this is not a very good model. No computer can have an input output device with infinitely many read-and-write cells. A computer is a finite automaton (see [6], p. 13) with some extension. We can set up the finite automaton by the notations in [6]. Let the machine be $M=<Q, S, S', d, q_0, F>$. Here the set of states Q is described by the situations in the memory and registers, and the operations that are to take, at that situation. The input and output symbols are just $S=S'=\{0,1\}$ if we consider a very short time unit. If the tape is broader we consider them as finite combinations of 0's and 1's. The transaction function d is to tell how the states of the machine are changing and the outputs are giving with a suitable input. Finally q_0, F are the starting state and final states. We also need a time counting procedure to go simultaneously with the running procedure of the machine. The differences between the finite automaton and the finite machine is that a finite automaton can not run without an input but a finite machine can; and a finite machine can go to an infinite loop in which the machine will not stop but a finite automaton will stop as soon as it finishes reading the input. To avoid the machine going to an infinite loop we can set up an external detection system that will check if two running states are identical in the running procedure. If it occurs the machine stops. The explanations above tell us that a computer is really an extended finite automaton with all its advantages and shortcomings. The advantages are that the machine is relatively simple in comparing with the Turing machine and we do not have to worry about having infinitely many states and going to an undecidable situation. The shortcomings are that the machine can do only finitely many things and the machine could accept only regular languages which are sectional identical and so are simple relatively. We will discuss this machine in another paper.

§ 4. The differentiation and integration are finite procedures

Originally the concepts of differentiation and integration are defined by limits. Taking a limit for example $\lim\limits_{x\to 0}\frac{\sin x}{x}$ is an infinite procedure. To compute the value p is also by taking the limit of the boundaries of polygons with 2^n sides. The derivation is defined by taking the limit of $d(f)$ over dx. The integration is defined by taking the limit of areas of some regions. And so on.

But the situation in a computer is very different. There is no function table for the function $\sin x$ in the computer. Whenever you call the value of the function for example $y=\sin(3)$ the computer computes the value for you through a procedure that is stored in the computer. To compute the value of differentiation or integration is also similar. If you want to define a new function for example $f(x)=d(\sin^2 x+x^2)/dx$, it is very possible you will need a procedure. You may ask how to obtain the procedure to compute the function $\sin x$. Originally the computing formula may come from a book that is proved using infinite procedures. For example the following formula

$$\sin x=x-\frac{x^3}{3!}+\frac{x^5}{5!}-\cdots+\frac{(-1)^n x^{2n+1}}{(2n+1)!}+\cdots$$

or its optimization can be used to compute the function sin x in a computer. Although the formulas were proved by infinite procedures it is possible we can establish a new theory to prove all of them through entirely finite procedures. Similar to the formula above we can use Taylor's formulas to write finite procedures for computing the differentiations and integrations.

Now suppose that you have a differential equation which is very crucial in constructing a reservoir. You take the equation to a computer to get the solution. The programmer will write a program to compute the solution for you through the procedures in the computer. Of course all of them are finite procedures. When you are not satisfied with the precision of data, the programmer will revise the program to obtain some better data. Finally everything is done. Your reservoir is built in good quality. You will not pay any attention to the source formulas coming from infinite procedures or finite procedures.

§ 5. The living period for all human beings is finite

The debate about the existence of infinity is still continuing. People trend to believe that the life period of all human beings is finite; the existence periods of the earth and the universe are also finite. It is impossible for us to put infinite things together. The debaters would say that although we cannot put infinite things together by ourselves but theoretically we can put them together in the abstract thinking. We are still not having any reasons to convince them not to do so. But it is certainly not necessary to use a troublesome concept to establish our computer science.

§ 6. Conclusion

In our opinion the computer science is a practical science. Although a lot of ideas and concepts in the history of building computers were from mathematics but we do not need to repeat the trace that was going through by the mathematicians. It is not necessary for us to use a Turing machine with a tape of infinite length as the model of a computer. We also do not need to suggest that our database have infinitely many lines (There are a lot of papers published at this direction). Some computer scientists would say that without infinity we do not know how to do our research. We do not think so. There are still a lot of unknown areas in finite computational procedures. The following are our suggestions.

1. To do '+1' consumes resources and time although it might be very little. Therefore we can not do it infinity many times. From this reason we can not put infinitely many natural numbers together. There is not an infinite set. The natural numbers form a finite set.

2. The length of a decimal is finite. The length of decimal p is finite. The lengths of them depend on the resource in a computer.

3. The number of real numbers is finite. How many of them can be used also depends on the resource in a computer.

4. Functions sin x, cos x and some other functions are computed by finite procedures.

5. Differentiation and integration are computed by finite procedures.

6. A Turing machine with a tape of infinite length is not the correct model of a computer. A computer is a machine with finitely many states.

References

[1] Luo L. A talk on computer science. Shuxue Tongbao, 1996, (9): 24—27(Chinese).

[2] Fraenkel A. A and Hillel R L. Foundations of Set Theory. Amsterdam: North-Holland, 1958.

[3] Gödel K. The Consistency of the Axiom of Choice and of the Generalized Continuum Hypothesis with the Axioms of Set Theory. Ann. Math. Studies 3 (Princeton Univ. Press, Princeton, N. J.), 2nd edit, 1951.

[4] Kleene S C. Introduction to matamathematics. Amsterdam: North-Holland, 1952.

[5] Chang C C and Keisler H J. Model Theory. Amsterdam: North-Holland, 3rd ed., 1990.

[6] Hopcroft J E and Ullman J D. Introduction to Automata theory Languages and Computation. Reading: Addison-Wesly, 1979.

Acta Mathematica Sinica, English Series, 2006, 22(3): 865～872

没有等号的有限模型论①

The Theory of Finite Models without Equal Sign

Abstract In this paper, it is the first time ever to suggest that we study the model theory of all finite structures and to put the equal sign in the same situation as the other relations. Using formulas of infinite lengths we obtain new theorems for the preservation of model extensions, submodels, model homomorphisms and inverse homomorphisms. These kinds of theorems were discussed in Chang and Keisler's Model Theory, systematically for general models, but Gurevich obtained some different theorems in this direction for finite models. In our paper the old theorems manage to survive in the finite model theory. There are some differences between into homomorphisms and onto homomorphisms in preservation theorems too. We also study reduced models and minimum models. The characterization sentence of a model is given, which derives a general result for any theory T to be equivalent to a set of existential-universal sentences. Some results about completeness and model completeness are also given.

Keywords **Finite models, Equal sign, Preservation theorems.**

---

① Received May 21, 2003, Accepted September 14, 2004.
This project is supported by the research foundation of Guangdong Women's Professional Technical College.

§ 0. Introduction

The general model theory has been studied for a long time. Chang and Keisler in [1] gave a detailed introduction. It seems that the model theory mainly is to study infinite models. For the model theory of finite structures some people would think that it is easy to deal with and not as important as the infinite case. The development of computer science makes us hold a different view. Database is finite, operating system is finite, even the computer itself is considered as a finite machine. The model theory of finite structures becomes more important. The reason for not giving a predefined equal sign is that in some cases in reality we know only the relations of the elements but don't know their identities. Or we can say that we identify the elements according to the relations amongst them. Another reason for studying model theory of finite structures independently is that many of the theorems in general model theory cannot be used in the finite cases, for example, Gödel's Incompleteness Theorem, Craig's Interpolation Theorem and Compactness Theorem. Without these theorems the model theory of finite structures looks quite different from the general model theory. The research also finds out that it is not always easy to prove a theorem in the model theory of finite structures as compared with the same theorem in the general model theory. In some cases the proofs are even more difficult. A partial reason for that is the lack of the theorems mentioned above. Gurevich in [2] and [3] gave a systematic discussion for the model theory of finite structures in connection with the theory of computer science. Then in [4] he investigated the preservation theorems for the model theory of finite structures. He gave counter-examples in the model theory of finite structures to some of the theorems for infinite models, and some of the others are still valid in the model theory of finite structures. In our paper we prove some different kinds of preservation theorems for finite structure theory. In general model theory the equal relation is supposed to be built into the system but we treat all relations equally. Our notations are as follows:

$$\mathfrak{A}=\langle A, c_1, c_2, \cdots, c_p, R_1, R_2, \cdots, R_q\rangle$$

is a structure with the universe A, p constants $c_1, c_2, \cdots, c_p$ and q relations $R_1, R_2, \cdots, R_q$. The universe A is always finite and the language $L=\langle A, c_1, c_2, \cdots, c_p, R_1, R_2, \cdots, R_q\rangle$ is also finite. The logical formulas in our paper are usually first order formfilas. We also use infinite conjunctions and disjunctions of first order formulas. In the latter case we will use capital Greek letters to denote a formula with infinite length. We adopt the idea from [4] to say that two formulas are equivalent if they are equivalent in all finite structures. They are logically equivalent if they are equivalent in all finite and infinite models. The notation $\mathfrak{A}\models T$ means that all first order sentences in T are true in $\mathfrak{A}$. T_1 and T_2 are theories. The notation $T_1\models T_2$ means that for any model $\mathfrak{A}$, $\mathfrak{A}\models T_1$ implies $\mathfrak{A}\models T_2$. T_1 is equivalent to T_2 if and only if $T_1\models T_2$ and $T_2\models T_1$. A theory T is consistent if and only if T has a (finite) model.

§ 1. Elementary Properties

As in the general model theory we can define elementary equivalence between models, elementary submodels and elementary extensions of models with almost the same definitions. Most of the theorems about these concepts in general model theory are still true in finite model theory. The others are similar. We provide some of them here and omit the proofs.

Theorem 1. 1 *Two models $\mathfrak{A}$ and $\mathfrak{B}$ are elementarily equivalent to each other if and only if the following two conditions are true:*

(1) *There is a constant keeping function F from A to B such that for any relation R in the language L with arity k and any k-tuple of elements $a_1, a_2, \cdots, a_k$ in A,*

$\mathfrak{A}\models R[a_1, a_2, \cdots, a_k]$ *if and only if* $\mathfrak{B}\models R[Fa_1, Fa_2, \cdots, Fa_k]$.

(2) *There is a constant keeping function G from B to A such that*

for any relation R in the language L with arity k and any k-tuple of elements $b_1, b_2, \cdots, b_k$ *in B*,

$$\mathfrak{B} \models R[b_1, b_2, \cdots, b_k] \text{ if and only if } \mathfrak{A} \models R[Gb_1, Gb_2, \cdots, Gb_k].$$

Proof (an outline) By the functions F and G we can find

$$\mathfrak{A} = \mathfrak{A}_0 \supseteq \mathfrak{A}_1 \supseteq \mathfrak{A}_2 \supseteq \cdots, \mathfrak{B} = \mathfrak{B}_0 \supseteq \mathfrak{B}_1 \supseteq \mathfrak{B}_2 \supseteq \cdots,$$

such that $\mathfrak{A}_i \equiv \mathrm{B}_{i+1}$ and $\mathfrak{B}_i \equiv \mathrm{A}_{i+1}$. Since they are finite models the two sequences will be the same models in finitely many steps.

Definition 1.1 *Model* $\mathfrak{A}$ is an elementary submodel of model $\mathfrak{B}$ *if* $A \subseteq B$ and for any first order formula $\varphi(x_1, x_2, \cdots, x_k)$ and for any k-tuple $(z_1, z_2, \cdots, z_k)$ form $A \cup \{c_1, c_2, \cdots, c_p\}$ the sentence $\varphi(z_1, z_2, \cdots, z_k)$ is true in $\mathfrak{A}$ if and only if the same sentence is true and in $\mathfrak{B}$.

Theorem 1.2 *Model* $\mathfrak{A}$ *is an elementary submodel of* $\mathfrak{B}$ *(or Model* $\mathfrak{B}$ *is an elementary extension of* $\mathfrak{A}$*) if and only if there is a constant keeping function F from B onto A such that for any relation R in the language L with arity k and any k-tuple of elements* $b_1, b_2, \cdots, b_k$ *in B*,

$$\mathfrak{B} \models R[b_1, b_2, \cdots, b_k] \text{ if and only if } \mathfrak{A} \models R[Fb_1, Fb_2, \cdots, Fb_k].$$

We also give the definitions of reduced models and minimum models.

Definition 1.2 Two elements a and b are said to satisfy the same relation in a model $\mathfrak{A}$ of language $L = \langle A, c_1, c_2, \cdots, c_p, R_1, R_2, \cdots, R_q \rangle$ with respect to elements $a_1, a_2, \cdots, a_n$ if for all relations $R(x_1, x_2, \cdots, x_k)$, all k-tuples $(z_1, z_2, \cdots, z_k)$ from $\{a_1, a_2, \cdots, a_n\} \cup \{c_1, c_2, \cdots, c_p\}$ whenever the relation $R(z_1, z_2, \cdots, z_{i-1}, a, z_{i+1}, z_{i+2}, \cdots, z_k)$ is true, the relation $R(z_1, z_2, \cdots, z_{i-1}, b, z_{i+1}, \cdots, z_k)$ is also true and vise versa.

In a model $\mathfrak{A}$ if two elements a and b satisfy the same relation with respect to all elements of the model $\mathfrak{A}$, then we can reduce one of them to form a new model $\mathfrak{B}$. The new model is an elementary submodel of the original model $\mathfrak{A}$.

Definition 1.3 A model in which any two elements do not satisfy the same relation with respect to all elements of the model is call a reduced model.

Definition 1.4 A model $\mathfrak{A}$ of a theory T is called a minimum model of T if $\mathfrak{A}$ itself is a reduced model and it does not contain any proper submodel of T.

Theorem 1.3 *For any model $\mathfrak{A}$ there is a reduced model $\mathfrak{B}$ such that $\mathfrak{B} \prec \mathfrak{A}$. The reduced model $\mathfrak{B}$ is unique up to isomorphism.*

Theorem 1.4 *A theory T is complete if and only if T has only one reduced model up to isomorphism.*

We can define model completeness as in the general model theory, but here the theorem is different.

Theorem 1.5 *If a theory T is complete then T is model complete.*

Proof If $\mathfrak{A} \subset \mathfrak{B}$ are T models, then both of them can be reduced to the same reduced model.

Theorem 1.6 *For any model $\mathfrak{A}$ of language $L = \langle A, c_1, c_2, \cdots, c_p, R_1, R_2, \cdots, R_q \rangle$ there is a characterization formula φ such that for any model $\mathfrak{B}$, $\mathfrak{B} \models \varphi$ if and only if $\mathfrak{B}$ is elementarily equivalent to $\mathfrak{A}$. The characterization formula is an existential universal sentence.*

Proof Suppose that model $\mathfrak{A}$ has n elements $\{a_1, a_2, \cdots, a_n\}$. We define a conjunction formula

$$\underset{y_1 = a_{i_1}, a_{i_2}, \cdots, y_m = a_{i_m}}{\chi}(y_1, y_2, \cdots, y_m)$$

with free variables $(y_1, y_2, \cdots, y_m)$ where y_1 represents $a_{i_1}, a_{i_2}, \cdots, y_m$ represents a_{i_m}. The formula χ is a conjunction of all positive and negative relations $\varepsilon R(z_1, z_2, \cdots, z_k)$ where the k-tuple $(z_1, z_2, \cdots, z_k)$ runs over all possible combinations (including repetitions) of symbols in

$$Z = \{y_1, y_2, \cdots, y_m\} \cup \{a_1, a_2, \cdots, a_n\} \cup \{c_1, c_2, \cdots, c_p\}.$$

The symbol ε is chosen according to the truth of the relation $R(z_1, z_2, \cdots, z_k)$ in $\mathfrak{A}$. We give the following definition:

$$\underset{y_i = a_{i_1}, a_{i_2}, \cdots, y_n = a_{i_m}}{\chi}(y_1, y_2, \cdots, y_m) = \bigwedge_{\substack{R(z_1, z_2, \cdots, z_k) \in L \\ (z_1, z_2, \cdots, z_k) \subset Z}} \Big(\bigwedge_{\mathfrak{A} \models R(z_1, z_2, \cdots, z_k)} R(z_1, z_2, \cdots, z_k) \cdot$$

$$\wedge \cdot \bigwedge_{\mathfrak{A} \not\models R(z_1, z_2, \cdots, z_k)} \neg R(z_1, z_2, \cdots, z_k) \Big).$$

Using formula χ we can define the conjunctive formula of the model $\mathfrak{A}$.

$$\chi_{\mathfrak{A}}(x_1, x_2, \cdots, x_n) = \underset{x_1 = a_1, x_2 = a_2, \cdots, x_n = a_n}{\chi}(x_1, x_2, \cdots, x_n).$$

Formula $\chi_{\mathfrak{A}}(x_1, x_2, \cdots, x_n)$ is true if and only if elements $x_1, x_2, \cdots, x_n$ have the same positive and negative relations as elements $a_1, a_2, \cdots, a_n$ in model $\mathfrak{A}$.

Next we define a formula ψ_i for an element y_i to have the same relation as the element a_i with respect to all a's.

$$\psi_i(y_i, x_1, x_2, \cdots, x_n) = \chi_{y_i = a_i, x_1 = a_1, \cdots, x_n = a_n}(y_i, x_1, x_2, \cdots, x_n).$$

Putting together $\psi_i(y_i, x_1, x_2, \cdots, x_n)$, for $i = 1, 2, \cdots, n$ and replacing y for y_i we obtain a disjunction $\bigvee_{i=1}^{n} \psi_i(y_i, x_1, x_2, \cdots, x_n)$. The above formula is true if and only if y satisfies the same relation as one of the elements in $a_1, a_2, \cdots, a_n$, with respect to all a's.

If $y_{i_1}, y_{i_2}, \cdots, y_{i_k}$ have the same relations as elements $a_{i_1}, a_{i_2}, \cdots, a_{i_k}$ with respect to all elements of $\mathfrak{A}$ we want the list $(y_{i_1}, y_{i_2}, \cdots, y_{i_k})$ to satisfy the same relations as $(a_{i_1}, a_{i_2}, \cdots, a_{i_k})$. In the following the arity k is considered as the maximum of the arity of all relations in the language L and the k-tuple $(y_{i_1}, y_{i_2}, \cdots, y_{i_k})$ runs through all possible strings of length k in $(y_1, y_2, \cdots, y_n)$. The formula is as follows:

$$\bigwedge_{\{y_{i_1}, y_{i_2}, \cdots, y_{i_k}\} \subset \{y_1, y_2, \cdots, y_n\}} \left(\bigwedge_{j=1}^{k} \psi_{i_j}(y_{i_j}, x_1, x_2, \cdots, x_n) \rightarrow \chi_{y_{i_1} = a_{i_1}, y_{i_2} = a_{i_2}, \cdots, y_{i_k} = a_{i_k}}(y_{i_1}, y_{i_2}, \cdots, y_{i_k}) \right)$$

Now we are ready to define our main sentence φ. The meaning of φ is that there exist n elements $x_1, x_2, \cdots, x_n$ such that they satisfy the same relations as $a_1, a_2, \cdots, a_n$ and for any y, y satisfies the same relation as one of the x's and for any k-tuple $(y_{i_1}, y_{i_2}, \cdots, y_{i_k})$ chosen from $\{y_1, y_2, \cdots, y_n\}$ if each y_{i_j} satisfies the same relations as a_{i_j}, then the k-tuple $(y_{i_1}, y_{i_2}, \cdots, y_{i_k})$ satisfies the same relations as $(a_{i_1}, a_{i_2}, \cdots, a_{i_k})$

$$\varphi = \exists x_1, x_2, \cdots, x_n \Big(\chi_{\mathfrak{A}}(x_1, x_2, \cdots, x_n) \cdot \wedge \cdot \forall y \Big(\bigvee_{i=1}^{n} \psi_i(y, x_1, x_2, \cdots, x_n) \Big) \cdot \wedge \cdot \forall y_1, y_2, \cdots y_n \Big(\bigwedge_{\{y_{i_1}, y_{i_2}, \cdots, y_{i_k}\} \subset \{y_1, y_2, \cdots, y_n\}} \Big(\bigwedge_{j=1}^{k} \psi_{i_j}(y_{i_j}, x_1, x_2, \cdots, x_n) \rightarrow \chi_{y_{i_1} = a_{i_1}, y_{i_2} = a_{i_2}, \cdots, y_{i_k} = a_{i_k}}(y_{i_1}, y_{i_2}, \cdots, y_{i_k}) \Big) \Big) \Big).$$

We will prove that for any model $\mathfrak{B}$, $\mathfrak{B}\models\varphi$ if and only if $\mathfrak{B}\equiv\mathfrak{A}$. Let $\mathfrak{B}$ be a model of φ and $B=\{b_1, b_2, \cdots, b_m\}$ be the universe of $\mathfrak{B}$. We can set up a many-to-many and onto mapping from B to A satisfying $F(b_i)=a_j$ if b_i has the same relations as a_j with respect to $a_1, a_2, \cdots, a_n$. For any relation R in the language L with arity k and for any k-tuple $(b_{i_1}, b_{i_2}, \cdots, b_{i_k})$ in B, $\mathfrak{B}\models R[b_{i_1}, b_{i_2}, \cdots, b_{i_k}]$ if and only if $\mathfrak{A}\models R[Fb_{i_1}, Fb_{i_2}, \cdots, FB_{i_k}]$. With this property we can prove that the two models $\mathfrak{A}$ and $\mathfrak{B}$ are elementarily equivalent to each other.

Observe that there is not a quantifier in formulas ψ and χ. Therefore ψ is an existential universal sentence.

The following two theorems are of special interest for finite model theory. We give only a simple proof for Theorem 1. 7 and omit the proof of Theorem 1. 8.

Theorem 1. 7 *Any theory T is equivalent to a countable disjunction of first order existential universal sentences.*

Proof For every model $\mathfrak{A}$ with n elements we can write the characterization sentence $\varphi_{\mathfrak{A}}$ as in the proof of Theorem 1. 6 such that $\varphi_{\mathfrak{A}}$ is true in a model $\mathfrak{B}$ if and only if $\mathfrak{B}$ is elementarily equivalent to $\mathfrak{A}$. Let $\sum$ be the disjunction of $\varphi_{\mathfrak{A}}$ over all models of T. $\sum$ is equivalent to T.

Theorem 1. 8 *Any theory T is equivalent to a countable set of first order universal existential sentences.*

Proof (It is the dual argument of Theorem 1. 7)

§ 2. Preservation Theorems

There are a lot of theorems in general model theory in preserving properties among models. Using preservation theorems Chang and Keisler in [1, p. 123] gave a list of mathematical theories which preserve certain kinds of properties. In that book the following theorems are given:

Theorem 3. 2. 2[MT] *A theory T is preserved under submodels if and only if T has a set of universal axioms.*

Theorem 3.2.3[MT] *A theory T is preserved under union of chains if and only if T has a set of universal existential axioms.*

Theorem 3.2.4[MT] *A consistent theory T is preserved under homomorphisms if and only if T has a set of positive axioms.*

Can any of the preservation theorems be true in the theory of finite models? The answer is negative for many cases. Gurevich in [4] gave counter-examples to obtain the following results for finite models:

(i) There is a first order sentence φ such that φ is preserved by extensions of models but it cannot be equivalent to a first order existential sentence.

(ii) There is a first order sentence φ such that φ is preserved by submodels but it cannot be equivalent to a first order universal sentence.

(iii) There is a first order sentence $\varphi(P)$ such that $\varphi(P)$ is monotone in P but it cannot be equivalent to a first order sentence positive in P.

There are two differences between Gurevich's results and the results in [1]. Gurevich's theorems work for finite models only, while the theorems in [1] work for all finite and infinite models. Also Gurevich is discussing the possibility of a first order sentence to be equivalent to another first order sentence while the theorems in [1] are using a set of (possibly) infinitely many sentences to be equivalent to another set of infinitely many sentences. We will show in this section that [MT] Theorem 3.2.2 survives in finite model theory. We also give similar theorems for the preservation of homomorphisms of models.

Theorem 2.1 *A theory T is preserved under extensions of models if and only if T is equivalent to a countable disjunction of existential sentences.*

Proof The right-to-left implication is trivial so we have to prove only the left-to-right implication. Assume T is preserved under extensions. For every finite model $\mathfrak{A}$ in language $L=\langle c_1, c_2, \cdots, c_p, R_1, R_2, \cdots, R_q\rangle$ with n elements, we can write an existential sentence φ which describes all relations and negative relations among all elements. Suppose that the n elements are $A=\{a_1,a_2,\cdots,a_n\}$. We first define a conjunctive formula

$$\underset{x_1=a_1,x_2=a_2,\cdots,x_n=a_n}{\chi}(x_1,x_2,\cdots,x_n)$$

with free variables $x_1,x_2,\cdots,x_n$, where x_1 represents $a_1,a_2,\cdots$, and x_n represents a_n. The formula χ is a conjunction of all positive and negative atomic formulas for all relations $\varepsilon R(z_1,z_2,\cdots,z_k)$, where the k-tuple $(z_1, z_2,\cdots,z_k)$ runs over all possible combinations (including repetitions) of

$$Z=\{y_1,y_2,\cdots,y_m\}\cup\{a_1,a_2,\cdots,a_n\}\cup\{c_1,c_2,\cdots,c_p\}.$$

The symbol ε is chosen according to the truth of relation $R(z_1,z_2,\cdots, z_k)$ with the corresponding k-tuple in model $\mathfrak{A}$. To specify the details we give the following definition:

$$\underset{x_1=a_1,x_2=a_2,\cdots,x_n=a_n}{\chi}(x_1,x_2,\cdots,x_n)=$$

$$\bigwedge_{\substack{R(z_1,z_2,\cdots,z_k)\in L\\(z_1,z_2,\cdots,z_k)\subset Z}}\Big(\bigwedge_{\mathfrak{A}\models R(z_1,z_2,\cdots,z_k)}R(z_1,z_2,\cdots,z_k)\cdot\wedge\cdot\bigwedge_{\mathfrak{A}\not\models R(z_1,z_2,\cdots,z_k)}\neg R(z_1,z_2,\cdots,z_k)\Big).$$

The maximum existential conjunctive sentence is defined as follows:

$$\varphi(\mathfrak{A})=\exists x_1,x_2,\cdots,x_n\underset{x_1=a_1,x_2=a_2,\cdots x_n=a_n}{\chi}(x_1,x_2,\cdots,x_n).$$

Let Σ be the disjunction of the above sentence $\varphi(\mathfrak{A})$ over all models of the theory T.

$$\Sigma=\bigvee_{\mathfrak{A}\models T}\varphi(\mathfrak{A}).$$

Every model $\mathfrak{A}$ of the theory T is a model of the formula Σ because $\mathfrak{A}$ satisfies at least a term of the disjunction of Σ. Conversely, every model $\mathfrak{B}$ of the formula Σ satisfies a disjunct of Σ which is written according to a model $\mathfrak{A}$. Model $\mathfrak{B}$ is an extension of the reduced model of $\mathfrak{A}$. $\mathfrak{A}$ is a model of T. Therefore $\mathfrak{B}$ is a model of the theory T.

Theorem 2.2 *A theory T is preserved under submodels if and only if T is equivalent to a set of universal sentences.*

Proof The theory T is considered as a conjunction $\prod$ of (possibly infinitely many) sentences. We consider the theory $T\leftrightarrow\prod=\wedge_{1\leqslant i<\infty}\varphi_i$. The negation of it is $\Sigma=\vee_{1\leqslant i<\infty}\neg\varphi_i$. It is easy to see that Σ is preserved under the extension of models. In the proof of Theorem 2.1 the theory T is very general. We do not require the T to be a set of first order sentences or in any other special form. So we can use it for $\sum$. Hence $\sum$ is equivalent to a countable disjunction $\sum' = \bigvee_{1\leqslant i<\infty}\psi_i$, where each ψ_i is a first order existential sentence. Therefore the original theory T is equivalent to $\neg\sum' = \bigwedge_{1\leqslant i<\infty}\neg\psi_i$. The left-hand side is equivalent to a set of first order universal sentences.

A homomorphism is considered as a mapping h from a model $\mathfrak{A}$ into (including onto) a model $\mathfrak{B}$ such that for any relation R and any elements $a_1, a_2, \cdots, a_k \in \mathfrak{A}$, if $R(a_1, a_2, \cdots, a_k)$ is true in $\mathfrak{A}$, then $R(ha_1, ha_2, \cdots, ha_k)$ is true in $\mathfrak{B}$. In [1, p. 126], a result of Lyndon's characterizes the theories preserving onto homomorphisms. We first discuss the preservation property of into homomorphisms in finite models. In our theorem we construct a countable disjunction of first order positive existential sentential sentences to be equivalent to the original theory.

Theorem 2.3 *A theory T is preserved under into homomorphisms if and only if T is equivalent to a countable disjunction of positive existential sentences.*

Proof For every finite model $\mathfrak{A}$ with n elements we can write a first order existential sentence φ which describes all positive relations (but not negative relations) among all elements

$$\varphi = \exists x_1, x_2, \cdots, x_n \bigwedge_{\substack{R(z_1, z_2, \cdots, z_k)\in L \\ \{z_1, z_2, \cdots, z_k\}\subset Z}} \Big(\bigwedge_{\mathfrak{A}\models R(z_1, z_2, \cdots, z_k)} R(z_1, z_2, \cdots, z_k) \Big),$$

where x_1 represents a_1, x_2 represents a_2, $\cdots$, x_n represents a_n and

$$Z = \{x_1, x_2, \cdots, x_n\} \cup \{a_1, a_2, \cdots, a_n\} \cup \{c_1, c_2, \cdots, c_p\}.$$

Let $\sum$ be the disjunction of the above sentences over all models of the theory T. Every model $\mathfrak{B}$ of the theory T is a model of the formula $\sum$ because $\mathfrak{B}$ satisfies at least a disjunct of $\sum$ which is written according to the model $\mathfrak{B}$ itself. Conversely if model $\mathfrak{B}$ satisfies a disjunct of $\sum$ which is written according to the model $\mathfrak{A}$, Model $\mathfrak{A}$ can be reduced to a model $\mathfrak{C}$. $\mathfrak{B}$ also satisfies the positive existential sentence written according to model $\mathfrak{C}$. Hence $\mathfrak{B}$ is a homomorphism image of $\mathfrak{C}$. $\mathfrak{C}$ is a model of T.

Therefore $\mathfrak{B}$ is a model of the theory T because T is preserved under into homomorphisms.

Theorem 2.4 *A theory T is preserved under inverse into homomorphisms if and only if T is equivalent to a set of negative universal sentences.*

Proof Similar to the proof of Theorem 2.2.

For the case of onto homomorphisms our result is weaker. We will use a countable disjunction of countable conjunctions of first order positive sentences to be equivalent to the original theory. We first give a lemma.

Lemma 2.1 *A theory T in the language $L=\langle A, c_1, c_2, \cdots, c_p, R_1, R_2, \cdots, R_q\rangle$ is preserved under onto homomorphisms. Let Δ be the set of all positive sentences in L. If $\mathfrak{A}$ is a model of T and every sentence $\delta\in\Delta$ that holds in $\mathfrak{A}$ holds in the model $\mathfrak{B}$, then $\mathfrak{B}$ is a model of T.*

Proof Suppose that the model $\mathfrak{A}$ has n elements $\{a_1, a_2, \cdots, a_n\}$. We define a conjunctive formula

$$\chi_{y_1=a_{i_1}, y_2=a_{i_2}, \cdots, y_n=a_{i_m}}(y_1, y_2, \cdots, y_m)$$

with free variables $(y_1, y_2, \cdots, y_m)$, where y_1 represents a_{i_1}, y_2 represents $a_{i_2}, \cdots, y_m$ represents a_{i_m}. The formula χ is a conjunction of all positive relations $R(z_1, z_2, \cdots, z_k)$ where k-tuple $(z_1, z_2, \cdots, z_k)$ runs over all possible symbols in

$$Z=\{y_1, y_2, \cdots, y_m\}\cup\{a_1, a_2, \cdots, a_n\}\cup\{c_1, c_2, \cdots, c_p\}.$$

We give the following definition:

$$\chi_{y_1=a_{i_1}, y_2=a_{i_2}, \cdots, y_n=a_{i_m}}(y_1, y_2, \cdots, y_m)=\bigwedge_{\substack{R(z_1, z_2, \cdots, z_k)\in L \\ \{z_1, z_2, \cdots, z_k\}\subset Z}}\Big(\bigwedge_{\mathfrak{A}\models R(z_1, z_2, \cdots, z_k)} R(z_1, z_2, \cdots, z_k)\Big).$$

Using formula χ we can define the positive conjunctive formula of the model $\mathfrak{A}$.

$$\chi_{\mathfrak{A}}(x_1, x_2, \cdots, x_n)=\chi_{x_1=a_1, x_2=a_2, \cdots, x_n=a_n}(x_1, x_2, \cdots, x_n).$$

Formula $\chi_{\mathfrak{A}}(x_1, x_2, \cdots, x_n)$ is true if and only if the n-tuple $(x_1, x_2, \cdots, x_n)$ has at least the same positive relations as $(a_1, a_2, \cdots, a_n)$ in the model $\mathfrak{A}$.

Let $\mathfrak{B}=\langle B, L\rangle$, where the universe is $B=\{b_1, b_2, \cdots, b_m\}$ with $m\leqslant n$. Choose a set of new variables $Z=\{z_1, z_2, \cdots, z_l\}$, where $l=m-n$. Let

Ω be the set of all functions from Z to the universe of $\mathfrak{A}$. $\Omega=\{\omega\mid\omega$ is a function of Z to $A\}$. We define a formula ψ_ω with free variables $\{z_1, z_2,\cdots,z_l\}$ saying that the l-tuple $(z_1,z_2,\cdots,z_l)$ has at least the same positive relations as $(\omega z_1,\omega z_2,\cdots,\omega z_l)$ with respect to all elements of $\mathfrak{A}$

$$\psi_\Omega(z_1,z_2,\cdots,z_l)=\underset{z_1=\omega z_1,z_2=\omega z_2,\cdots,z_l=\omega z_l}{\chi}(x_1,x_2,\cdots,x_n).$$

Putting together $\psi_\Omega(z_1,z_2,\cdots,z_l)$ for $\omega\in\Omega$ we obtain a formula: $\Psi=\bigvee_{\omega\in\Omega}\psi_\Omega(z_1,z_2,\cdots,z_l)$. Formula Ψ is true if and only if $(z_1,z_2,\cdots, z_l)$ satisfies at least the same positive relations as one of the l-tuples of elements $(\omega z_1,\omega z_2,\cdots,\omega z_l)$ with respect to all elements of $\mathfrak{A}$.

Now we define our main formula φ

$\varphi=\exists x_1,x_2,\cdots,x_n\chi_{\mathfrak{A}}(x_1,x_2,\cdots,x_n)\cdot\wedge\cdot\forall y_1,y_2,\cdots,y_l\Psi(y_1,y_2,\cdots,y_l)$.

φ is a positive sentence which holds in $\mathfrak{A}$. From the condition of the theorem $\mathfrak{B}\models\varphi$. By the above explanations and the structure of sentence φ, we know that there are elements $b_1,b_2,\cdots,b_n\in\mathfrak{B}$ satisfying at least the same positive relations as $a_1,a_2,\cdots,a_n\in\mathfrak{B}$. Now let $d_1,d_2,\cdots,d_l\in\mathfrak{B}$ be the elements in $\mathfrak{B}$ different from $b_1,b_2,\cdots,b_n$. By the formula Ψ we know that there is a function $\omega\in\Omega$ such that $d_1,d_2,\cdots,d_l$ satisfy at least the same positive relations as the l-tuple $(\omega z_1,\omega z_2,\cdots,\omega z_l)$ with respect to all a's.

Divide the elements of $\mathfrak{B}$ into subsets $B_1,B_2,\cdots,B_n$ in which $b_i\in B_i$ and every element in B_i satisfies at least the same positive relations as a_i with respect to all a's. Construct another model $\mathfrak{D}$ such that for every element $a_i\in\mathfrak{A}$, $\mathfrak{D}$ has a subset D_i with $|B_i|$ different copies of elements having the same (positive and negative) relations as the element $a_i\in\mathfrak{A}$. Although it is possible $b_i=b_j$ for $i\neq j$, but D_i and D_j are disjoint. From Theorem 1.2, $\mathfrak{D}$ is an elementary extension of $\mathfrak{A}$. Hence $\mathfrak{D}$ is a model of T. Define a function F so that it maps all the elements of D_i to element b_i in B. We can check that F is a homomorphism from $\mathfrak{D}$ onto $\mathfrak{B}$. Because the theory T is preserved under onto homomorphisms, $\mathfrak{B}$ is therefore a model of T.

Theorem 2.5 *A theory T is preserved under onto homomorphisms if and only if T is equivalent to a countable disjunction of countable* *conjunction of positive existential sentences.*

Proof For every T model $\mathfrak{A}$ with n elements we can write a countable conjunction of positive sentence $\Phi = \bigwedge_{0 \leqslant l < \infty} \varphi_l$ as in the proof of Lemma 2. 1 such that Φ is true in a model $\mathfrak{B}$ if and only if $\mathfrak{B}$ is an image of an onto homomorphism of an elementary extension of $\mathfrak{A}$. Let $\sum$ be the disjunction of the above formula Φ over all models of T. Every model of T satisfies at least a disjunct of $\sum$ which is written according to itself. Conversely, every model $\mathfrak{B}$ of the sentence $\sum$ satisfies a disjunct which is written according to a model $\mathfrak{A}$. $\mathfrak{B}$ is an image of an onto homomorphism of an elementary extension of $\mathfrak{A}$. Therefore $\mathfrak{B}$ is a model of T.

Theorem 2. 6 *A theory T is preserved under inverse onto homomorphisms if and only if T is equivalent to a set of countable disjunctions of negative universal sentences.*

Proof Similar to the proof Theorem 2. 2.

In Section 3, an example is given which shows that a sentence φ is preserved under extensions of models but it can not be equivalent to an existential sentence. Of course from our Theorem 2. 1 the sentence φ is equivalent to a countable disjunction of existential sentences.

§ 3. Examples

We give an example to show how we work differently in the finite model theory and in the general model theory.

Example 3. 1 The language L has two relations $x<y$ and $x\,S\,y$ expressing the relations 'x is smaller than y' and 'x is succeeded by y', respectively, and two constants m and M representing the minimum and maximum elements of the linear order.

The symbol '$x \sim y$' is defined as the simplification of the following:

$x \sim y$: $\forall z((x<z \rightarrow y<z) \cdot \wedge \cdot (z<x \rightarrow z<y))$.

Then we define the following sentences:

(1) $\forall xyz(x<y \wedge y<z :\rightarrow x<z)$,

(2) $\forall x \neg (x<x)$,

(3) $\forall xy\,(x<y\rightarrow\neg\,(y<x))$,

(4) $\forall x\,(\neg(x<m))$,

(5) $\forall x\,(\neg(M<x))$,

(6) $\forall xy\,(x<y\vee y<x\vee x\sim y)$,

(7) $\forall xy\,(xSy\rightarrow x<y)$,

(8) $\forall xyz\,(xSy\wedge xSz:\rightarrow.\ y\sim z)$,

(9) $\forall xyz\,(xSy\wedge x<z:\rightarrow:y\sim z\vee y<z)$,

(10) $\forall x\,(x\sim M\vee\exists y(xSy))$.

We need strong axioms here because there is not an equal sign in our language. It is not necessary to discuss the independence of the axioms. The sentences (1)~(6) give a linear order with two end elements. (7)~(10) give a successor relation which is consistent with the linear order. The sentence φ is now given as the conjunction of all the above 10 sentences $\varphi=\bigwedge_{i=1}^{10}(i)$. The theory T consists of only one sentence $\{\varphi\}$. It is not difficult to see that the theory T is model complete but not complete.

If we change our sentence to the following: $\psi=\bigwedge_{i=1}^{9}(i)\rightarrow(10)$, then the sentence ψ is preserved by extensions of models but it can not be equivalent to a first order existential sentence. This is the original example given by Gurevich in [4].

References

[1] Chang C C. Keisler H J. Model Theory. North-Holland, Amsterdam, 3rd ed. , 1989.

[2] Gurevich Y. Toward logic tailored for computational complexity. Computation and proof theory (Ed. M. Richter et al.), Springer Lecture Notes in math. , 1984, 1104: 175—216.

[3] Gurevich Y. Logic and the challenge of computer science. Current Trends in Theoretical Computer Science (Ed. E. Börger), Computer Science Press, Indianapolis, 1988, 1—57.

[4] Gurevich Y. On finite model theory. Feasible Mathematics, (Ed. S. R. Buss et al.), A Mathematical Sciences Institute Workshop, Ithaca, New York, 1989, 211—219.

[5] Lyndon R C. An interpolation theorem in the predicate calculus. *Pacific J. Math.* , 1959, 9: 155—164.

数学研究，2009，42(2)：126～137

计算实数函数的图灵机的稳定性①

The Stability of Turing Maching in Computing Real Functions

Abstract The computable function over the set of all real numbers is a very important concept. There are two approaches in defining computable real functions. The first approach is to define the indexes of computable real numbers first. A computable function $y=f(x)$ defined on all computable real numbers only if there is a(partial) recursive function over the indexes which maps the index of x to the index of y. The study of functions on real numbers relies on the property of functions on natural numbers. The second approach of defining computable real functions is based on approximations. A real function is computable if it is both sequentially computable and effectively uniformly continuous. The condition is too strong preventing a lot of very useful functions to be computable. According to the definition predicates "$<$"and "$=$" are not computable because they are represented by discontinuous functions. In this paper we discuss the stability of Turing machines and give a more general definition for computable real functions based on stable Turing machines. Our definition does not use recursive functions over natural numbers. According to our definition the normally used functions especially some useful discontinuous functions on real numbers are computable.

① Received date：2008-09-25.

Our definition is more convenient in discussing properties of real computable functions.

Keywords **Computable real function, stability, Turing machine.**

§ 1. Introduction

Computable functions over the set of nature numbers have been studied for a long period of time. There are two main models in treating the computable functions. They are Turing machine and the recursive function. Since the nature numbers are discrete objects and countable, the treatment of their Computable functions are relatively easier. For the computable functions over the set of real numbers the treatment is more difficult because the set is continuous and having uncountably many elements. Researchers designed many ways to discuss the computable real functions. Turing in [18] used infinite decimal sequences to define computable numbers. A real number is computable if its digit sequence can be produced by an algorithm or a Turing machine. The algorithm takes an integer n as input and produces the n-th digit of the real number's decimal expansion as output. Some people use recursive functions or primitive recursive functions instead of the algorithm or Turing machine in the definition. Although the set of real numbers is uncountable, the set of computable numbers is countable and thus most real numbers are not computable. The computable numbers can be counted by assigning a Gödel number to each Turing machine. This gives a function from the natural numbers to the computable real numbers. Although the computable numbers are an ordered field, the set of Gödel numbers corresponding to computable numbers is not itself computably enumerable, because it is not possible to effectively determine which Gödel numbers correspond to Turing machines that produce computable reals. In order to produce a computable real, a Turing machine must compute a total function, but the corresponding decision problem is undecidable.

There are two approaches in defining computable real functions.

The first approach is to define the indexes of computable real numbers first. An index of a real number is an integer which represents the Gödel number of a machine that computes the real number. A computable function $y=f(x)$ defined on all computable real numbers only if there is a (partial) recursive function over the indexes which maps the index of x to the index of y. This definition succcssively avoids continuity and uncountability of the real numbers because the set of computable real numbers is countable. This approach, with some variations, has been considered in the Russian school of constructive analysis, represented by Markov[11], Shanin[15], Tseitin[16], Kushner[9], and others. It has also been considered by Moschovakis[12], Aberth[1] and by many others. Closely related is the constructive approach to analysis presented by Bishop and Bridges[3].

Since the computable real numbers are generated by a Turing machine or a program or any kinds of algorithm, the definition is very difficult to use. For example, if we want to add two real numbers $\frac{2}{3}$ and $\frac{3}{7}$, we must find two Turing machines A, B to produce the two digit sequences. The output of the two machines are 0. (6) and 0. (428571). Let a, b be the Gödel numbers of A, B which are usually astronomy numbers. After that, we need to find a number c which is the Gödel numbers of a machine C. The task of C is to print out 1. (095238). Since we can prove (which seems to be very complicated) that the function from a, b to c is recursive. Therefore addition is computable. The definition is also incomplete because all the incomputable real numbers do not join the operation which is unfair. For example: a very simple function which adds an incomputable real number to tile set of all computable real numbers is obvious computable. Actually we know how to add two real numbers regardless of whether they are computable or not. The addition should be defined for entire real space.

The second approach of defining computable real functions is based on approximations. A function $f: R \to R$ is sequentially computable if, for every computable sequence $\{x_i\}_{i=1}^{\infty}$ of real numbers, the sequence $\{f(x_i)\}_{i=1}^{\infty}$ is also computable. A function $f: R \to R$ is effectively uniformly continuous if there exists a recursive function $d: N \to N$ such that, if $|x-y| < \frac{1}{d(n)}$, then $|f(x)-f(y)| < \frac{1}{n}$. A real function is computable if it is both sequentially computable and effectively uniformly continuous. These definitions can be generalized to functions of more than one variable or functions only defined on a subset of R^n. The generalizations of the latter two need not be restated. A suitable generalization of the first definition is: Let D be a subset of R^n a function $D \to R$ is sequentially computable if, for every n-tuple of $(x_{i1})_{i=1}^{\infty}$, $(x_{i2})_{i=1}^{\infty}$, $\cdots$, $(x_{in})_{i=1}^{\infty}$ of computable sequences of real numbers such that $(\forall i)(x_{i1})_{i=1}^{\infty}$, $(x_{i2})_{i=1}^{\infty}, \cdots, (x_{in})_{i=1}^{\infty} \in D$ the sequence $\{f(x_i)\}_{i=1}^{\infty}$ is also computable. This kind of definitions are considered by Grzegorezyk[4], Klaua[7], Pour-El and Richards[13], Weihrauch[19], Ko[8], and by many others. Moschovakis[12] discussed, computable functions on metric spaces. Rogers[14] defined the effective computability. Weihrauch[18] developed computable analysis. Hertling[5] compared two types of computable real functions.

There are still a lot of definitions considered by different researchers: A real number is computable if and only if there is a computable Dedekind cut D converging to it. The function D is unique for each irrational computable number (although of course two different programs may provide the same function). Minsky[11] in chapter §9 defined "the Computable real numbers" Aberth[2] describes the development of the calculus over the computable number field. Weihrauch[19] in §1.3.2 introduces the definition by nested sequences of intervals converging to the singleton real. Other representations are discussed in §4.1.

In the second approach the definition of computable real functions has two conditions. 1. It is sequentially computable. 2. It is effectively uniformly continuous. In the first condition not only the elements of the two sequences are computable but the two sequences are also computably given. It seems that the functions are working between sequences which are very difficult to control. In the second condition the functions are required to be effectively uniformly continuous. The conditions are too strong preventing a lot of very useful functions to be computable. Through a very indirect proof the arithmetical operations on computable numbers are computable in the sense that whenever real numbers a and b are computable then the following real numbers are also computable: $a+b$, $a-b$, ab, and a/b if b is nonzero. According to the above definitions the order relation on the computable numbers is not computable. There is no Turing machine which on input A (the index of a Turing machine approximating the number a) outputs "YES" if $a>0$ and "NO" if $a\leqslant 0$. The reasons: suppose the machine described by A keeps outputting 0 as ε approximations. It is not clear how long to wait before deciding that the machine will never output an approximation which forces a to be positive. Thus the machine will eventually have to guess that the number will equal 0, in order to produce an output; the sequence may later become different from 0. This idea can be used, to show that the machine is incorrect on some sequences if it computes a total function. A similar problem occurs when the computable real numbers are represented as Dedekind cuts. The same holds for the equality relation: the equality test is not computable. This is also easily to see because both "greater" and "equal" relations are represented by discontinuous functions. We are not satisfied for the relations "$<$" and "$=$" being incomputable. An important property of Turing's definition is that the arithmetic operations are not uniformly computable in this representation. This can be seen most easily with addition: when adding decimal numbers, in order to produce one digit it may be necessary to look arbitrarily far to the right to determine if there is a carry to the current location. It is true, however, that if two real numbers have representations according to Turing's definitions then so will their sum, product, difference, and quotient. This lack of uniformity is one reason that the contemporary definition of computable numbers uses ε approximations rather than decimal expansions.

In the two approaches the definitions of computable real functions are using Gödel numbers, limits, sequences, Dedkind cuts and some other tools. So therefore the definitions are very difficult to use. They also do not react the properties of the functions. The implementation of the functions even is not considered. We would like to ask the following questions:

1. Is it possible to define the computable real functions without using the recursive functions over natural numbers?

2. Is it possible to define the computable real functions, so that they include normally used functions especially some useful discontinuous functions?

3. Is it possible to define the computable real functions, so that we can use the definitions to discuss some properties of the functions?

In this paper we discuss the stability of Turing machines in computing real functions and give a more general definition for the computable real function based on the stable Turing machine. Our definition positively answers the above three questions.

§ 2. The stability of a Turing machine

The Turing machines we used are basically the same kind of machines defined by Hopcroft and Ullman in [6]. A Turing machine is a sextuplets $M=(Q,\Sigma,\Gamma,\delta,q_0,B)$. Q is a set of finitely many states. Σ is a set of finitely many input and output symbols. Γ is a set of finitely many allowable tape symbols. B is the blank symbol which is a symbol of Γ. δ is a set of finitely many instructions which tells the machine how to move. Formally δ is a partial function defined as $Q\times\Gamma\rightarrow Q\times\Gamma\times\{L, R\}$. q_0 is the starting state. Only one component is different from the original Turing machine because we do not need final states here. Since the input and output sequences are written in infinitely many cells we must allow the Turing machine to run without halting. If the machine provides the correct resulting symbols one by one in finitely many steps for a fixed cell and never changes it again thereafter, the computation is successful. On the other hand, if the machine stops in the middle or it goes back and forth among finitely many cells, the computation is failed and the value of the function is undefined. If the above situation appears in deciding the membership of a language, the member is rejected. Because the input and output tapes are with infinite lengths, it can be used to compute the real functions with infinitely many digits. It makes possible to compute the accurate value for the digits of a real number which are computed only in approximations originally. We give the following definitions.

Definition 1 A Turing machine with k input tapes and an output tape is called stable if every cell of the output tape is visited for at most finitely many steps and leaves an output symbol in each cell at the last visit.

Definition 2 A function of real numbers with k arguments is computable if it can be computed by a stable Taring machine with k input tapes and an output tape.

Some people may argue that if the numbers on the input tapes are not computable how can you give the numbers in the input tapes. In our opinion it is not the responsibility of the function to give the input. The responsibility of the function is when you give the input it computes the output for you.

§ 3. Real functions computed by stable Turing machines

Because the usage of Turing machines in computing real functions is totally new we give the details of two Turing machines to compute the real addition and multiplication functions. The two Turing machines are computing real numbers between 0 and 1. They can easily be modified to compute any pair of real numbers.

Theorem 1 The addition of two real numbers between 0 and 1 is computable by a stable Turing machine.

Proof We use a multiple-track Turing machine to compute the function and binary infinite digits to represent the real numbers. The input numbers are printed on two input tapes in the form of $a_0 \cdot a_1a_2\cdots$ on the first tape and $c_0 \cdot c_1c_2\cdots$ on the second tape. Because the numbers are between 0 and 1, $a_0=c_0=0$. For the other input symbols $a_i, c_i=0$ or 1. Before the machine starts the output tape looks like $B \cdot BB\cdots$. The machine M has 3 scanning heads which move one cell to the right or one cell to the left simultaneously.

Let $M=(Q,\Sigma,\Gamma,\delta,q_0,B)$ be the Turing machine, where $Q=\{q_0, q_1, q_2, q_3\}$, $\Sigma=\{0,1,\cdot\}$ and $\Gamma=\{0,1,\cdot,X,B\}$. The definition of δ operator is divided into 5 groups. Each move is in the form of $\delta(q_i,s_1,s_2,s_3)=(q_j,t_1,t_2,t_3,A)$, where s_1,s_2,s_3 are the read in symbols and t_1,t_2,t_3 are the write-out symbols on the 3 tapes and $A=R$ or L showing the machine will go right or left for the next move.

(i) Starting moves:

$\delta(q_0,0,0,B)=(q_1,0,0,0,R)$; $\delta(q_1,\cdot,\cdot,\cdot)=(q_1,\cdot,\cdot,\cdot,R)$;

(ii) Adding moves: $\delta(q_1,0,0,B)=(q_1,0,0,0,R)$; $\delta(q_1,0,1,B)=(q_1,0,1,1,R)$; $\delta(q_1,1,0,B)=(q_1,1,0,1,R)$; $\delta(q_1,1,1,B)=(q_2,1,1,X,L)$;

(iii) Carrying moves:

$\delta(q_2,a,c,0)=(q_3,a,c,1,R)$; $\delta(q_2,a,c,1)=(q_2,a,c,0,L)$;

(iv) Recovering moves:

$\delta(q_3,a,c,0)=(q_3,a,c,0,R)$; $\delta(q_3,a,c,X)=(q_1,a,c,0,R)$;

(v) Special moves: In the case that the carrying needs to step over the decimal point:

$\delta(q_2,\cdot,\cdot,\cdot)=(q_2,\cdot,\cdot,\cdot,L)$; $\delta(q_3,\cdot,\cdot,\cdot)=(q_3,\cdot,\cdot,\cdot,R)$.

We will show the following:

1. Every digit on the third tape is stable after finitely many moves.

2. The result on the third tape is the sum of the real numbers on the first and second tapes.

At the starting point the digit before the decimal point is 0. Let the current symbols on the third tape be $0 \cdot d_1\cdots d_k01\cdots1$. There are m 1's at the tail. If the $(k+m+2)$th digit produces a carry then the machine puts a 0 on the $(k+m+2)$th digit, changes the m 1's to 0 and changes the $(k+1)th$ *digit to* 1. After that all the front $(k+m+1)$th digits will not be changed again because the later computations can change at most up to the 0 at the $(k+m+2)$th digit. It is possible the digit before the decimal point may be changed to 1. Every time a 0 is changed to 1 it will produce a new 0 which will prevent the changes of the decimals on the left of the 0. Therefore every digit on the third tape is stable after finitely many moves. For the correctness of the computation let $a=a_0 \cdot a_1a_2\cdots a_n\cdots$, $c=c_0 \cdot c_1c_2\cdots c_n\cdots$ and $d=d_0 \cdot d_1d_2\cdots d_n\cdots$ be the 3 real numbers on 3 tapes. Changing all digits after the nth digit of the 3 numbers we obtain $a^{(n)}=a_0 \cdot a_1a_2\cdots a_n00\cdots$, $c^{(n)}=c_0 \cdot c_1c_2\cdots c_n00\cdots$ and $d^{(n)}=d_0 \cdot d_1d_2\cdots d_n00\cdots$ We see that

$$\lim_{n\to\infty} a^{(n)}=a, \lim_{n\to\infty} c^{(n)}=c, \lim_{n\to\infty} d^{(n)}=d, \lim_{n\to\infty}(a^{(n)}+c^{(n)}-d^{(n)})=0.$$

Therefore we have $a+c=d$.

Theorem 2 The multiplication of two real numbers between 0 and 1 is computable by a stable Turing machine.

Proof We use a multi-tape Turing machine to compute the multiplication function. Let $M=(Q,\Sigma,\Gamma,\delta,q_0,B)$ be the Turing machine, where $Q=\{q_0,q_2,\cdots,q_9\}$, $\Sigma=\{0,1,\cdot\}$ and $\Gamma=\{0,1,\cdot,X,Y,B\}$. The input numbers are printed on two input tapes in the form of $a_0 \cdot a_1a_2\cdots$ on the first tape and $c_0 \cdot c_1c_2\cdots$ on the second tape. The output tape is given as $B \cdot BB\cdots$ at the beginning. Because the numbers are between 0 and 1, $a_0=c_0=0$. The machine M has 3 scanning head which move one cell to the left or to the right, or keep stationary independently. Although the heads move freely we still arrange the three tapes matching the decimals on the same vertical line and the heads start at the first digit before decimal point. We say the first head, the second head and the third head to represent them. We use the following 4 moves to start the machine which passes the decimal points and prints the product of the first digits (after decimal points, hereinafter) of the input tapes to the second digits on the output tape.

q_0	0 0 B	=	q_0	0 0 0	R R R

q_0	• • •	=	q_1	• • •	R R R

q_1	a c B	=	q_2	a c 0	S S R

q_2	a c B	=	q_3	a c ac	R R R

Then the machine starts a new round of the operation. In general suppose that the machine finishes the $(n-1)$-th round of the operation, in the n-th round, the machine adds all products of $a_ib_j\ (i+j=n)$ to the n-th digit of the output tape. This is done in the following moves. To represent more moves in the same table we use a, c and d to represent 0 or 1 on the input and output tapes, f is a function which maps $X=f(0)$, $Y=f(1)$, and Z represents X or Y.

q_3	a c B	=	q_4	$f(a)$ $f(c)$ 0	S L S

q_4	a c d	=	q_4	a c d	S L S

q_4	a • d	=	q_5	0 • d	S R S

In the above 3 moves the machine puts on the road markers X, Y and the second head goes left to search for the decimal point and turns back but the other heads keep stationary. Then the machine starts adding the product of two scanning cells of the input tapes to the cell on the output tape. The first head goes left, the second head goes right and the third head keeps stationary.

	Z			Z	L
q_5	c	$=$	q_e	c	R
	d			$(d+ac)\ mod\ 2, (a=f^{-1}(Z))$	S

In this move q_e depends on ac and d. If $ac=0$ or $d=0$, then the addition does not produce a carry. Therefore $q_e=q_5$. The machine repeats to work in q_5. If $ac=1$ and $d=1$. then the addition produces a carry and $q_e=q_6$. In the carrying procedure the first and second heads keep stationary without any changes and the third head works on the carry. We only write down the moves of the third head and skip the other two rows.

Carrying moves:

$\delta(q_6, 0)=(q_7, X, L)$; $\delta(q_7, 1)=(q_7, 0, L)$; $\delta(q_7, 0)=(q_8, 1, R)$;

Recovering moves: $\delta(q_8, d)=(q_8, d, R)$; $\delta(q_8, X)=(q_5, 0, S)$; (When the second head is not scanning X or Y. The other case will be given latter.)

After the carrying procedure the machine resumes to work on q_5 again to add the next term in $a_{n-1}c_1+a_{n-2}c_2+\cdots+a_1c_{n-1}\rightarrow d_n$.

When the second head reaches a road marker X or Y the adding procedure is working on the last term $\cdots+a_1c_{n-1}\rightarrow d_n$. So we need the moves:

If $ac=0$ or $d=0$, then the addition does not produce a carry. The move is:

	a			a	R
q_5	Z	$=$	q_9	Z	S
	d			$(d+ac)\ mod\ 2, (c=f^{-1}(Z))$	S

If $ac=1$ and $d=1$, then the addition produces a carry.

	a			a	R
q_5	Z	$=$	q_6	Z	S
	d			0	S

The machine should do the work in q_6 again. When the machine comes back in q_8. The moves are:

	a			a	R
q_8	Z	$=$	q_9	Z	S
	X			0	S

	Z_1			Z_1	R
q_8	Z_2	$=$	q_9	Z_2	S
	X			0	S

In q_9 the first head should go back to the point that they are on the same vertical line with the second head. The moves are:

	a			a	R
q_9	Z	$=$	q_9	Z	S
	d			d	S

	Z_1			$f^{-1}(Z_1)$	R
q_9	Z_2	$=$	q_3	$f^{-1}(Z_2)$	R
	d			d	R

Now the machine goes back to work on a new round of operations in q_3. In this way the machine is working continuously. We also show that the computation is stable and correct.

1. Every digit on the third tape is stable after finitely many moves.

2. The result on the third tape is the product of the real numbers on the first and second tapes.

We define an order relation in the set $D_n=\{0\cdot d_1\cdots d_n \mid d_i=0 \text{ or } 1\}$. Let $d=0\cdot d_1\cdots d_n$, and $d'=0\cdot d'_1\cdots d'_n$. The two rational numbers $d<d'$ if and only if there is an i, such that $d_1=d'_1, d'_2, \cdots, d_{i-1}=d'_{i-1}$, and $d_i<d'_i$. Let $e=0\cdot d_1d_2\cdots d_n\cdots$ and $e'=0\cdot d'_1d'_2\cdots d'_n\cdots$ be the situations appearing in the computation, where the prefixes of e, e' are d, d' respectively, and $d<d'$, then the appearance of e is before the appearance of e' and it never turns back. Let $d^{(1)}, d^{(2)}, \cdots, d^{(m)}, \cdots$ $(d^{(i)}\in D)$ be all appearances as prefixes in the computation order, we have $d^{(1)}\leqslant d^{(2)}\leqslant\cdots\leqslant$

$d^{(m)} \leqslant \cdots D^n$ is a finite set, therefore the infinite list must be all equal after a suitable m. This proves that the front n digits on the third tape is stable after finitely many moves. Since the n can be any large, hence every digit on the third tape is stable after finitely many moves. For the correctness of the computation, let $a = a_o \cdot a_1 a_2 \cdots a_n \cdots$, $c = c_0 \cdot c_1 c_2 \cdots c_n \cdots$ be the 2 real numbers on the input tapes. Changing all digits after the nth digit of the 2 numbers we obtain $a^{(n)} = a_0 \cdot a_1 a_2 \cdots a_n 00 \cdots$, $c^{(n)} = c_0 \cdot c_1 c_2 \cdots c_n 00 \cdots$ It is easy to see that

$$\lim_{n \to \infty} a^{(n)} = a, \lim_{n \to \infty} c^{(n)} = c.$$

In the computation the machine does the following:

1st round adding $a_1 c_1$ to the 2nd digit of the output tape;

2nd round adding $a_1 c_2 + a_2 c_1$ to the 3rd digit of the output tape;

3rd round adding $a_1 c_3 + a_2 c_2 + a_3 c_1$ to the 4th digit of the output tape …

nth round adding $a_1 c_n + a_2 c_{n-1} + \cdots + a_n c_1$ to the $(n+1)$th digit of the output tape…

Let the sum up to the nth round be $\sum_n$, we have

$$\sum_n \leqslant a^{(n)} \times c^{(n)} \leqslant \sum_{2n}.$$

$\lim_{n \to \infty} \sum_n = \lim_{n \to \infty} \sum_{2n} = d$. Therefore $\lim_{n \to \infty} (a^{(n)} \times {}^{(n)}) = d$. Hence $a \times c = d$.

§ 4. predicates "$>$" and "$=$" are computable

As we discussed in 2 the predicates "$>$" and "$=$" are not computable in the ε definition, but in our definition, they are computable, we will give the frame-work of the Turing machine for computing the two predicates.

For computing the predicate $x > 0$ the Turing machine has an input tape and an output tape. The machine keeps reading the input and moves right and prints 0's on the output tape simultaneously. Whenever it finds a symbol 1 it goes back and prints a 1 on the first cell of the out tape and passes the decimal point and prints infinitely many 0's on the output tape.

Similarly, for computing the predicate $x = y$ the Turing machine has two input tapes and an output tape. The machine keeps reading the two input symbols in the same digit position. If they are equal the machine prints a 1 in the first cell before the decimal point of the output tape. and moves right and prints 0's on the output tape. Whenever it finds two unequal symbols it goes back and prints a 0 on the first cell of the out tape and passes the decimal point, and prints infinitely many 0's on the output tape.

References

[1] Aberth O. Computable analysis. New York: McGraw-Hill, 1980.

[2] Aberth O. Analysis in the Computable Number Field. Journal of the Association for Computing Machinery, 1968, 15(2): 276—299.

[3] Bishop E, Bridges S D. Constructive Analysis. Berlin, Heidelberg: Springer-Verlag, 1985.

[4] Grzegorezyk A. On the definitions of computable real functions. Fund. Math., 1957, 44: 61—71.

[5] Peter Hertling. Banach-Mazur Computable Functions on Metric Spaces. CCA, 2000: 69—81.

[6] Hopcroft E J, Ullman D J. Introduction to Automata Theory Languages and Computation. Reading: Addison-Wesly. 1979.

[7] Klaua D. Konstruktive Analysis. Berlin: VEB Deutscher Verlag der Wiss, 1961.

[8] Ko K-I. Complexity Theory of Real Functions. Boston: Birkhäuser, 1991.

[9] Kushner A B. Lectures on Constructive Mathematical Analysis. Rhode Island; volume 60 of Translations of Mathematical Monographs. Amer. Math. Soc., Providence, 1984.

[10] Markov A A. On the continuity of constructive functions. Russian, Nauk: Uspekhi Mat., 1954, 9(3(61)): 226—230.

[11] Marvin Minsky. Computation: Finite and Infinite Machines. Prentice-Hall Inc Englewood Cliffs, NJ No ISBN Library of Congress Card Catalog, 1967, (42): 67—123.

[12] Moschovakis N Y. Recursive Metric Spaces. Fund. Math., 1964, 15: 215—238.

[13] Pour-El B M, Richards I J. Computability in Analysis and Physics. Berlin: Springer-Verlag, Heidelberg, 1989.

[14] Rogers H Jr. Theory of Recursive Functions and Effective Computability.

New York: McGrow-Hill Book Company, 1967.

[15] Shanin A N. Constructive real numbers and constructive function spaces. Math. Monographs Amer. Math. Soc. 1968, 21.

[16] Tseitin S G. Algorithmic operators in constructive metric spaces. Amer. Math. Soc. Transl. , 1967, 64(2): 1—80; Russian: Trudy Mat. Inst. Steklov, 67(1062).

[17] Turing A. On computable numbers, with an application to the "Entscheidungs problem". Proceedings of the London Mathematical Society, 1936, 42 (2): 230—265.

[18] Weihrauch K. Computable analysis: an introduction. New York: Springer-Verlag, Inc. Secaucus N J, 2000.

[19] Weihrauch K. Computability. Berlin: Springer-Veflag, 1987.

图灵机计算实数函数的稳定性

摘要 定义在全体实数上的可计算函数是一个很重要的概念，在这以前定义可计算的实数函数有两个途径. 第一个途径是首先要定义可计算实数的指标. 想要确定实数函数 $y=f(x)$ 是不是可以计算就要看是否存在一个自然数的(部分)递归函数将可计算实数 x 的指标对应到可计算实数 y 的指标. 这样一来对实数函数的研究依赖于对自然数函数的研究. 第二个定义可计算的实数函数的途径是以逼近为基础的. 一个实数函数是可以计算的，如果它既是序列可计算的同时也是一致连续的. 用这个途径来定义可计算实数函数使用的条件过强，以至于很多有用的实数函数成为不可计算的实数函数. 例如"$<$"和"$=$"的命题函数就是不可以计算的，因为它们是不连续的命题函数，本文讨论了图灵机的稳定性并且给出了一个基于稳定图灵机的可计算实数函数的定义，我们的定义不需要用到自然数的(部分)递归函数. 根据我们的定义很多常用实数函数特别是一些不连续的常用实数函数都是可以计算的. 用我们的定义来讨论可计算实数函数的性质比原来的定义要方便得多.

关键词 可计算实效函数，稳定性，图灵机.

前沿科学,2009,3(12):26～32

用 ω-图灵机计算实数函数①

Computing Functions on Real Numbers with Stable ω-Turing Machines

Abstract The computable function over the set of all read numbers is a very important concept. There are two approaches in defining computable real functions. The first approach is to define the indexes of computable real numbers first. An index of a computable real numbers is the Gödel number of a Turing machine which computes the real number. A function $y=f(x)$ defined on all computable real numbers is computable if and only if there is a (partial) recursive function over the indexes which maps the index of x to the index of y. In this way the study of functions on real numbers relies on the property of functions on natural numbers. The second approach of defining computable real functions is based on approximations. A real function is computable if and only if it is both sequentially computable and effectively uniformly continuous. The condition is too strong preventing a lot of very useful functions to be computable. According to the definition, predicates "$<$" and "$=$" are not computable because their characteristic functions are discontinuous functions.

In this paper we discuss the stability of ω-Turing machines and give a definition for ω-computable functions defined on the set of all infinite

① 收稿日期:2009-09-03;修订日期:2009-11-13.

strings of a finite alphabet. based on stable ω-Turing machines. We also prove that the composite functions of ω-computable functions are still ω-computable. Our definition does not use Gödel numbers and recursive functions over natural numbers. According to our definition many commonly used functions especially some useful discontinuous functions on real numbers are ω-computable. The conditions for a function to be ω-computable are easier to be verified. Under this definition the halting predicate of normal Turing machines is ω-computable.

Keywords stability of ω-Turing machines, ω-computable functions.

§ 1. Introduction

Computable functions over the set of all real numbers have been studied for a long period of time. Researchers designed many ways to discuss the computable real functions. There are two approaches in defining computable real functions. The first approach is to compute the indexes of computable real numbers first. Turing in [18] used infinite decimal sequences to define computable numbers. A real number is computable if its digit sequence can be produced by an algorithm or a Turing machine. The algorithm takes an integer n as input and produces the n-th digit of the real number's decimal expansion as output, An index of a real number is an integer which represents the Gödel number of a machine that computes the real number. A computable function $y=f(x)$ defined on all computable real numbers only if there is a (partial) recursive function over the indexes which map the index of x to the index of y. This approach, with some variations, has been considered in the Russian school of constructive analysis, represented by Markov [11], Shanin [16], Tseitin[17], Kushner[9]. It has also been considered by Moschovakis[13], Aberth[1]and by many others. Closely related method is the constructive approach of analysis presented by Bishop and Bridges[3]. Since the computable real numbers are generated by a Turing machine or a program or any kinds of algorithms, the definition is very difficult to use. The definition is also incomplete because all the incomputable real numbers do not join the operation which is unfair. For example: a very simple function which adds an incomputable real number to the set of all computable real numbers is obvious computable. Actually we know how to add two real numbers regardless of whether they are computable or not. The addition should be defined for the entire real space.

The second approach of defining computable real functions is based on approximations. A function $f:\mathbf{R}\rightarrow\mathbf{R}$ is sequentially computable if, for every sequence $(x_i: i=1,2,\cdots)$ of computable real numbers, the sequence $(f(x_i): i=1,2,\cdots)$ is also computable. A function $f:\mathbf{R}\rightarrow\mathbf{R}$ is effectively uniformly continuous if there exists a recursive function $d:\mathbf{N}\rightarrow\mathbf{N}$ such that, if $|x-y|<\frac{1}{d(n)}$, then $|f(x)-f(y)|<\frac{1}{n}$. A real function is computable if it is both sequentially computable and effectively uniformly continuous. These definitions can be generalized to functions of more than one variable or functions only defined on a subset of $\mathbf{R}^n$. A suitable generalization of the first definition is: Let $\mathbf{D}$ be a subset of $\mathbf{R}^n$. A function $\mathbf{D}\rightarrow\mathbf{R}$ is sequentially computable if, for every n-tuple of $(x_{1i}, x_{2i}, \cdots, x_{ni}): i=1,2,\cdots$ of computable sequences of real numbers such that $(\forall i)\ (x_{1i}, x_{2i}, \cdots, x_{ni}): i=1,2,\cdots \in \mathbf{D}$, the sequence $f(x_{1i}, x_{2i}, \cdots, x_{ni}): i=1,2,\cdots$ is also computable. This kind of definitions is considered by Grzegorczyk [4], Klaua [7], Pour-El and Richards [14], Weihrauch [20], Ko [8], and by many others. Moschovakis in [13] discussed computable functions on metric spaces. Rogers in [15] defined the effective computability. Weihrauch in [19] developed computable analysis. Hertling in [5] compared two types of computable real functions.

There are still a lot of definitions considered by different researchers: A real number is computable if and only if there is a computable Dedekind cut D converging to it. Minsky in [12] chapter §9 defined "the Computable real numbers". Aberth in [2] describes the development of the calculus over the computable number field. Weihrauch [19] §1.3.2 introduces the definition by nested sequences of intervals converging to the singleton real. Other representations are discussed in 4. 1.

In the second approach the definition of computable real functions has two conditions. 1. It is sequentially computable. 2. It is effectively uniformly continuous. In the first condition not only the elements of the two sequences are computable but the two sequences are also computably given. In the second condition the functions are required to be effectively uniformly continuous. The condition is too strong preventing a lot of very useful functions to be computable. According to the above definitions the order and equality relations on the computable numbers is not computable because they are represented by discontinuous functions. We are not satisfied for the relations "$<$" and "$=$" being incomputable.

In the two approaches the definitions of computable real functions are using Gödel numbers, limits, sequences, Dedkind cuts and some other tools. So therefore the definitions are very difficult to use in studying the properties of the functions. For example, it is very difficult to see that the addition function satisfies associative law according to the definition. The implementation of the functions is not even considered. We would like to ask the following questions:

(1) Is it possible to define the computable real functions without using Gödel numbers and the recursive functions over natural numbers?

(2) Is it possible to define the computable real functions, so that they include normally used functions especially some useful discontinuous functions?

(3) Is it possible to define computable function on the set of all real numbers regardless the numbers being computable or not?

In paper [10] we discussed the stability of Turing machines in computing real functions and give a more general definition for the computable real function based on the stable Turing machine. In this paper since the usage of Turing machines is different from the old definitions we call them ω-Turing machines and the functions computed are called ω-computable functions. Our definition does not use the recursive functions over natural numbers. According to our definition normally used functions especially some useful discontinuous functions are computable. Our definition of computable functions is easier to verify comparing the old definition which uses the recursive function over Gödel numbers. In a lot of the quoted references the computable functions are defined in a space with certain kind of measure because they use limits. Our definition didn't emphasize the input symbols. Actually we can take any finite set of symbols to form the infinite strings and it is not necessary to define any kind of measure among them. We discuss the computable functions over the set of all infinite strings on a finite set of alphabet. The real numbers are special cases of this kind of strings.

§ 2. The definition and stability of the ω-Turing machine

The ω-Turing machines we used are basically the same kind of machines defined by Hopcroft and Ullman in [6].

Definition 2. 1 An ω-Turing machine is a sextuplets M=$(Q, \Sigma, \Gamma, \delta, q_0, B)$. Q is a set of finitely many states. Σ is a set of finitely many input and output symbols. Γ is a set of finitely many allowable tape symbols. B is the blank symbol which is a symbol of Γ. δ is a set of finitely many instructions which tells the machine how to move. Formally δ is a partial function defined as $Q\times\Gamma\rightarrow Q\times\Gamma\times\{\mathrm{L},\mathrm{R}\}$. q_0 is the starting state.

There are some differences between ω-Turing machine and the normal Turing machine.

(i) An ω-Turing machine does not have a final state but a normal Turing machine has. An ω-Turing machine never stops.

(ii) Since the input and output sequences are written in infinitely many cells we must allow the ω-Turing machine to run without halting. If the machine provides a correct resulting symbol for every fixed cell on the output tape in finitely many steps and never changes it again thereafter, the computation is successful. On the other hand, if the output head stops or it goes back and forth among finitely many cells, the computation is failed and the value of the function is undefined.

(iii) Because the input and output tapes are with infinite lengths. The ω-Turing machine can be used to compute the real functions with infinitely many digits. It is possible to compute the accurate value for all digits of the output number which are computed only in approximations with other tools.

(iv) Since an ω-Turing machine works for infinitely many steps it is not considered as a finite method in computing functions. Anyway infinite strings and real numbers are not finite objects either. It is more natural to study the real functions with infinite methods. Although the running time of an ω-Turing machine is infinite but its controlling box has only finitely many instructions.

(v) The computing power of an ω-Turing machine with one tape is equivalent to an ω-Turing machine with multi-head multi-tape and even multi-head on one tape. The proof is similar to the proof of Theorem 7. 2 in [6].

Definition 2. 2 An ω-Turing machine M with k input tapes, a working tapes and an output tape is called stable if every cell of the output tape is visited by M for at most finitely many steps and leaves an output symbol in each cell at the last visit.

Definition 2. 3 A function of real numbers with arity k is ω-computable if it can be computed by a stable ω-Turing machine with k input tapes a working tapes and an output tape.

Definition 2. 4 The ω-computable functions of arity 0 are called ω-computable numbers.

§ 3. Some ω-computable real functions

It is not difficult to design ω-Turing machines to compute real addition, subtraction, multiplication, and none-zerodivision functions. In [10] we give the details of ω-Turing machines computing real addition, multiplication the predicates "<" and "=".

We give two functions to show some new ω-computable functions on the set of infinite strings. One of them is to manipulate the strings the other is the halting problem.

Theorem 3.1 Let Σ be a finite set of input symbols. Let $\boldsymbol{a}=(a_{11},\cdots,a_{1k},\cdots,a_{n1},\cdots,a_{nk},\cdots)$ be an infinite string on Σ. $\pi:\Sigma\to\Sigma$ be a function on finite input symbols, then the function $f(a)=(a_{1\pi(1)},a_{1\pi(2)},\cdots,a_{1\pi(k)},\cdots,a_{n\pi(1)},a_{n\pi(2)},\cdots,a_{n\pi(k)},\cdots)$, is a computable function on infinite strings of Σ.

Proof The ω-Turing machine computing f is not difficult to design.

Some people used this kind of structures to construct real functions in the interval (0,1) which are useful in some problems. We now apply the construction to the real numbers. Let $\Sigma=\{0,1,\cdots,9\}$ be the set of input symbols. If we change two consecutive digits of the decimal $a=0.1000\cdots00\cdots$ we obtain $b=f(a)=0.0100\cdots00\cdots$ But as a real number a it has another decimal expression $b=0.0999\cdots99\cdots$ and the same function would change b to $c=f(b)=0.9099\cdots99\cdots=0.9100\cdots00\cdots$ Therefore the value of function is not uniquely determined. Therefore we have to use only one form of the decimal as input. The ω-Turing machine works like the following: The input must have a 1 digit. Starting from the first appearance of 10 in the decimal $0.\cdots10\cdots$ the machine guesses that there might all be 0's after that and goes back to change the previous 1 to 0 and works forward as if the decimal becomes $0.\cdots099\cdots$ The operation is to switch every pair of consecutive digits. If the input head sees another 1 in the later part of the input tape the machine goes back to restore the previous section and does the operation again, In this way the machine obtains only one $f(a)$ for every a.

Our definition of the ω-Turing machines actually has more computing power. We will show this in the next theorem.

Theorem 3.2 The halting predicate of the Turing machines is ω-computable.

Proof The ω-Turing machine M has three tapes. The first tape is for writing the input Turing machine m' and the input x of machine m'

which occupies only a finite area. This can be done by certain kind of coding of m'. The second tape is the working tape for the machine m'. The third tape is the output for the ω-Turing machine M. The machine M simulates the machine m' working on input x and prints 0's of decimal 0. 00…0…on the output tape in one 0 per step of m'. If the machine m' halts on input x, the machine M goes back to the beginning to print 1. 0…0…

§ 4. Composite functions on ω-computable functions

In this section we prove that the composite function of ω-computable functions is also ω-computable. A similar theorem for normal Turing machine computable functions is not difficult to prove. The situation for ω-Turing machine computable functions is different but we still have a similar result.

Theorem 4. 1 Suppose that $z=f(y_1, y_2, \cdots, y_n)$ and $y_i=g_i(x_1, x_2, \cdots, x_m), i=1,2,\cdots,n$ are ω-computable functions, then the composite function $z=h(x_1, x_2, \cdots, x_m)=f(g_1(x_1, x_2, \cdots, x_m), \cdots, g_n(x_1, x_2, \cdots, x_m))$ is also ω-computable.

Proof Let ω-Turing machines $M, M_i, i=1,2,\cdots,n$ compute functions $f, g_i, i=1,2,\cdots,n$ respectively and machine $\overline{M}$ computes the composite function h. We use M_{1-n} to represent machines $\{M_i, i=1,2,\cdots, n\}$. The output tapes of M_{1-n} are the input tapes of $\overline{M}$. The main difficulty is that at a time when machine M reads the input from the output tapes of M_{1-n} to produce its own output, the result may not be usable because some of the machines in M_{1-n} may come back latter to change their output and we do not know when would the output of M_{1-n} are stabilized. To overcome this difficulty we must setup a checking mechanism whenever an output head of M_i is going backward, the computation of M must go back to a certain point allowing machine M_i to reproduce its output. The ω-Turing machine $\overline{M}$ works as the following.

(i) $\overline{M}$ has m input tapes for printing the input strings $x_1, x_2, \cdots, x_m$.

(ii) All input tapes and working tapes of M_{1-n} are working tapes of $\overline{M}$.

(iii) Each M_i has a scanning head on each input tape of $\overline{M}$ which copies the input of $\overline{M}$ to its own input tape at the first visit of that cell and after that it works only on its own input cells.

(iv) The output tapes of M_{1-n} are the input tapes of M. There are two heads on each of these tapes. One of them is called writing head w_i whose duty is to print the output of M_i. The other is called reading head r_i whose duty is to read the input for M.

(v) $\overline{M}$ has a working tape T to keep track of all writing and reading heads and to coordinate their activities according to the following rules.

(v. 1) Suppose that we arrange all n output tapes of M_{1-n} matching together. At the beginning all writing and reading heads are on the same vertical line at the first cell of each tape.

(v. 2) At the beginning, all working tapes are filled in with blank symbols. There is an input copying procedure whenever one of the M_{1-n}'s needs a new input all input heads of M_{1-n}'s must copy the input of $\overline{M}$ to their own input tapes and then return to their own operations. Therefore the lengths of none-blank symbols on all input tapes of M_{1-n}'s are the same.

(v. 3) On all output tapes of M_{1-n} all reading head $r_1, r_2, \cdots, r_n$ must not go to the right of any writing head $w_1, w_2, \cdots, w_n$. Since all $2n$ such heads are having records on working tape T. It is possible to arrange.

(v. 4) M_{1-n} has priority of working to produce output. The working order according to the position and the subscript of their writing heads. The leftmost writing head works first. If there are more than one leftmost writing heads, then the one with the smallest subscript works first.

(v. 5) There are two conditions for M to start working.

(v. 5. 1) M_{1-n} produce enough output and the distance between the vertical lines of rightmost reading head and the leftmost writing head has at least two cells.

(v. 5. 2) M must recognize the symbols on all of its reading heads. It is possible the output of machines M_{1-n} has some symbols which machine M doesn't recognize and refuses to work. In this case we can arrange M to keep waiting until M_{1-n} come back to give recognized output. For doing this we arrange that all working symbols of M_{1-n} and M in $\Gamma_i-(\Sigma\cup\{B\})$, $i=1,2,\cdots,n$ and $\Gamma-(\Sigma\cup\{B\})$ are different so that M keeps waiting when it meets any symbols in $\Gamma_i-(\Sigma\cup\{B\})$, $i=1,2,\cdots,n$.

(v. 5. 3) When the above two conditions are fulfilled, M works continuously until the distance between the vertical lines of rightmost reading head and the leftmost writing head has only one cell.

(vi) Following (v. 5. 3) according also (v. 4) a machine M_i starts to work. After M_i working for one step there are three possibilities. The writing head w_i may go to the right, left or keep stationary.

(vi. 1) w_i goes to the right. In this case if there are still some other head w_j on the previous vertical line then M_j works. If there is no such a head w_j on the previous vertical line $\overline{M}$ works as in (v. 5) because the distance between the vertical lines of rightmost reading head and the leftmost writing head has two cells.

(vi. 2) w_i keeps stationary. In this case M_i works for one more step.

(vi. 3) w_i goes to the left. In this case machine M goes to a reverse procedure that is described in (vii).

(vii) $\overline{M}$ has a working tape S which stores all working records of M. When in the situation of (vi. 3) to leave rooms for M_i to change its output, M does the reverse operation δ^{-1} according to the records on S until the rightmost reading head withdraws back one cell left to the leftmost writing head. At the same time the records on S is also changed to fit in the situation. To prove that $\overline{M}$ can give the correct output for $z=h(x_1,x_2,\cdots,x_m)=f(g_1(x_1,x_2,\cdots,x_m),\cdots,g_n(x_1,x_2,\cdots,x_m))$ we use the stability of M and M_{1-n}. Let n_0 be a fixed natural number. According to the stability of M_{1-n} there must be a finite number of steps for machines M_{1-n} to produce correct output on all n output tapes up to n_0 digits. Between two steps of M_{1-n} it is possible machine M may run for finitely many steps. Therefore after finitely many steps machine M can work on correct input with n_0 digits. Since n_0 can be any number therefore M can work on correct input with any length. The correct output thus can be given finally. ■

Referrences

[1] Aberth O. Comtutable analysis. McGraw-Hill, New York, 1980.

[2] Aberth O. Analysis in the computable number fielf. Journal of the Association for Computing Machinery (JACM), 1968, 15(2): 276—299.

[3] Bishop E and Bridges D S. Constructive Analysis. Springer-Verlag, Berlin, Heidelberg, 1985.

[4] Grzegorczyk A. On the definitions of computable real functions. Fund. Math., 1957, 44: 61—71.

[5] Peter Hertling: Banach-Mazur Computable Functions on Metric Spaces. CCA 2000: 69—81.

[6] Hopcroft J E and Ullman J D. Introduction to Automata theory Languages and Computation. Reading: Addison-Westy, 1979.

[7] Klaua D. Konstruktive Analysis. VEB Deutscher Verlag der Wiss., Berlin, 1961.

[8] Ko K-I. Complexity Theory of Real Functions, Birkh" {a}user. Boston, 1991.

[9] Kushner B A. Lectures on Constructive Mathematical Analysis. volume 60 of Translations of Mathematical Monographs. American Math. Soc., Providence, Rhode Island, 1984.

[10] Luo L. The stability of Turing machine in computing real functions. Journal of Math. Study, 2009, 42(2): 126—131.

[11] Markov A A. On the continuity of constructive functions. Uspekhi Mat. Nauk, 1954, 9(3(61)): 226—230 (Russian).

[12] Marvin Minsky 1967, Computation: Finite and Infinite Machines. Prentice-Hall, Inc. Englewood Cliffs, NJ. No ISBN. Library of Congress Card Catalog No. 67—12342.

[13] Moschovakis Y N. Recursive Metric Spaces. Fund. Math., 1964, 15: 215—238.

[14] Pour-El M B and Richards J I. Computability in Analysis and Physics. Springer-Verlag, Berlin, Heidelberg, 1989.

[15] Rogers Jr H. Theory of Kecursive Functions and Effective Computability. McGrow-Hill Book Company, New York, 1967.

[16] Shanin N A. Constructive real numbers and constructive function spaces.

volume 21 of Translations of Math. Monographs. Amer. Math. Soc. Providence, R. I. ,1968.

[17] Tseitin G S. Algorithmic operators in constructive metric spaces. Amer. Math. Soc. Transl. , 1967 64(2): 1－80, Trudy Mat. Inst. Steklov. 67 (1062),(Russian).

[18] Turing A. On computable numbers, with an application to the "Entscheidungs problem". Proceedings of the London Mathematical Society, 1936, 42 (2):230－265.

[19] Weihrauch K. Computable analysis: an introduction. Springer-Verlag New York, Inc. , Secaucus, NJ, 2000.

[20] Weihrauch K. Computability. Springer-Verlag, Berlin, 1987.

用 ω-图灵机计算实数函数

摘要 实数的可计算函数是一个非常重要的概念，定义可计算实数函数有两个途径，第一个途径是先定义可计算实数的指标. 一个可计算实数的指标是一个计算该实数的 Turing 机的 Gödel 数. 一个定义在全体可计算实数上的函数 $y=f(x)$是可计算的，当且仅当存在一个在全体可计算实数的指标上的(部分)递归函数将 x 的指标对应到 y 的指标，这样对实数函数的研究依赖于自然函数的性质. 第二个定义可计算实数的途径是基于逼近. 一个实数函数是可计算的，当且仅当它既是序列可计算的也是有效地一致连续的，这个条件太强使得很多非常有用的实数函数不能成为可计算函数. 根据这个定义"＜"和"＝"都是不可计算的，因为它们的特征函数是不连续的.

在这篇文章里我们讨论 ω-Turing 机的稳定性并且定义了在一个有限字母表的全体无限序列上的 ω-可计算函数，我们也证明了 ω-可计算函数的复合函数也是 ω-可计算的. 我们的定义没有用到 Gödel 数或者递归函数. 根据我们的定义很多常用函数特别是一些有用的不连续函数是 ω-可计算函数. 一个函数是否 ω-可计算的验证较为容易，根据我们的定义普通 Turing 机的停机命题是 ω-可计算的.

关键词 ω-Turing 机的稳定性，ω-可计算函数.

Acta Mathmatica Sinica, English Series, 2002, 18(3): 531～538

非标准数论的新定理①

The Generalization of the Chinese Remainder Theorem

Abstract We generalize the Chinese Remainder Theorem, use it to study number theory models, compare and analyse several number theory theorems in non-standard number theory models.

Keywords **Non-standard number, THN model, Chinese Remainder Theorem.**

§ 0. Introduction

Vaught's conjecture concerning the number of countable models is one of the central problems in the research of model theory. It states that the number of non-isomorphic countable models of a complete countable first-order theory is cither $\leqslant\omega$ or $=2^{\omega}$. During the last 40 years, several mathematicians studied this conjecture and obtained partial results.

Morley defines $\mathcal{L}_{\alpha}$ through extending types step by step in [1], which marks a big progress in this direction. Therefore, the number and characteristics of types of a countable theory are important concepts when we discuss this theory. For this reason, our group launches a series

① Received August 18, 1999, Accepted February 5, 2001
与于丽荣合作.

of research for different mathematicial theories. This paper is a part of the unified project (see Luo [2]). Throughout the paper $\mathcal{L}=\langle S,+,\cdot,0\rangle$, THN is the complete theory of the standard natural numbers on language $\mathcal{L}$. Because THN has 2^{ω} types and a countable model can realize at most countably many types, so as far as we are concerned, Vaught's conjecture has been solved for THN, but this will not affect our search for its types, and the property of types still plays an important role in studying THN models. Through constructing definable types Schmerl in [3] proves that if $L_t(N)$ is used to denote the lattice formed by the collection of the elementary substructure of N then for any countable, non-standard model M of THN, those finite lattices, which can be realized as $L_t(N/M)$ where N is a cofinal extension of M, depend only on Th (M). Very little effort is made in studying the inner structures of THN models.

Because THN is undecidable and does not admit quantifier elimination, we will start with the Chinese reminder theorem, making use of remainders to characterize a type to some extent. The number of prime divisors is increased from a finite number to infinity, and the dividends stretch from standard numbers to non-standard numbers. In this way, we will get the 'generalized Chinese remainder theorem', which is undoubtedly another key to the research of non-standard numbers.

First from the Chinese remainder theorem we arrive at the generalized Chinese remainder theorem and with it we make the remainder sequence a characteristic for every non-standard number, then we verify the number of countable THN models. Also we prove the existence of several non-standard numbers which have some special properties. In the second part, we will show that there exists an ω-saturated THN model. We will analyse the number of its elements, the characteristics and special elements, and compare it with the standard THN model. The last part is used to study the inner structure of a THN model. In the following, let N be the set of all natural numbers.

§ 1. Generalized Chinese Remainder Theorem

Lemma 1.1(Chinese remainder theorem) *Let* $m_1, m_2, \cdots, m_n \in \mathbf{N}$ *and* $(m_i, m_j)=1$ $(i \neq j)$. *Then for all* $r_1, r_2, \cdots, r_n \in \mathbf{N}$, *the system of congruences*,

$$\begin{cases} x \equiv r_1 (\bmod m_1), \\ \cdots \\ x \equiv r_n (\bmod m_n), \end{cases}$$

has an integral solution that is uniquely determined modulo $m_1, m_2, \cdots, m_n$. [4, p. 34].

So we can see, if we fix a group of prime a numbers $p_1, p_2, \cdots, p_k$, for arbitrary $r_1, r_2, \cdots, r_k \in \mathbf{N}$ $(0 \leqslant r_i < p_i)$ (*) there is a unique n: $0 \leqslant n < p_1, p_2, \cdots, p_k$, which satisfies $n = q_i p_i + r_i (i_1 = 1, 2, \cdots, k)$. (**)

On the other hand, given a number $n \in \mathbf{N}$ there must be a unique sequence $r_1, r_2, \cdots, r_k$ satisfying (*) and (**). It is thus concluded that there is a one-to-one correspondence between $(r_1, r_2, \cdots, r_k)$ (*) and n $(0 \leqslant n < p_1, p_2, \cdots, p_k)$ (**). For a fixed k, Lemma 1.1 can be written as a first order sentence. Therefore it is still true in non-standard models. Moreover, we can generalize this Lemma 1.1 to a sequence of infinitely many prime numbers.

Theorem 1.2(generalized Chinese remainder theorem, first form) *Let* $(p_1, p_2, \cdots)$ *be a sequence of standard prime numbers, and* $(r_1, r_2, \cdots)$ *be a sequence of natural numbers satisfying* (*), *taking it as a group of remainders. It will be shown that*

$$\Gamma(x) = \{(\exists q_i)(x \equiv q_i p_i + r_i) \mid i = 1, 2, \cdots\}$$

is consistent with THN.

Proof We first let $\mathcal{L}' = \mathcal{L} \cup \{c\}$, where c is a constant symbol not in $\mathcal{L}$. Then $\Gamma(c)$ is a set of sentences in $\mathcal{L}'$. We consider the set $\Sigma = \text{THN} \cup \Gamma(c)$. Any finite subset Σ' of Σ will involve a finite number of sentences of the form $(\exists q_i)(c \equiv q_i p_i + r_i)$ where $i = j_1, j_2, \cdots$ or j_n. In the set of usual natural numbers, as the Chinese reminder theorem states, there is an n $(0\leqslant n<p_{j_1}p_{j_2}\cdots p_{j_n})$ such that $(\exists q_i)(c\equiv q_ip_i+r_i)$ $i=j_1, j_2,\cdots,j_n$. So $\langle N,+,\cdot,0,n\rangle$ is a model of $\sum'$, where n is an interpretation of c in N. By the Compactness theorem ([5, p. 67]) $\sum$ has a model $\mathfrak{A}'=\langle\mathfrak{A},c\rangle$. The reduction to $\mathcal{L}$ of this model is a THN model which satisfies $\Gamma(x)$. Therefore $\Gamma(x)$ is consistent with T.

Taking advantage of the generalized remainder theorem, we can describe the characteristics of non-standard numbers in detail.

In the following we fix the sequence $(p_1, p_2,\cdots)$ as the list of all standard prime numbers in the natural order.

Theorem 1.3 *The number of countable* THN *models is* 2^{ω}.

Proof By Theorem 1.2 we know that every $\Gamma(x)$ is THN-consistent and can be extended to a maximal consistent set of formulas (whose proof is similar to the proof concerning consistent sets of sentences on [5, p. 26]). This maximal consistent set of formulas is a type and can be realized by a non-standard number. In this way, every sequence $(r_1, r_2,\cdots)$ $(0\leqslant r_i<p_i)$ corresponds to a type and different sequences correspond to different types. Therefore the number of types in number theory must be at least equal to that of such sequences. The number is at least $\prod_{i<\omega}p_i=2^{\omega}$. On the other hand, as a countable theory, the number of THN types is at most 2^{ω}. Therefore THN has 2^{ω} types. Every type can be realized by a countable THN model because of its THN-consistency. As we know a countable model can realize at most countably many types, so the number of non-isomorphic countable models is 2^{ω}.

By Theorem 1.2, an arbitrary infinite remainder sequence $(r_1, r_2,\cdots)$ correponds to a number in a suitable model. On the other hand, an arbitrary number corresponds to such a remainder sequence which we call 'the remainder sequence of this non-standard number'.

We notice that when $r_i=0$ $(i=1,2,\cdots)$ in Theorem 1.2, $\Gamma(x)$ becomes $\Gamma(x)=\{\exists q_i(x\equiv p_iq_i)\mid i=1,2,\cdots\}$ which is THN-consistent. There is a THN model $\mathfrak{B}$ which realizes $\Gamma(x)$. Let the element which realizes $\Gamma(x)$ in $\mathfrak{B}$ be q. Then q can be divided by any prime number:

$$2|q, 3|q, 5|q, \cdots \quad ①$$

For $2|q$, there is a q_1 such that $q=2 \cdot q_1$ and for $3|q$ there is a q_2 such that $q_1=3 \cdot q_2$; and $5|q$ gives us a q_3 such that $q_2=5 \cdot q_3$, …If $x=q_0$, we will have:

$$q_1|q_0, q_2|q_1, q_3|q_2, \cdots \quad ②$$

① and ② are both properties that standard numbers don't have. Studying this kind of proper and analysing the differences between standard and non-standard numbers are the reasons why we generalize the Chinese remainder theorem. We are trying to use it to find the properties of non-standard numbers and describe their characteristics.

Theorem 1.4 *There exists a* THN *model which has an element that cannot be divided by any finite prime number.*

Proof Evidently, as in the proof of Theorem 1.3, given a remainder sequence with $r_i=1, i=1,2,\cdots$ we can obtain a consistent set of formulae, say $\Gamma(x)$. The THN model which realizes $\Gamma(x)$ satisfies the condition in Theorem 1.4. The existence of such a model is proved.

Theorem 1.5 *There exists a non-standard number which is a prime number.* (*Therefore it cannot be divided by any finite prime number*)

Proof First we let $\Gamma(x)=\{x>1, x>2, x>3, \cdots\}$ *and* $\mathcal{L}'=\mathcal{L}\cup\{c\}$. We consider $\sum=T\cup\Gamma(c)\cup\sigma(c)$, where $\sigma(c)$ shows that c is a prime number, that is, $\sigma(x)$: $\forall a\ (a|x \rightarrow a\equiv x \vee a\equiv 1)$. Any finite subset $\sum'$ of $\sum$ will involve a finite subset $\Gamma'(c)$ of $\Gamma(c)$. We denote it as $c>n_i (1\leqslant i\leqslant m)$, where m is a finite number. Let

$$n=\max_{1\leqslant i\leqslant m}$$

Then any prime number c which is larger than n will realize $\sum'$.

By the compactness theorem, $\sum$ has a model $\mathfrak{A}'=(\mathfrak{A}, c)$. The reduct $\mathfrak{A}$ of $\mathfrak{A}'$ to $\mathcal{L}$ gives a THN model which realizes $\Gamma(x)\cup\sigma(x)$. The element c has an interpretation a in $\mathfrak{A}$ and a is a non-standard number which satisfies the theorem.

§ 2. Saturated Model of Number Theory

From the discussion above there are 2^ω THN types. Because the types of standard numbers are in any THN model, the key point is the types of non-standard numbers. Let $\{\Gamma_i(x_i) : i \in I\}$ be all types of THN (where I is an index set with 2^ω elements).

Definition *In a language $\mathcal{L}=(F,R,C)$ with uncountably many constants (the numbers of functions and relations are assumed to be countable). A type of $x_1,x_2,\cdots,x_n$ is defined as a complete consistent set of formulae in which the number of constants appearing in the formula is at most countable.*

This definition is helping us to avoid the number of types of $\mathcal{L}$ exceeding 2^ω.

Definition A model $\mathfrak{A}$ is called ω-saturated if any countable set of formula $\sum(x)$ consistent with Th($\mathfrak{A}_A$) can be realized in $\mathfrak{A}_A$.

This definition is stronger than the same definition in [5]. Our definition is more conveinient when we discuss an uncountable model.

Theorem 2.1 *There exists an ω-saturated* THN *model with 2^ω elements.*

Proof For any THN model $\mathfrak{A}$, let $\{\Gamma(x_i)\}_{i\in I}$ denote the set of all types consistent with Th($\mathfrak{A}_A$).

From Theorem 1.3 we know that THN has 2^ω types. So if $\mathfrak{A}$ is a countable THN model, $\{\Gamma(x_i)\}_{i\in I}$ has 2^ω types. If $\mathfrak{A}$ is an uncountable THN model of power $\leqslant 2^\omega$, because THN is a countable theory, we know that $\mathfrak{A}_A$ has at most 2^ω new constants. And by our new definition every type contains at most countably many constants, so $2^\omega \cdot 2^\omega \cdot \cdots \cdot 2^\omega = 2^\omega$. $\{\Gamma(x_i)\}_{i\in I}$ still has 2^ω types.

Let $\mathfrak{A}_0=\mathfrak{A}$ be a countable THN model.

1. For $0\leqslant\alpha<2^\omega$ we define $\mathfrak{A}_{\alpha+1}$ from $\mathfrak{A}_\alpha$.

Let $C=\{c_a : a\in A_\alpha\}$ and $\mathcal{L}_\alpha=\mathcal{L}\cup C$. We expand $\mathfrak{A}_\alpha$ to the model $\mathfrak{A}_{A_\alpha}=(\mathfrak{A},a)_{a\in A_\alpha}$ for $\mathcal{L}_{A_\alpha}$ by interpreting each new constant c_a with element a. Let I be an index set of power 2^ω and $D=\{d_i : i\in I\}$ be a set of new constants not in C.

Consider the set $\sum=\mathrm{Th}((\mathfrak{A}_\alpha)_{A_\alpha})\cup\{\Gamma(d_i)\}_{i\in I}$ where $\{\Gamma(d_i)\}_{i\in I}$ is the set of all types consistent with $\mathrm{Th}((\mathfrak{A}_\alpha)_{A_\alpha})$. For any finite set $\{\sigma_1(d_1),\sigma_2(d_2),\cdots,\sigma_n(d_n)\}\subset\{\Gamma(d_i)\}_{i\in I}$ where $\sigma_i(d_i)\in\Gamma(d_i)$, $i=1,2,\cdots,n$, the existential sentence $(\exists x_1,x_2,\cdots,x_n)(\sigma_1(x_1)\wedge\sigma_2(x_2)\wedge\cdots\wedge\sigma(x_n))$ is equivalent to $(\exists x_1)\sigma_1(x_1)\wedge(\exists x_2)\sigma_2(x_2)\wedge\cdots\wedge(\exists x_n)\sigma_n(x_n)$. $\exists x_i\sigma(x_i)$ holds in a suitable model of $\mathrm{Th}((\mathfrak{A}_\alpha)_{A_\alpha})$. Hence it also holds in $(\mathfrak{A}_\alpha)_{A_\alpha}$ because $\exists x_i\sigma(x_i)$ doesn't have new constants in D. Therefore $(\exists x_1,x_2,\cdots,x_n)(\sigma_1(x_1)\wedge\sigma_2(x_2)\wedge\cdots\wedge\sigma_n(x_n))$ holds in $\mathfrak{A}_\alpha$. By the compactness theorem we know that $\sum$ is consistent and $\sum$ has a model $\mathfrak{A}'_{A_\alpha}$. Because $\{\Gamma(c_i)\}_{i\in I}$ has 2^ω types, $|A'|\geqslant 2^\omega$. On the other hand, $|\mathcal{L}_\alpha|=2^\omega$, by the downward Löwenheim-Skolem-Tarski theorem, $\mathfrak{A}'_{A_\alpha}$ can be reduced to a model $(\mathfrak{A}_{\alpha+1})_{A_\alpha}$ with exactly 2^ω elements. The reduction $\mathfrak{A}_{\alpha+1}$ to $\mathcal{L}$ is a THN model. By [5, Proposition 3.3.1], $\mathfrak{A}_{\alpha+1}$ is an elementary extension of $\mathfrak{A}_\alpha$.

2. For a limit ordinal number α we define

$$\mathfrak{A}_\alpha=\bigcup_{\beta<\alpha}\mathfrak{A}_\beta.$$

Finally we define

$$\mathfrak{B}=\bigcup_{\alpha<2^\omega}\mathfrak{A}_\alpha.$$

By the elementary chain theorem, $\mathfrak{A}_0\prec\mathfrak{A}_1\prec\cdots\prec\mathfrak{A}_\alpha\prec\cdots\prec\mathfrak{B}$. Therefore $\mathfrak{B}$ is a THN model.

It remains to prove that $\mathfrak{B}$ is ω-saturated. Let $\sum(x)$ be a countable set of formulas consistent with $\mathrm{Th}(\mathfrak{B}_B)$. Extend $\sum(x)$ to a type $\Gamma(x)$ in $\mathrm{Th}(\mathfrak{B}_B)$. Suppose $Y=\{y_{i_1},y_{i_2},\cdots,y_{i_n},\cdots\}$ is the set of all new constants appearing in $\sum(x)$. Then there exists an $\mathfrak{A}_\alpha$ such that $Y\subset A_\alpha$. By the construction of $\mathfrak{B}$, we know that $\Gamma(x)$ can be realized in $\mathfrak{A}_{\alpha+1}$. And so it can be realized in $\mathfrak{B}_B$. Thus we have proved that $\mathfrak{B}$ is ω-saturated.

Theorem 2.2 (generalized Chinese remainder theorem, second form) *Let $\mathfrak{A}$ be an ω-saturated* THN *model. Suppose that $(p_1,p_2,\cdots)$ is a sequence of standard or non-standard prime numbers and $(r_1,r_2,\cdots)$ is a* *sequence of standard or non-standard numbers satisfying* $0 \leqslant r_i < p_i$ *for* $0 \leqslant i < \omega$ *(taking it as a group of remainders). It will be shown that*

$$\Gamma(x) = \{(\exists q_i)(x \equiv q_i p_i + r_i) \mid i = 1, 2, \cdots\}$$

is consistent with T*, where* T *is the theory of* $\langle N, S, +, \cdot, 0 \rangle$. *There is an element* a *realizing* $\Gamma(x)$ *in* $\mathfrak{A}$.

Proof We first let $\mathcal{L}' = \mathcal{L} \cup \{c\}$, where c is a constant symbol not in $\mathcal{L}$. Then $\Gamma(c)$ is a set of sentences in $\mathcal{L}'$. We consider the set $\Sigma = T \cup \Gamma(c)$. Any finite subset Σ' of Σ will involve a finite number of sentences of the form $(\exists q_i)(c \equiv q_i p_i + r_i)$ where $i = j_1, j_2, \cdots$ or j_n. In the set of usual natural numbers, as the Chinese reminder theorem states, there is an n $(0 \leqslant n < p_{j_1} p_{j_2} \cdots p_{j_n})$ such that $(\exists q_i)(c \equiv q_i p_i + r_i)$ $i = j_1, j_2, \cdots, j_n$. So $\langle N, +, \cdot, 0, n \rangle$ is a model of Σ' where n is an interpretation of c in N. By the compactness theorem ([5, p. 67]) Σ has a model $\mathfrak{A}' = \langle \mathfrak{A}, c \rangle$. The reduction to $\mathcal{L}$ of this model is a THN model which satisfies $\Gamma(x)$. Therefore $\Gamma(x)$ is consistent with T.

Theorem 2. 3(generalized Chinese remainder theorem, third form) *Let* $m_1, m_2, \cdots, m_n, \cdots \in \mathbf{N}$ *and* $(m_i, m_j) = 1$ $(i \neq j)$. *Then for all* $r_1, r_2, \cdots r_n \cdots \in \mathbf{N}$, *the system of congruences*,

$$\begin{cases} x \equiv r_1 \pmod{m_1} \\ \cdots \\ x \equiv r_n \pmod{m_n}, \\ \cdots \end{cases}$$

has an integral solution.

Proof (This is similar to the proof of the above theorem.)

By Theorem 1. 4 there exists a non-standard number that cannot be divided by any finite prime number. It can be a prime, or not. For the convenience of studying its properties, we give the following definition:

Definition A number that cannot be divided by any finite prime number is called a quasiprime number. If p and $p+2$ are both prime, $(p, p+2)$ is called a pair of twin prime numbers.

In number theory it has been conjectured that there are infinitely many pairs of twin prime numbers $(p, p+2)$. Now we have:

Theorem 2.4 *There exists infinitely many pairs of twin quasi-prime numbers.*

Proof First we construct a pair of twin quasi-prime numbers. We choose a special remainder sequence $(1,2,1,1,\cdots)$ where $r_1=1, r_2=2, r_i=1$ $(i=3,4,\cdots)$. Using this sequence we construct a set of formulas $\Gamma(x)=\{(\exists q_i)(x \equiv q_i p_i + r_i) \mid i=1,2,\cdots\}$. From the proof of Theorem 2.2, $\Gamma(x)$ is THN-consistent. In $\mathfrak{B}$ there is a non-standard number p realizing $\Gamma(x)$. p is evidently a quasi-prime number and because of the special values of its remainder sequence $p+2$ is also a quasi-prime number. Hence $(p, p+2)$ is a pair of twin quasi-prime numbers. Similarly, using the sequence $(0,0,\cdots)$ where $r_i=0, i=1,2,\cdots$ as another remainder sequence we obtain a non-standard number q. Then letting $p_1=p+q, p_2=p+2\cdot q, p_3=p+3\cdot q,\cdots$, we get infintely many pairs of twin quasi-prime numbers:

$$(p, p+2), (p_1, p_1+2), (p_2, p_2+2), (p_3, p_3+2), \cdots$$

Theorem 2.5 *In an ω-saturated number theory model there exists an element a such that $2^n \mid a, (n=1,2,\cdots)$ and the only prime divisor of a is 2.*

Proof We consider the set $\Gamma(x)=\{\exists a_1(x=2\cdot a_1), \exists a_2(x=2^2\cdot a_2), \exists a_3(x=2^3\cdot a_3),\cdots\}$ and $\sigma(x)$: $\forall q(\forall a(a \mid q \rightarrow a \equiv q \vee a \equiv 1) \wedge q \mid x \rightarrow 2 \equiv q)$. Let $\Sigma = T \cup \Gamma(c) \cup \sigma(c)$. Any finite subst of Σ will involve finite sentences of $\Gamma(c)$ and these sentences can be realized in the standard number theory model. By the Compactness theorem, Σ has a model. Therefore $\Gamma(x)$ is T-consistent and can be extended to a type which corresponds to a special element that can be divided infinitely many times by 2 and the only prime divisor of which is 2. Because ω-saturated models can realize all types, they therefore realize $\Gamma(x)$. This tells us that in ω-saturated models such a condition exists: One number has infinitely many identical divisors.

What is worth referring to is that, although the element above satisfies the following conditions:

(i) $\forall p(\sigma(p) \wedge p|a \rightarrow p=2)$ ($\sigma(p)$ means p is a prime number),

(ii) $a \neq \underbrace{2 \cdot 2 \cdots 2}_{n \text{ times}}$ (n is a finite number),

its properties are very different from that of 2^{ω} in set theory. Because in set theory we know that $2^{\omega}=2^{\omega} \cdot 2^{\omega}$. Here we will not have $a \cdot a=a$. Because in our THN model, first-order THN theory requires $a \cdot a=a \rightarrow a=1$ which leads to a contradiction.

There is a famous Archimedean property in standard number theory. That is, for any a, b there is a number n such that $na>b$ (where $na=\underbrace{a+a+\cdots+a}_{n \text{ times}}$). This property will no longer hold in ω-saturated THN models. Obviously, when a is a standard number and b is a non-standard number, we will have $na<b$ (for $n=1,2,\cdots$):

Theorem 2.6 *ω-saturated* THN *models are non-Archimedean.*

Theorem 2.7 (*Euclidean property: Any number $n>1$ can be expressed as a product of finitely many prime numbers uniquely determined up to the order of the factors*) *ω-saturated* THN *models have a non-Euclidean property.*

Theorem 2.8 *In ω-saturated* THN *models there is always a non-standard prime number after any non-standard number.*

An ω-saturated model is just an elementary extension of the standard THN model. Therefore the first-order theory of the standard THN model still holds in any ω-saturated THN model.

§ 3. The Hierarchy of the Non-standard Numbers

Let m, n be two finite numbers and a, b be two non-standard numbers which satisfy $a+m>b, b+n>a$. Then a, b lie in the same segment, similar to Z. We consider every segment as a point, then all these points form a dense order set without endpoints. The proof of this can be found in many common books about non-standard numbers.

In the whole universe of an ω-saturated THN model, for two standard numbers a, b, by the Archimedean property we have $\exists mn$ $(na>b \wedge$

$mb>a$), but a non-standard number a is much larger than any standard number b, that is, $\forall n\ (nb<a)$. How about the condition between two non-standard numbers?

All non-standard numbers are said to be infinitely large as compared with standard numbers. But how large are they? Different non-standard numbers may be different. We know there is some definition about infinity of the same order in the calculus of fluxion:

$$a\sim b\Leftrightarrow \exists m, n\ (na>b \wedge mb>a).$$

Here a, b can be two limits or an infinite series.

Thinking in the same way, we define the degree of non-standard numbers:

Definition *The elements a and b are called in the same degree (denoted as $a \sim b$) if and only if there exist polynomials f, g with finite integer coefficients such that $f(a)>b \wedge g(b)>a$.*

(Attention: Here 'integer coefficients' are not elements of the ω-saturated model but integers in the usual sense; actually, they can be reduced to formal language.)

In fact, '$\sim$' is an equivalent relation, which partitions the elements in the ω-saturated model into infinitely many classes. In one class the elements are of the same degree. According to their degrees, these classes form a linear order. Because an ω-saturated THN model realizes all types, its inner structure is intricate. Now we will prove the order formed by the degrees above is similar to the order of real numbers.

Theorem 3 *The degrees of all non-standard elements in an ω-saturated* THN *model form a dense linear order set without endpoints.*

Proof Let Σ be the set of all polynomials with integer coefficients and suppose a and b are two arbitrary elements in an ω-saturated THN model $\mathfrak{A}$ and a is in a smaller degree than b. We will prove that there is a degree between that of a and that of b. We consider the set $\Gamma(x)=\{x>f(a), g(x)<b \mid \text{for all } f, g\in\Sigma\}$. Any finite subset of $\Gamma(x)$ will involve finitely many formulae. We might as well suppose that they are $x>$

$f_1(a), x>f_2(a), \cdots, x>f_n(a), g_1(x)<b, g_2(x)<b, \cdots, g_m(x)<b$. Let $f_i(a)$ be the maximal number in $\{f_1(a), f_2(a), \cdots, f_n(a)\}$. So $f_i(a)+1$ can satisfy $g_j(x)<b(j=1,2,\cdots,m)$. Therefore $\Gamma(x)$ is finitely satisfiable in $\mathfrak{A}$. And $\Gamma(x)$ is satisfiable by the compactness theorem. So $\Gamma(x)$ can be extended to a type and realized by an element d. The degree of d is between those of a, b.

This is because $d>f(a)$ (for all $f\in\Sigma$) and $f(a)\geqslant a$ give us $d>a$. So $f(d)\geqslant d>a$, that is, $f(d)>a$. Therefore $d>f(a)\wedge f(d)>a$ (for all $f\in\Sigma$) and the degree of d is larger than that of a.

On the other hand, by $g(d)<b$ (for all $g\in\Sigma$) and $d\leqslant g(d)$ we have $d<b$. Then $d<b\leqslant g(b)$. So $g(d)<b\wedge d<g(b)$ (for all $g\in\Sigma$) shows that d is in a degree smaller than b.

Then the density is proved.

Similarly, for any element a in the ω-saturated THN model, the sets $\{x>f(a)\mid \text{for all } f\in\Sigma\}$, $\{g(x)<a\mid \text{for all } g\in\Sigma\}$ are both THN-consistent. The corresponding elements of the types extended from them respectively belong to such degrees: one is larger than that of a, the other is smaller. Because a is an arbitrary element, we know the order of these degrees of elements form a dense order set without endpoints.

References

[1] Morley M D. The number of countable models. J. Symbolic Logic, 1970, 35: 14—18.

[2] Luo L B. Vaught's Conjecture on Unitary Relations. Acta Math. Sin., New Series, 1998, 14(2): 277—284.

[3] Schmerl J H. Substructure Lattices of Models of Peano Arithmatic. Logic Colloquim'84, North-Holland, 225—243.

[4] Ireland K., Rosen M. A Classical Introduction to Modern Number Theory. Springer-Verlag, 1972.

[5] Chang C C, Keisler H J. Model Theory, North-Holland. 3rd ed., 1990.

[6] Takeuti G, Zaring W M. Introduction to Axiom. Set Theory. Springer-Verlag, 1982.

论文和著作目录

Bibliography of Papers and Works

论文目录

[序号] 作者. 论文题目. 杂志名称,年份,卷(期):起页—止页.

[1] 王世强,罗里波. 用牛顿法求实根界限的精确性. 数学通报,1955,(7):18—20.

[2] 王世强,罗里波. 集合与一一对应. 数学通报,1956,(5):12—15;(6):14—18.

[3] 罗里波,王世强. 有限结合系与有限群(Ⅰ). 数学进展,1957,3(2):268—270.

[4] 罗里波. 强不可接近基数上 P(K)的插入定理. 数学学报,1980,23(2):177—182.

[5] 罗里波. 关于代数系统自同构群的一个问题. 数学学报,1980,23(4):500—505.

[6] 罗里波. 模型的并、积与齐次模型. 北京师范大学学报(自然科学版),1980,(3—4):31—40.

[7] 罗里波. 一类自由群内方程. 科学通报,1980,25(6):287—288.

[8] Luo Libo. A class of equations in free groups. Chinese Science Bulletin,1980,25(6):532—533.

[9] 罗里波. 自由群内方程的讨论. 中国科学,1981,(4):521—528 (Luo Libo. A discussion on the equations in free groups. Scientia Sinica,

1982,25(2):161—170).

[10] 罗里波. 可换群中无限生成元直和项消去条件的探讨. 数学学报, 1981,24(3): 472—480.

[11] Luo Libo. On the number of countable homogeneous models. Journal of Symbolic Logic, 1983, 48(3): 539—541.

[12] Luo Libo. The τ-theory for free groups is undecidable. Journal of Symbolic Logic, 1983, 48(3): 700—703.

[13] 王世强,罗里波,翁稼丰. 群的部分初等等价性(Ⅰ). 北京师范大学学报(自然科学版),1986,(3):7—12.

[14] Luo Libo. Computational complexity of Abelian groups. Annals of Pure and Applied Logic, 1988, 37(2): 205—248.

[15] Luo Libo. The computational complexity of positive formulas in real addition. 名古屋商科大学论集(Bulletin of the Faculty of Nagoya University of Commerce and Bussiness Administration), 1991, 36(1): 191—204.

[16] Luo Libo. Functions and functionals on finite systems. Journal of Symbolic Logic, 1992, 57(1): 118—130.

[17] 罗里波. 计算机科学发展漫谈. 数学通报,1996,(9):24—27.

[18] 罗里波. 多个一元关系上的 Vaught 猜想. 数学学报,1998,41(3): 617—622 (Luo Libo. Vaught's conjecture on unitary relations. Acta Mathematica Sinica, New Series, 1998, 14(2): 277—284).

[19] Luo Libo. *P*-time algorithms in number theory. 南京大学学报(数学半年刊),1998,15(1):100—107.

[20] Yu Lirong, Luo Libo. New theorems in non-standard number theory. Acta Mathematica Sinica, New Series, 2002, 18(3): 531—538.

[21] 刘吉强,廖东升,罗里波. 完全二叉树的量词消去. 数学学报,2003, 46(1):95—102.

[22] 罗里波. 无原子布氏代数理论的计算复杂性. 数学研究,2004,37 (2):144—154.

[23] 罗里波,龚成清,蒋桂梅,陈永遥. 利用计算机计算古典数论问题. 数学通报,2004,(8):40—43.

[24] 于丽荣,罗里波. 中国剩余定理在非标准数论中的推广. 数学学报,2005,48(5):1 029—1 034.

[25] 罗里波. 康托尔实数的局限性. 数学研究,2008,41(1):72—78.

[26] 李志敏,罗里波,李祥. 完全二叉树理论的计算复杂度. 数学学报,2008,51(2):311—318.

[27] 罗里波. 非良基集合论模型悖论. 北京师范大学学报(自然科学版),2009,45(3):221—225.

[28] Luo Libo. The stability of Turing machine in computing real functions. 数学研究,2009,42(2):126—137.

[29] Luo Libo. Computing functions on real numbers with stable ω-Turing machines. 前沿科学,2009,3(4):26—32.

著作

[序号] 著者. 书名. 出版地:出版社,出版年份.

[1] 吴兴玲,黄成泉,罗里波. 计算复杂性与算法分析. 西安:电子科技大学出版社,2010.

[2] 罗里波. 模型论及其在计算机科学中的应用. 北京:北京师范大学出版社,2012.

后　　记

Postscript by the Chief Editor

北京师范大学数学科学学院系统地开展几个现代数学方向的科学研究，是从王世强老师在 20 世纪 50 年代初进行数理逻辑的研究工作，发表了一批论文开始的. 50 年代中期，严士健老师在华罗庚教授的指导下，进行了环上的线性群、辛群的自同构的研究，首次得到了它们的完整形式，还用自己提出的方法得到了 n 阶模群的定义关系. 刘绍学老师于 1956 年在莫斯科大学获得了副博士学位，他对结合环、李环、若当环和交错环做了统一的处理，获得完整的结果. 回国以后，在国内带动了环论的研究. 1958 年 2 月，孙永生老师在莫斯科大学完成了他的副博士学位论文《关于乘子变换下的函数类利用三角多项式的最佳逼近》，结果深刻，当时受到数学界前辈陈建功的称赞. 回国后，又解决了余下的困难问题. 这 4 位老师在数学科学学院被戏称为"四大金刚". 正是由于他们开创性的工作，使得数学科学学院的科学研究逐渐形成了一支具有相当学术素养的队伍；有一批确定的研究方向，形成了自己的风格和传统；获得了丰富而系统的、达到世界学科前沿的科研成果，其中有一些已经达到世界先进水平；在国内具有一定的学术地位，在国际上有一定的知名度；对国家的数学发展做出了一定的贡献. 之所以能取得这样的成绩，是经过几代人的探索和努力，遭受了诸多的困苦和磨难. 1984 年 5 月 29 日，王梓坤老师被国务院任命为北京师范大学校长并到学院工作，大大加强了学院概率论学科的力量. 这 5 位老师均是学院的学术带头人，为学院的学科建设和人才培养，花费了毕生的精力，做出了重大的贡献. 5 位老师均在 1981 年被批准为首批博士生导师(我校理科首批博士生导师还有 5 位：黄祖洽、刘若庄、陈光旭、汪堃仁和周廷儒老师)，此次批准的博士生导师的数量，提高了学院在学校中的地位，且此举对学院在全国数学界的地位奠定了重要基础，开创了近30多年来的良好局面. 由于5位老师在学院学科建设中的重要地位和学术贡献，将他们的论著整理出来，作为《北京师范大学数学家文库》系列出版，是一件意义重大的事情. 在北京师范大学出版社的大力支持下，该文集系列已经在2005年由北京师范大学出版社出版.

在5部数学文集出版之后，2005年12月25日，学校隆重举行了北京师范大学数学系成立90周年庆祝大会暨王世强、孙永生、严士健、王梓坤和刘绍学教授文集首发式. 5部文集的出版在国内数学界产生了很好的影响. 5部文集的作者按年龄排序为：王世强、孙永生、严士健、王梓坤和刘绍学，除了王世强教授在1927年出生外，其余4位均在1929年出生，广泛一点，在20世纪20年代出生. 考虑学院在30年代出生的博士生导师们，按批准为博士生导师先后的顺序为：陆善镇、汪培庄、王伯英、李占柄、刘来福、陈公宁、罗里波. 由于他们出生在1936～1939年，按年龄从大到小排序为：李占柄、罗里波、汪培庄、王伯英、刘来福、陈公宁和陆善镇. 他们均为学院的发展和建设作出了重要贡献，出版他们的文集是学院的基本建设. 因此，学院将在近几年内陆续出版他们的文集，并由院党委书记李仲来教授任主编.《文集》的结构为：照片、序、论文选、发表的论文和著作目录、后记.

《罗里波文集》是这套文库的第11部. 该文集的出版得到了北京师范大学出版社的大力支持，在此一并表示衷心的感谢.

华罗庚教授说："一个人最后余下的就是一本选集."(龚升论文选集，中国科学技术大学出版社，2008) 这些选集的质量反映了我们学院某一学科，或几个学科，或学科群的整体学术水平. 而将北京师范大学数学科学学院著名数学家、数学教育家和科学史专家论文进行整理和选编出版，是学院学科建设的一项重要的和基础性的工作，是学院的基本建设之一. 它对提高学院的知名度和凝聚力，激励后人，有着重要的示范作用. 当然，这项工作还在继续做下去，搜集和积累数学科学学院各种资料的工作还在继续进行.

主编　李仲来

2012-03-14